Praise for *Prime Obsession*

"Rewarding."
—*The Washington Post Book World*

"A historical adventure . . . full of interesting stories."
—*The Economist*

"An intellectual tour de force and an excellent read."
—*The Washington Times*

"A riveting account of Riemann's times."
—*American Scientist*

"Derbyshire elegantly explores a vexing topic."
—*Science News* magazine

"Derbyshire is a talented expositor determined to make the reader understand some serious mathematics."
—*The New Criterion*

"An informative, comprehensive, well-written account of the unsolved problem that most mathematicians regard as the most important open problem in the field."
—Keith Devlin, author of *The Millenium Problems*

John Derbyshire is a mathematician and linguist by education, a systems analyst by profession, and a celebrated writer in his spare time. His work appears frequently in *National Review* and *The New Criterion*. Born and raised in England, he has made his home in the United States for the past fifteen years. He currently lives in Huntington, New York, with his wife and two children.

Bernhard Riemann and the Greatest Unsolved Problem in Mathematics

John Derbyshire

A PLUME BOOK

PLUME
Published by Penguin Group
Penguin Group (USA) Inc., 375 Hudson Street, New York, New York 10014, U.S.A.
Penguin Group (Canada), 90 Eglinton Avenue East, Suite 700, Toronto, Ontario,
Canada M4P 2Y3 (a division of Pearson Penguin Canada Inc.)
Penguin Books Ltd., 80 Strand, London WC2R 0RL, England
Penguin Ireland, 25 St. Stephen's Green, Dublin 2, Ireland (a division of Penguin Books Ltd.)
Penguin Group (Australia), 250 Camberwell Road, Camberwell, Victoria 3124, Australia
(a division of Pearson Australia Group Pty. Ltd.)
Penguin Books India Pvt. Ltd., 11 Community Centre, Panchsheel Park,
New Delhi – 110 017, India
Penguin Group (NZ), 67 Apollo Drive, Rosedale, North Shore 0745, Auckland,
New Zealand (a division of Pearson New Zealand Ltd.)
Penguin Books (South Africa) (Pty.) Ltd., 24 Sturdee Avenue, Rosebank,
Johannesburg 2196, South Africa

Penguin Books Ltd., Registered Offices: 80 Strand, London WC2R 0RL, England

Published by Plume, a member of Penguin Group (USA) Inc. This is an authorized reprint of a hardcover edition published by Joseph Henry Press. For information address Joseph Henry Press, 500 Fifth Street, NW, Washington, D.C. 20001.

First Plume Printing, June 2004
10 9

The Library of Congress has catalogued the Joseph Henry Press edition as follows:

Derbyshire, John.
Prime obsession : Bernhard Riemann and the greatest unsolved problem in mathematics / John Derbyshire.
p. cm.
Includes index.
ISBN 0-309-08549-7 (hc.)
ISBN 978-0-452-28525-5 (pbk.)
1. Numbers, Prime. 2. Series. 3. Riemann, Bernhard, 1826–1866. I. Title.
QA246.D47 2003
512'.72—dc21

2002156310

Printed in the United States of America

For Rosie

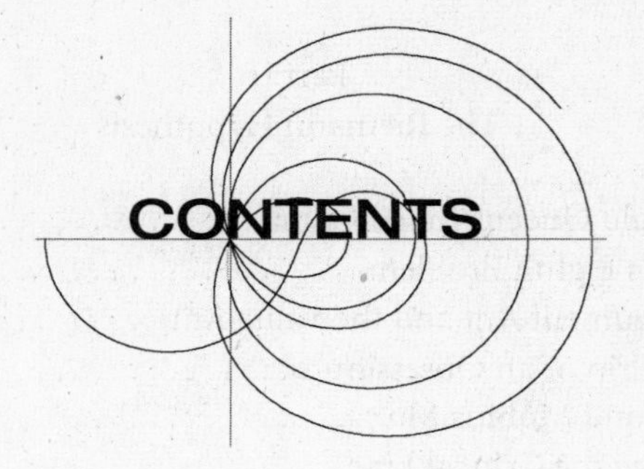
CONTENTS

Part II
The Riemann Hypothesis

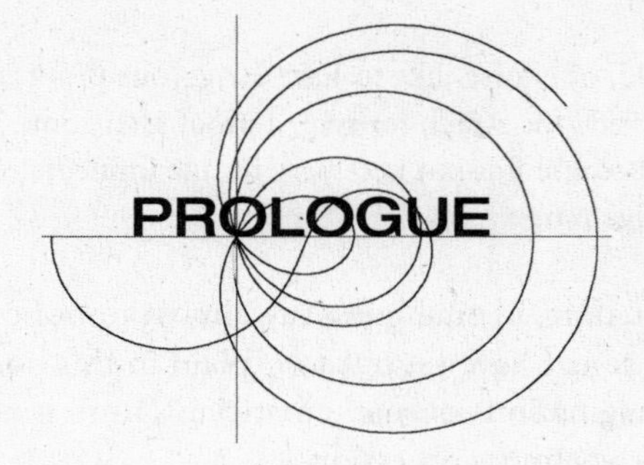

In August 1859, Bernhard Riemann was made a corresponding member of the Berlin Academy, a great honor for a young mathematician (he was 32). As was customary on such occasions, Riemann presented a paper to the Academy giving an account of some research he was engaged in. The title of the paper was: "On the Number of Prime Numbers Less Than a Given Quantity." In it, Riemann investigated a straightforward issue in ordinary arithmetic. To understand the issue, ask: How many prime numbers are there less than 20? The answer is eight: 2, 3, 5, 7, 11, 13, 17, and 19. How many are there less than one thousand? Less than one million? Less than one billion? Is there a *general rule or formula* for how many that will spare us the trouble of counting them?

Riemann tackled the problem with the most sophisticated mathematics of his time, using tools that even today are taught only in advanced college courses, and inventing for his purposes a mathematical object of great power and subtlety. One-third of the way into the paper, he made a guess about that object, and then remarked:

> One would, of course, like to have a rigorous proof of this, but I have put aside the search for such a proof after some fleeting vain attempts because it is not necessary for the immediate objective of my investigation.

That casual, incidental guess lay almost unnoticed for decades. Then, for reasons I have set out to explain in this book, it gradually seized the imaginations of mathematicians, until it attained the status of an overwhelming obsession.

The Riemann Hypothesis, as that guess came to be called, remained an obsession all through the twentieth century and remains one today, having resisted every attempt at proof or disproof. Indeed, the obsession is now stronger than ever since other great old open problems have been resolved in recent years: the Four-Color Theorem (originated 1852, proved in 1976), Fermat's Last Theorem (originated probably in 1637, proved in 1994), and many others less well known outside the world of professional mathematics. The Riemann Hypothesis is now the great white whale of mathematical research.

The entire twentieth century was bracketed by mathematicians' preoccupation with the Riemann Hypothesis. Here is David Hilbert, one of the foremost mathematical intellects of his time, addressing the Second International Congress of Mathematicians at Paris in August 1900:

> Essential progress in the theory of the distribution of prime numbers has lately been made by Hadamard, de la Vallée Poussin, von Mangoldt and others. For the complete solution, however, of the problems set us by Riemann's paper "On the Number of Prime Numbers Less Than a Given Quantity," it still remains to prove the correctness of an exceedingly important statement of Riemann, viz. . . .

There follows a statement of the Riemann Hypothesis. A hundred years later, here is Phillip A. Griffiths, Director of the Institute for Advanced Study in Princeton, and formerly Professor of Mathematics at Harvard University. He is writing in the January 2000 issue of *American Mathematical Monthly*, under the heading: "Research Challenges for the 21st Century":

> Despite the tremendous achievements of the 20th century, dozens of outstanding problems still await solution. Most of us would probably agree that the following three problems are among the most challenging and interesting.
>
> *The Riemann Hypothesis.* The first is the Riemann Hypothesis, which has tantalized mathematicians for 150 years. . . .

An interesting development in the United States during the last years of the twentieth century was the rise of private institutes for mathematical research, funded by wealthy math enthusiasts. Both the Clay Mathematics Institute (founded by Boston financier Landon T. Clay in 1998) and the American Institute of Mathematics (established in 1994 by California entrepreneur John Fry) have targeted the Riemann Hypothesis. The Clay Institute has offered a prize of one million dollars for a proof or a disproof; the American Institute of Mathematics has addressed the Hypothesis with three full-scale conferences (1996, 1998, and 2002), attended by researchers from all over the world. Whether these new approaches and incentives will crack the Riemann Hypothesis at last remains to be seen.

Unlike the Four-Color Theorem, or Fermat's Last Theorem, the Riemann Hypothesis is not easy to state in terms a nonmathematician can easily grasp. It lies deep in the heart of some quite abstruse mathematical theory. Here it is:

The Riemann Hypothesis
All non-trivial zeros of the zeta function
have real part one-half.

To an ordinary reader, even a well-educated one, who has had no advanced mathematical training, this is probably quite incomprehensible. It might as well be written in Old Church Slavonic. In this book, as well as describing the history of the Hypothesis, and some of the personalities who have been involved with it, I have attempted to bring this deep and mysterious result within the understanding of a general readership, giving just as much mathematics as is needed to understand it.

* * * * *

The plan of this book is very simple. The odd-numbered chapters (I was going to make it the *prime*-numbered, but there is such a thing as being *too* cute) contain mathematical exposition, leading the reader, gently I hope, to an understanding of the Riemann Hypothesis and its importance. The even-numbered chapters offer historical and biographical background matter.

I originally intended these two threads to be independent, so that readers who don't like equations and formulae could read only the even-numbered chapters while readers who did not care for history or anecdote could just read the odd-numbered ones. I did not quite manage to hold to this plan all the way through, and I now doubt that it can be done with a subject so intricate. Still, the basic pattern was not altogether lost. There is much more math in the odd-numbered chapters, and much less in the even-numbered ones, and you are, of course, free to try reading just the one group or the other. I hope, though, that you will read the whole book.

I have aimed this book at the intelligent and curious but nonmathematical reader. That statement, of course, raises a number of questions. What do I mean by "nonmathematical?" How much math knowledge have I assumed my readers possess? Well, everybody knows *some* math. Probably most educated people have at least an inkling of what calculus is all about. I *think* I have pitched my book to the level of a person who finished high school math satisfactorily and perhaps went on to a couple of college courses. My original goal was, in fact, to explain the Riemann Hypothesis *without using any calculus at all.* This proved to be a tad over-optimistic, and there is a very small quantity of very elementary calculus in just three chapters, explained as it goes along.

Pretty much everything else is just arithmetic and basic algebra: multiplying out parentheses like $(a + b) \times (c + d)$, or rearranging equations so that $S = 1 + xS$ becomes $S = 1/(1 - x)$. You will also need a willingness to take in the odd shorthand symbols mathematicians use to spare the muscles of their writing hands. I claim at least this much: I don't believe the Riemann Hypothesis can be explained using math more elementary than I have used here, so if you don't understand the Hypothesis after finishing my book, you can be pretty sure you will never understand it.

* * * * *

Various professional mathematicians and historians of mathematics were generous with their help when I approached them. I am profoundly grateful to the following for their time, freely given, for their advice, sometimes not taken, for their patience in dealing with my repetitive dumb questions, and in one case for the hospitality of his home: Jerry Alexanderson, Tom Apostol, Matt Brin, Brian Conrey, Harold Edwards, Dennis Hejhal, Arthur Jaffe, Patricio Lebeuf, Stephen Miller, Hugh Montgomery, Erwin Neuenschwander, Andrew Odlyzko, Samuel Patterson, Peter Sarnak, Manfred Schröder, Ulrike Vorhauer, Matti Vuorinen, and Mike Westmoreland. Any gross errors in this book's math are mine, not theirs. Brigitte Brüggemann and Herbert Eiteneier helped plug the gaps in my German. Commissions from my friends at *National Review*, *The New Criterion*, and *The Washington Times* allowed me to feed my children while working on this book. Numerous readers of my online opinion columns helped me understand what mathematical ideas give the most difficulty to nonmathematicians.

Along with these acknowledgments goes an approximately equal number of apologies. The topic this book deals with has been under intensive investigation by some of the best minds on our planet for a hundred years. In the space available to me, and by the methods of exposition I have decided on, it has proved necessary to omit entire large regions of inquiry relevant to the Riemann Hypothesis. You will find not one word here about the Density Hypothesis, the approximate functional equation, or the whole fascinating issue—just recently come to life after long dormancy—of the moments of the zeta function. Nor is there any mention of the Generalized Riemann Hypothesis, the Modified Generalized Riemann Hypothesis, the Extended Riemann Hypothesis, the Grand Riemann Hypothesis, the Modified Grand Riemann Hypothesis, or the Quasi-Riemann Hypothesis.

Even more distressing, there are many workers who have toiled away valiantly in these vineyards for decades, but whose names are absent from my text: Enrico Bombieri, Amit Ghosh, Steve Gonek, Henryk Iwaniec (half of whose mail comes to him addressed as "Henry K. Iwaniec"), Nina Snaith, and many others. My sincere apologies. I did not realize, when starting out, what a vast subject I was taking on. This book could easily have been three times, or thirty times, longer, but my editor was already reaching for his chainsaw.

And one more acknowledgment. I hold the superstitious belief that any book above the level of hired drudge work—any book written with care and affection—has a presiding spirit. By that, I only mean to say that a book is *about* some one particular human being, who is in the author's mind while he works, and whose personality colors the book. (In the case of fiction, I am afraid that all too often that human being is the author himself.)

The presiding spirit of this book, who seemed often to be glancing over my shoulder as I wrote, whom I sometimes imagined I heard clearing his throat shyly in an adjoining room, or moving around discreetly behind the scenes in both my mathematical and historical chapters, has been Bernhard Riemann. Reading him, and reading about him, I developed an odd mixture of feelings for the man: great sympathy for his social awkwardness, wretched health, repeated bereavements, and chronic poverty, mixed with awe at the extraordinary powers of his mind and heart.

A book should be dedicated to someone living, so that the dedication can give pleasure. I have dedicated this book to my wife, who knows very well how sincere that dedication is. There is a sense, though, not to be left unremarked in a prologue, in which this book most properly belongs to Bernhard Riemann, who, in a short life blighted with much misfortune, gave to his fellow men so very, very much of everlasting value—including a problem that continues to vex them a century and a half after, in a characteristically diffident aside, he noted his own "fleeting vain attempts" to resolve it.

John Derbyshire
Huntington, New York
June 2002

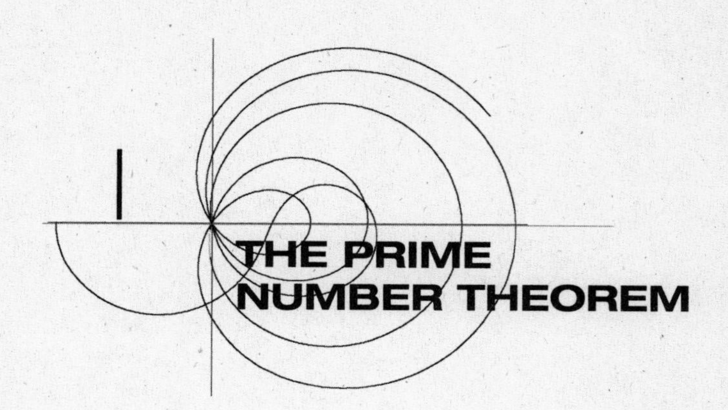

I

THE PRIME NUMBER THEOREM

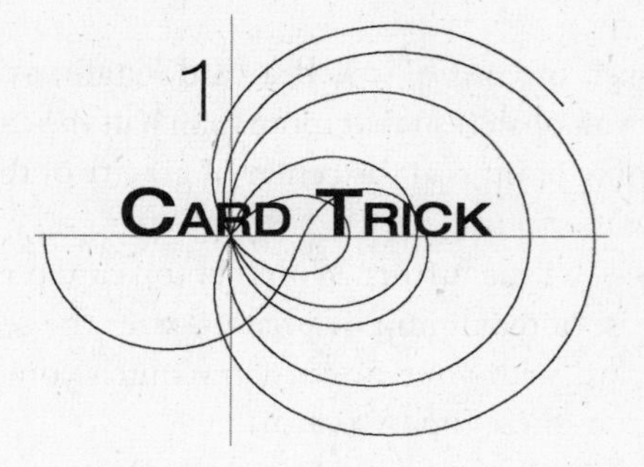

1 Card Trick

I. Like many other performances, this one begins with a deck of cards.

Take an ordinary deck of 52 cards, lying on a table, all four sides of the deck squared away. Now, with a finger slide the topmost card forward without moving any of the others. How far can you slide it before it tips and falls? Or, to put it another way, how far can you make it overhang the rest of the deck?

FIGURE 1-1

The answer, of course, is half a card length, as you can see in Figure 1-1. If you push it so that more than half the card overhangs, it falls. The tipping point is at the center of gravity of the card, which is halfway along it.

Now let's go a little further. With that top card pushed out half its length—that is, to maximum overhang—over the second one, push that second card with your finger. How much combined overhang can you get from these top two cards?

The trick is to think of these top two cards as a single unit. Where is the center of gravity of this unit? Well, it's halfway along the unit, which is altogether one and a half cards long; so it's three-quarters of a card length from the leading edge of the top card (see Figure 1-2). The combined overhang is, therefore, three-quarters of a card length. Notice that the top card still overhangs the second one by half a card length. You moved the top two cards as a unit.

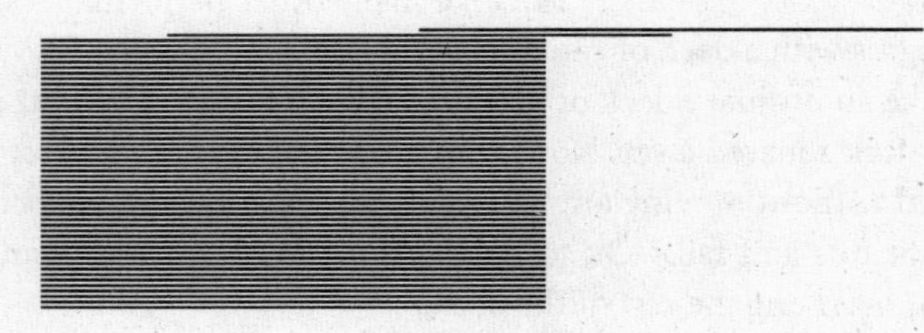

FIGURE 1-2

If you now start pushing the third card to see how much you can increase the overhang, you find you can push it just one-sixth of a card length. Again, the trick is to see the top three cards as a single unit. The center of gravity is one-sixth of a card length back from the leading edge of the third card (see Figure 1-3).

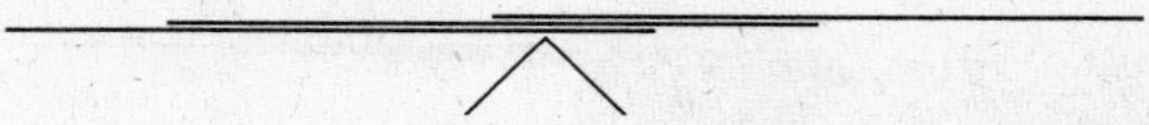

FIGURE 1-3

In front of this point is one-sixth of the third card, a sixth plus a quarter of the second card, and a sixth plus a quarter plus a half of the top card, making a grand total of one and a half cards.

$$\frac{1}{6}+\left(\frac{1}{6}+\frac{1}{4}\right)+\left(\frac{1}{6}+\frac{1}{4}+\frac{1}{2}\right)=1\frac{1}{2}$$

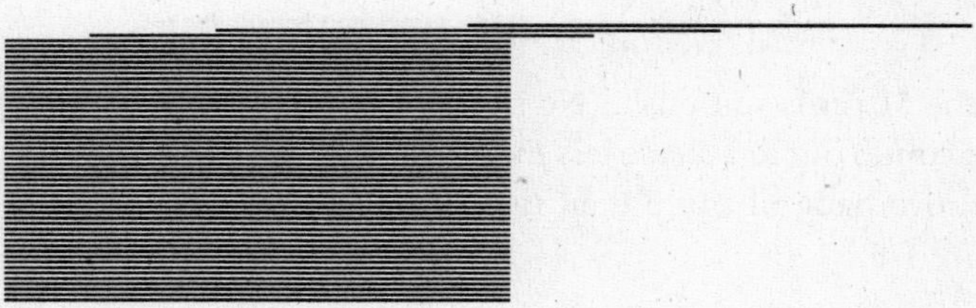

FIGURE 1-4

That's half of three cards—the other half being behind the tipping point. Here's what you have after pushing that third card as far as it will go (see Figure 1-4).

The total overhang now is a half (from the top card) plus a quarter (from the second) plus a sixth (from the third). This is a total of eleven twelfths of a card. Amazing!

Can you get an overhang of more than one card? Yes you can. The very next card—the fourth from the top—if pushed forward carefully, gives another one-eighth of a card length overhang. I'm not going to do the arithmetic; you can trust me, or work it out as I did for the first three cards. Total overhang with four cards: one-half plus one-quarter plus one-sixth plus one-eighth, altogether one and one-twenty-fourth card lengths (see Figure 1-5).

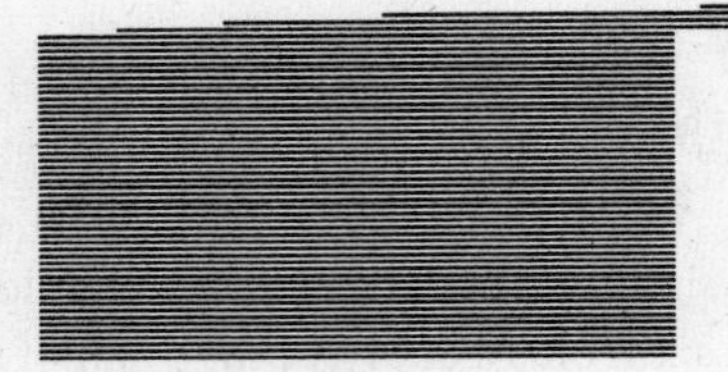

FIGURE 1-5

If you keep going the overhangs accumulate like this.

$$\frac{1}{2}+\frac{1}{4}+\frac{1}{6}+\frac{1}{8}+\frac{1}{10}+\frac{1}{12}+\frac{1}{14}+\frac{1}{16}+\cdots+\frac{1}{102}$$

for the 51 cards you push. (No point pushing the very bottom one.) This comes out to a shade less than 2.25940659073334. So you have a total overhang of more than two and a quarter card lengths! (See Figure 1-6.)

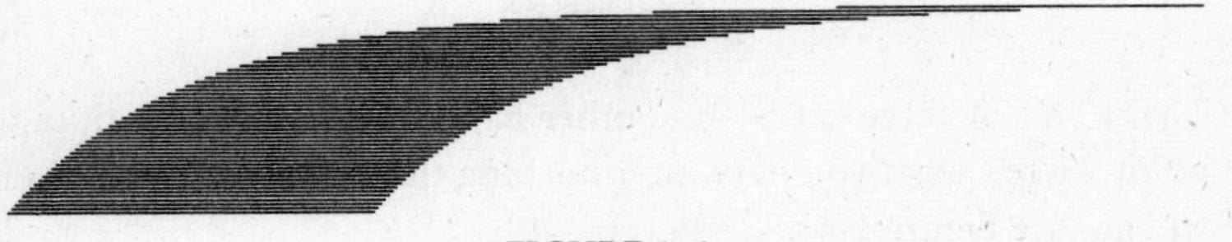

FIGURE 1-6

I was a college student when I learned this. It was summer vacation and I was prepping for the next semester's work, trying to get ahead of the game. To help pay my way through college I used to spend summer vacations as a laborer on construction sites, work that was not heavily unionized at the time in England. The day after I found out about this thing with the cards I was left on my own to do some clean-up work in an indoor area where hundreds of large, square, fibrous ceiling tiles were stacked. I spent a happy couple of hours with those tiles, trying to get a two and a quarter tile overhang from 52 of them. When the foreman came round and found me deep in contemplation of a great wobbling tower of ceiling tiles, I suppose his worst fears about the wisdom of hiring college students must have been confirmed.

II. One thing mathematicians like to do, and find very fruitful, is *extrapolation*—taking the assumptions of a problem and stretching them to cover more ground.

I assumed in the above problem that we had 52 cards to work with. We found that we could get a total overhang of better than two and a quarter cards.

Why restrict ourselves to 52 cards? Suppose we had more? A hundred cards? A million? A trillion? Suppose we had an *unlimited* supply of cards? What's the biggest possible overhang we could get?

First, look at the formula we started to develop. With 52 cards the total overhang was

$$\frac{1}{2}+\frac{1}{4}+\frac{1}{6}+\frac{1}{8}+\frac{1}{10}+\frac{1}{12}+\frac{1}{14}+\frac{1}{16}+\cdots+\frac{1}{102}$$

Since all the denominators are even, I can take out one-half as a factor and rewrite this as

$$\frac{1}{2}\left(1+\frac{1}{2}+\frac{1}{3}+\frac{1}{4}+\frac{1}{5}+\frac{1}{6}+\frac{1}{7}+\frac{1}{8}+\cdots+\frac{1}{51}\right)$$

If there were a hundred cards, the total overhang would be

$$\frac{1}{2}\left(1+\frac{1}{2}+\frac{1}{3}+\frac{1}{4}+\frac{1}{5}+\frac{1}{6}+\frac{1}{7}+\frac{1}{8}+\cdots+\frac{1}{99}\right)$$

With a trillion cards it would be

$$\frac{1}{2}\left(1+\frac{1}{2}+\frac{1}{3}+\frac{1}{4}+\frac{1}{5}+\frac{1}{6}+\frac{1}{7}+\frac{1}{8}+\cdots+\frac{1}{999999999999}\right)$$

That's a lot of arithmetic; but mathematicians have shortcuts for this kind of thing, and I can tell you with confidence that the total overhang with a hundred cards is a tad less than 2.58868875882, while for a trillion cards it is a wee bit more than 14.1041183904147 9.

These numbers are doubly surprising. The first surprise is that you can get a total overhang of more than 14 full card lengths, even though you need a trillion cards to get it. Fourteen card lengths is more than four feet, with standard playing cards. The second surprise, when you start thinking about it, is that the numbers aren't bigger. Going from 52 cards to 100 got us only an extra one-third of a card overhang (a bit less than one-third, in fact). Then going all the way to a trillion—a stack of a trillion standard playing cards would go most of the way from the Earth to the Moon—gained us only another $11\frac{1}{2}$ card lengths.

And if we had an unlimited number of cards? What is the absolute biggest overhang we could get? The remarkable answer is, there is no limit. Given enough cards, you could have an overhang of any size. You want an overhang of 100 card lengths? You'd need a stack of about 405,709,150,012,598 trillion trillion trillion trillion trillion trillion cards—a stack whose height would far, far exceed the bounds of the known universe. Yet you could get still bigger overhangs, and bigger, as big as you want, if you're willing to use unimaginably large numbers of cards. A million-card overhang? Sure, but the number of cards you need now is so huge it would need a fair-sized book just to print it in—it has 868,589 digits.

III. The thing to concentrate on here is that expression inside the parentheses

$$1+\frac{1}{2}+\frac{1}{3}+\frac{1}{4}+\frac{1}{5}+\frac{1}{6}+\frac{1}{7}+\cdots$$

This is what mathematicians call a *series*, addition of terms continuing indefinitely, where the terms follow some logical progression. Here the terms 1, $\frac{1}{2}$, $\frac{1}{3}$, $\frac{1}{4}$, $\frac{1}{5}$, $\frac{1}{6}$, $\frac{1}{7}$,... are the reciprocals of the ordinary counting numbers 1, 2, 3, 4, 5, 6, 7, ...

The series $1+\frac{1}{2}+\frac{1}{3}+\frac{1}{4}+\frac{1}{5}+\frac{1}{6}+\frac{1}{7}+\ldots$ is sufficiently important that mathematicians have a name for it. It is called the *harmonic series.*

What I have stated above amounts to this: by adding enough terms of the harmonic series, you can get a total as big as you please. The total has no limit.

A crude, but popular and expressive, way to say this is: the harmonic series adds up to infinity.

$$1+\frac{1}{2}+\frac{1}{3}+\frac{1}{4}+\frac{1}{5}+\frac{1}{6}+\frac{1}{7}+\cdots=\infty$$

Well-brought-up mathematicians are taught to sniff at expressions like that; but so long as you know the pitfalls of using them I think they are perfectly all right. Leonhard Euler, one of the half-dozen greatest mathematicians who ever lived, used them all the time with very fruitful results. However, the proper mathematical term of art is: *The harmonic series is divergent.*

Well, I have said this, but can I prove it? Everybody knows that in mathematics you must prove every result by strict logic. Here we have a result: the harmonic series is divergent. How do you prove it?

The proof is, in fact, rather easy and depends on nothing more than ordinary arithmetic. It was produced in the late Middle Ages by a French scholar, Nicole d'Oresme (*ca.* 1323-1382). D'Oresme pointed out that $\frac{1}{3}+\frac{1}{4}$ is greater than $\frac{1}{2}$; so is $\frac{1}{5}+\frac{1}{6}+\frac{1}{7}+\frac{1}{8}$; so is $\frac{1}{9}+\frac{1}{10}+\frac{1}{11}+\frac{1}{12}+\frac{1}{13}+\frac{1}{14}+\frac{1}{15}+\frac{1}{16}$; and so on. In other words, by taking 2 terms, then 4 terms, then 8, then 16 terms, and so on, you can group the series into an infinite number of blocks, every one of which is bigger than one-half. The entire sum must, therefore, be infinite. Don't be perplexed by the fact that the blocks get bigger very quickly. There is an awful lot of room in "infinity," and no matter how many blocks you take, the next block is well defined and waiting for you. There is always another one-half to be added; and that means that the total increases without limit.

D'Oresme's proof of the divergence of the harmonic series seems to have been mislaid for several centuries. Pietro Mengoli proved the result all over again in 1647, using a different method; then, forty years later, Johann Bernoulli proved it using yet another method; and shortly after that, Johann's elder brother Jakob produced a proof by a fourth method. Neither Mengoli nor the Bernoullis seem to have been aware of d'Oresme's fourteenth-century proof, one of the barely known masterpieces of medieval mathematics. D'Oresme's proof remains the most straightforward and elegant of all the proofs, though, and is the one usually given in textbooks today.

IV. The amazing thing about series is not that some of them are divergent, but that any of them are not. If you add together an infinity of numbers, you expect to get an infinite result, don't you? The fact that you sometimes don't can be easily illustrated.

Take an ordinary ruler marked in quarters, eighths, sixteenths, and so on (the more "so on" the better—I've shown a ruler marked in sixty-fourths). Hold a sharp pencil point at the very first mark on the ruler, the zero. Move the pencil one inch to the right. The pencil point is now on the one-inch mark and you have moved it a total of one inch (see Figure 1-7).

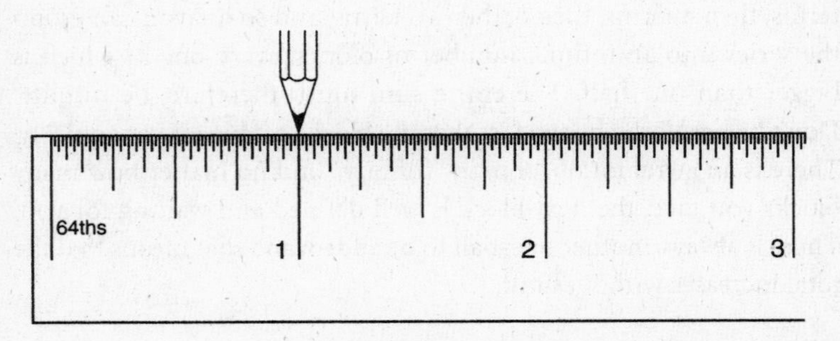

FIGURE 1-7

Now move the pencil half an inch further to the right (see Figure 1-8).

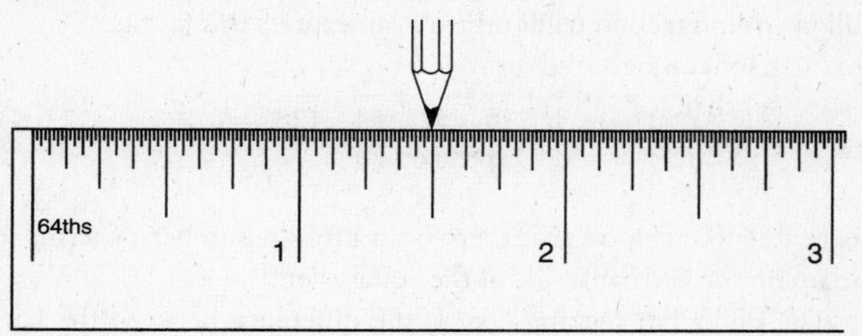

FIGURE 1-8

Now move the pencil a quarter-inch further to the right ... then an eighth-inch ... then a sixteenth ... then a thirty-second ... then a sixty-fourth. Your pencil is in the position shown in Figure 1-9

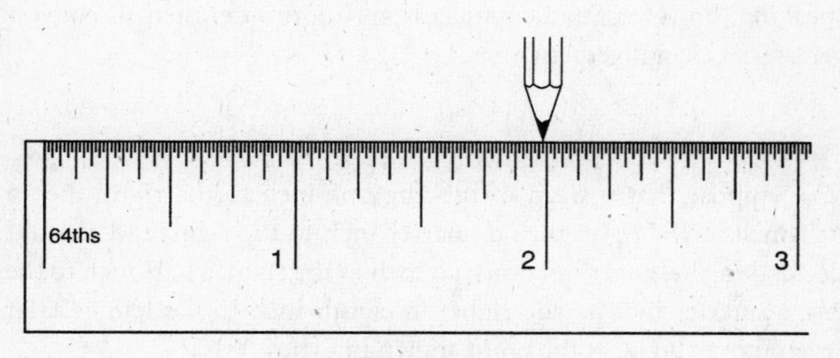

FIGURE 1-9

... and you have moved to the right a total distance of

$$1+\frac{1}{2}+\frac{1}{4}+\frac{1}{8}+\frac{1}{16}+\frac{1}{32}+\frac{1}{64}$$

which is, as you can see, $1\frac{63}{64}$. Clearly, if you could go on like this, halving the distance each time, you would get closer and closer to the two-inch mark. You would never quite reach it; but there is no limit to how close you could get. You could get to within a millionth of an inch of it; or a trillionth; or a trillion trillion trillion trillion trillion trillion trillion trillion trillionth. We can express this fact as

$$1+\frac{1}{2}+\frac{1}{4}+\frac{1}{8}+\frac{1}{16}+\frac{1}{32}+\frac{1}{64}+\frac{1}{128}+\cdots=2$$

Expression 1-1

where it is understood that there is an infinite number of terms to add up on the left-hand side of the equals sign.

The point I'm making here is the difference between the harmonic series and this new one. With the harmonic series I added up an infinite number of terms and got infinity. Here I am adding up an infinite number of terms and getting 2. The harmonic series is *divergent.* This one is *convergent.*

The harmonic series has its charms, and it stands at the center of the topic this book addresses—the Riemann Hypothesis. Generally speaking, however, mathematicians are more interested in convergent series than divergent ones.

V. Suppose that instead of moving one inch to the right, then a half-inch to the right, then a quarter-inch to the right, and so on, I decided to alternate directions: an inch to the right, a half-inch to the left, a quarter-inch to the right, an eighth-inch to the left.... After seven moves I'd be at the point shown in Figure 1-10.

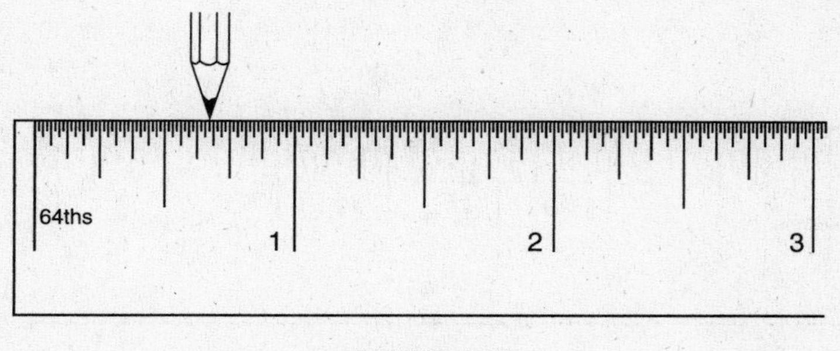

FIGURE 1-10

Since from the mathematical point of view a move to the left is just a negative move to the right, this is equivalent to

$$1-\frac{1}{2}+\frac{1}{4}-\frac{1}{8}+\frac{1}{16}-\frac{1}{32}+\frac{1}{64}$$

which is $\frac{43}{64}$. In fact, it's rather easy to show—I'll prove it in a later chapter—that if you keep on adding and subtracting to infinity you get

$$1-\frac{1}{2}+\frac{1}{4}-\frac{1}{8}+\frac{1}{16}-\frac{1}{32}+\frac{1}{64}-\frac{1}{128}+\cdots=\frac{2}{3}$$

Expression 1-2

VI. Now, suppose that instead of starting out with a ruler marked in halves, quarters, eighths, sixteenths, and so on, I have a ruler marked in thirds, ninths, twenty-sevenths, eighty-firsts, and so on. In other words, instead of halves, halves of halves, halves of halves of halves ... I have thirds, thirds of thirds, thirds of thirds of thirds, and so on. And suppose I do an exercise similar to the first one, move the pencil along one inch, then a third of an inch, then a ninth, then a twenty-seventh (see Figure 1-11).

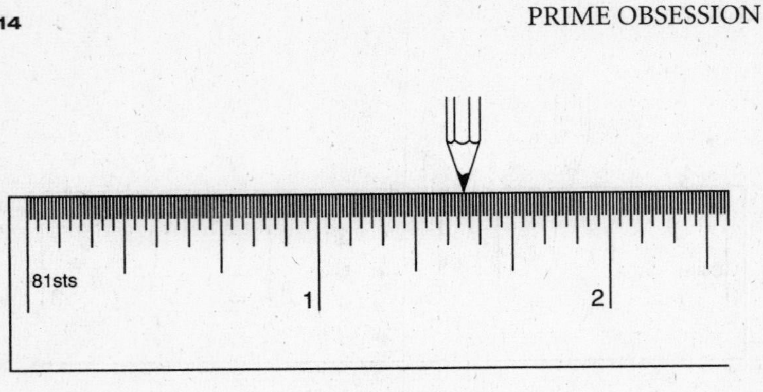

FIGURE 1-11

I don't think it's too hard to see that if you continue forever, you end up moving right a total $1\frac{1}{2}$ inches as shown in Expression 1-3. That is,

$$1+\frac{1}{3}+\frac{1}{9}+\frac{1}{27}+\frac{1}{81}+\frac{1}{243}+\frac{1}{729}+\frac{1}{2187}+\cdots=1\frac{1}{2}$$

Expression 1-3

And of course, I can do the alternating movement with this new ruler, too: right one inch, left a third, right a ninth, left a twenty-seventh, and so on (see Figure 1-12).

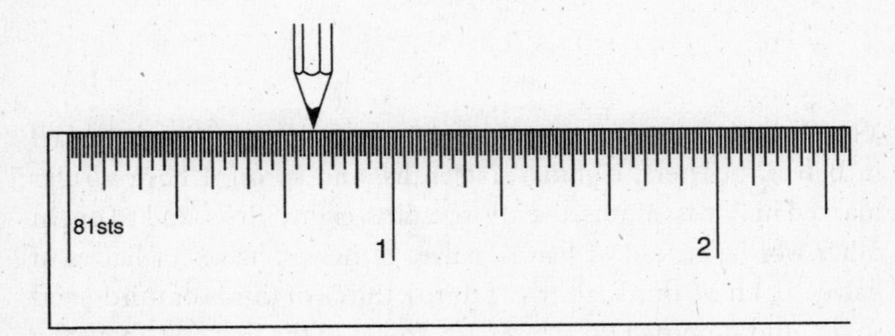

FIGURE 1-12

The math of Expression 1-4 is not so visually obvious, but it's a fact that

$$1-\frac{1}{3}+\frac{1}{9}-\frac{1}{27}+\frac{1}{81}-\frac{1}{243}+\frac{1}{729}-\frac{1}{2187}+\cdots=\frac{3}{4}$$

Expression 1-4

So here we have four *convergent* series, the first (Expression 1-1) creeps closer and closer to 2 from the left, the second (Expression 1-2) closes in on $\frac{2}{3}$ from left and right alternately, the third (Expression 1-3) creeps closer and closer to $1\frac{1}{2}$ from the left, the fourth (Expression 1-4) closes in on $\frac{3}{4}$ from left and right alternately. Before that, I showed one *divergent* series, the harmonic series.

VII. When reading math, it is important to know where in math you are—what region of this vast subject you are exploring. The particular zone these infinite series dwell in is what mathematicians call *analysis*. Analysis used, in fact, to be thought of as the study of the infinite, that is, the infinitely large, and of the infinitesimal, the infinitely small. When Leonhard Euler—of whom I shall write much more later—published the first great textbook of analysis in 1748, he called it *Introductio in analysin infinitorum*: "Introduction to the Analysis of the Infinite."

The notions of the infinite and the infinitesimal created serious problems in math during the early nineteenth century, though, and eventually they were swept away altogether in a great reform. Modern standard analysis does not admit these concepts. They linger on in the vocabulary of mathematics, and I shall make free use of the word "infinity" in this book. This usage, however, is only a convenient and imaginative shorthand for more rigorous concepts. Every mathematical statement that contains the word "infinity" can be reformulated without that word.

When I say that the harmonic series adds up to infinity, what I really mean is that given any number *S*, no matter how large, the sum of the harmonic series eventually exceeds *S*. See?—No "infinity." The whole of analysis was rewritten in this kind of language in the middle third of the nineteenth century. Any statement that can't be so rewritten is not allowed in modern mathematics. Nonmathematical people sometimes ask me, "You know math, huh? Tell me something I've always wondered, What is infinity divided by infinity?" I can only reply, "The words you just uttered do not make sense. That was not a mathematical sentence. You spoke of 'infinity' as if it were a number. It's not. You may as well ask, 'What is truth divided by beauty?' I have no clue. I only know how to divide numbers. 'Infinity,' 'truth,' 'beauty'—those are not numbers."

What is a modern definition of analysis, then? I think *the study of limits* will do for my purposes here. The concept of a limit is at the heart of analysis. All of calculus, for example, which forms the largest part of analysis, rests on the idea of a limit.

Consider the following sequence of numbers: $\frac{1}{1}$, $\frac{3}{2}$, $\frac{7}{5}$, $\frac{17}{12}$, $\frac{41}{29}$, $\frac{99}{70}$, $\frac{239}{169}$, $\frac{577}{408}$, $\frac{1393}{985}$, $\frac{3363}{2378}$, ... Each fraction is built from the one before by a simple rule: add top and bottom to get new bottom, add top and twice bottom to get new top. That sequence converges to the square root of 2. If you square $\frac{3363}{2378}$, for example, you get $\frac{11309769}{5654884}$, which is 2.000000176838287.... We say that the limit of the sequence is $\sqrt{2}$.

Here is another case: $\frac{4}{1}$, $\frac{8}{3}$, $\frac{32}{9}$, $\frac{128}{45}$, $\frac{768}{225}$, $\frac{4608}{1575}$, $\frac{36864}{11025}$, $\frac{294912}{99225}$,.... To get the N-th member of that sequence: if N is even, multiply the previous member by $\frac{N}{N+1}$, if N is odd, multiply the previous member by $\frac{N+1}{N}$. That converges to π. The last fraction shown is 2.972154... (this sequence converges very slowly). Here is yet another: 1^1, $\left(1\frac{1}{2}\right)^2$, $\left(1\frac{1}{3}\right)^3$, $\left(1\frac{1}{4}\right)^4$, $\left(1\frac{1}{5}\right)^5$, ... If you work them out, these come to $1, 2\frac{1}{4}$, $2\frac{10}{27}, 2\frac{113}{256}, 2\frac{1526}{3125}$,... a sequence that converges to a number close to 2.718281828459. This is an exceedingly important number—I shall use it later.

Notice that all of these are *sequences*, just strings of numbers separated by commas. They are not *series*, where the numbers are actually added up. From the point of view of analysis, however, a series is just a sequence in thin disguise. The statement "The series $1 + \frac{1}{2} + \frac{1}{4} + \frac{1}{8}$

$+ \frac{1}{16} + \frac{1}{32} + \ldots$ converges to 2" is mathematically equivalent to: "The sequence $1, 1\frac{1}{2}, 1\frac{3}{4}, 1\frac{7}{8}, 1\frac{15}{16}, 1\frac{31}{32}, \ldots$ converges to 2." The fourth term of the sequence is the sum of the first four terms of the series, and so on. (The term of art for this kind of sequence is *the sequence of partial sums.*) Similarly, of course, the statement, "The harmonic series diverges" is equivalent to: "The sequence $1, 1\frac{1}{2}, 1\frac{5}{6}, 2\frac{1}{12}, 2\frac{17}{60}, 2\frac{27}{60}, \ldots$ diverges," where the N-th term of the sequence is the previous term plus $\frac{1}{N}$.

This is analysis, the study of limits, of how a sequence of numbers can get closer and closer to a limiting number without ever quite reaching it. If I say the sequence goes on forever, I mean that no matter how many terms you write down, I can always write another. If I say it has the limit a, I mean that no matter how tiny a number x you pick, from some point on, every number in the sequence differs from a by less than x. If you choose to say: "The sequence is infinite," or: "The limit of the N-th term, when N goes to infinity, is a," you are free to do so, as long as you understand that these are just loose and convenient ways of speaking.

VIII. The traditional division of mathematics into subdisciplines is as follows.

- *Arithmetic*—The study of whole numbers and fractions. Sample theorem: If you subtract an odd number from an even number you get an odd number.
- *Geometry*—The study of figures in space—points, lines, curves, and three-dimensional objects. Sample theorem: The angles of a triangle on a flat surface add up to 180 degrees.
- *Algebra*—The use of abstract symbols to represent mathematical objects (numbers, lines, matrices, transformations), and the study of the rules for combining those symbols. Sample

theorem: For any two numbers x and y, $(x + y) \times (x - y) = x^2 - y^2$.

- *Analysis*—The study of limits. Sample theorem: The harmonic series is divergent (that is, it increases without limit).

Modern mathematics contains much more than that, of course. It includes set theory, for example, created by Georg Cantor in 1874, and "foundations," which another George, the Englishman George Boole, split off from classical logic in 1854, and in which the logical underpinnings of all mathematical ideas are studied. The traditional categories have also been enlarged to include big new topics—geometry to include topology, algebra to take in game theory, and so on. Even before the early nineteenth century there was considerable seepage from one area into another. Trigonometry, for example (the word was first used in 1595), contains elements of both geometry and algebra. Descartes had in fact arithmetized and algebraized a large part of geometry in the seventeenth century, though pure-geometric demonstrations in the style of Euclid were still popular—and still are—for their clarity, elegance, and ingenuity.

The fourfold division is still a good rough guide to finding your way around mathematics, though. It is a good guide, too, for understanding one of the greatest achievements of nineteenth-century math, what I shall later call "the great fusion"—the yoking of arithmetic to analysis to create an entirely new field of study, analytic number theory. Permit me to introduce the man who, with one single published paper of eight and a half pages, got analytic number theory off the ground and flying.

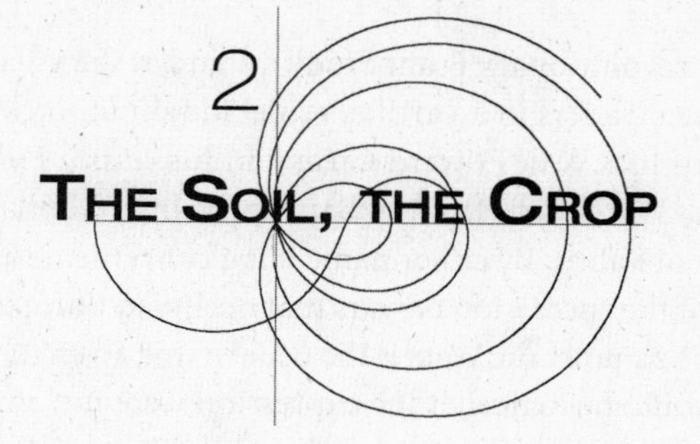

2 The Soil, the Crop

I. We do not know much about Bernhard Riemann. He left no record of his inner life, other than what can be deduced from his letters. His friend and contemporary, Richard Dedekind, was the only person close to him who wrote a detailed memoir; but that was a mere 17 pages and revealed little. What follows, therefore, cannot hope to capture Riemann, but I hope it will at least leave him more than a mere name in the reader's mind. I have reduced his academic career to a brief sketch in this chapter. I shall describe it in much more detail in Chapter 8.

First, let me set the man in his time and place.

II. Supposing that their Revolution had left the French disorganized and ineffective, and disturbed by its republican and antimonarchical ideals, France's enemies moved to take advantage of the situation. In 1792 a huge force of mainly Austrian and Prussian troops, but which included 15,000 emigré French, advanced on Paris. To their surprise, the army of revolutionary France took a stand at the village of Valmy, engaging the invaders in an artillery duel fought in thick fog on September 20 of that year. Edward Creasy, in his classic *Fifteen Decisive Battles of the World*, calls this the Battle of Valmy. Germans call it the Cannonade of Valmy. By either name it is a convenient marker for the beginning of the succession of wars that occupied Europe for the next 23 years. The Napoleonic Wars is the usual name given to these events; though it would be logical, if the expression were not already spoken for, to put them all under the heading First World War, since they included engagements in both the Americas and the Far East. When it all ended at last, with a peace treaty worked out at the Congress of Vienna (June 8, 1815), Europe settled into a long period, almost a century, of relative peace.

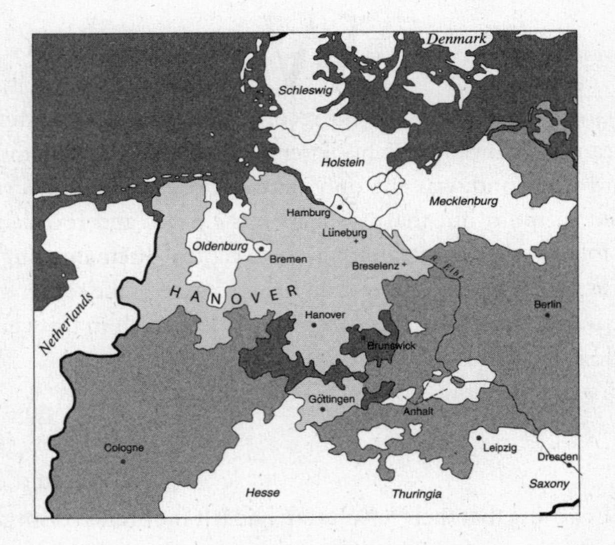

Northwest Germany after 1815. Note that Hanover (the state) is in two pieces; both Hanover (the city) and Göttingen belong to it. Prussia is in two large pieces and some smaller ones; both Berlin and Cologne are Prussian cities. Brunswick is in three pieces.

One consequence of the treaty was a modest tidying up of the German peoples in Europe. Before the French Revolution a German-speaking European might have been a citizen of Hapsburg Austria (in which case he was probably a Catholic) or of the Kingdom of Prussia (making him more likely a Protestant) or of any one of three hundred-odd petty principalities scattered across the map of what we now call Germany. He might also have been a subject of the king of France, or of the king of Denmark, or a citizen of the Swiss Confederation. "Tidying up" is a relative term—there was enough untidiness left over to occasion several minor wars, and to contribute to the two great conflicts of the twentieth century. Austria still had her empire (which included great numbers of non-Germans: Hungarians, Slavs, Romanians, Czechs, and so on); Switzerland, Denmark, and France still included German speakers. It was a good start, though. The three hundred-odd entities that comprised eighteenth-century Germany were consolidated into 34 sovereign states and 4 free cities, and their cultural unity was recognized by the creation of a German Confederation.

The largest German states were still Austria and Prussia. Austria's population was about 30 million, only 4 million of them German speakers. Prussia had about 15 million citizens, most of them German speakers. Bavaria was the only other German state with a population over 2 million. Only four others had more than a million: the kingdoms of Hanover, Saxony, and Württemberg, and the Grand Duchy of Baden.

Hanover was something of an oddity in that, although a kingdom, its king was hardly ever present. The reason for this was that, for complicated dynastic reasons, he was also king of England. The first four of what English people call the "Hanoverian kings" were all named George,[1] and the fourth was on the throne in 1826, when the central character in the story of the Riemann Hypothesis first appeared.

III. Georg Friedrich Bernhard Riemann was born on September 17, 1826, in the village of Breselenz in the eastern salient of the Kingdom of Hanover. This part of the kingdom is known as Wendland, "Wend" being an old German word for the Slavic-speaking peoples they encountered. Wendland was the furthest west reached by the great Slavic advance of the sixth century. The name "Breselenz" itself derives from the Slavic word for "birch-tree." Slavic dialects and folklore survived into modern times—the philosopher Leibnitz (1646–1716) promoted research into them—but from the late Middle Ages onward German immigrants moved into Wendland and by Riemann's time the population was pretty solidly German.

Wendland was, and still is, something of a backwater. With only 110 inhabitants per square mile, it is the most thinly populated district in its modern region, Lower Saxony. There is little industry and few large towns. The mighty Elbe—it is about 250 yards wide here—flows just 7 miles from Breselenz and was the principal connection with the world beyond until modern times. In the nineteenth century sailing ships and barges carried timber and agricultural produce down to Hamburg from Central Europe, returning with coal and industrial goods. During the recent decades of division, the Wendland stretch of the Elbe was part of the border between East and West Germany, a fact that did nothing to help local development. It is a flat, dull countryside of farm, heath, marsh, and thin woodland, prone to flooding. There was a serious flood in 1830 that must have been the first great external event of Bernhard Riemann's childhood.[2]

Riemann's father, Friedrich Bernhard Riemann, was a Lutheran minister and a veteran of the wars against Napoleon. He was already middle-aged when he married Charlotte Ebell. Bernhard was their second child and seems to have been especially close to his older sister, Ida—he named his own daughter after her. Four more children followed, a boy and three girls. With today's standard of living, which of course we take for granted, it is difficult to imagine the hardships that faced a country parson, well into his middle years, with a wife and six children to support, in a poor and undeveloped region of a middling country in the early nineteenth century. Of the six Riemann children, only Ida lived a normal life span. The others all died young, probably in part from poor nutrition. Riemann's mother, too, died young, before her children were grown.

Poverty aside, it needs an effort of imagination for us, living and working in a modern economy, to grasp the sheer difficulty of finding a job in those times and circumstances. Outside large cities the middle class barely existed. There was a scattering of merchants, parsons, schoolteachers, physicians, and government officials. Everyone else who did not own land was a craftsman, a domestic servant, or a peasant. The only respectable employment for women was as governesses; otherwise they relied on their husbands or male family members for support.

When Bernhard was still an infant, his father took up a new position as minister in Quickborn, a few miles from Breselenz, and closer to the great river. Quickborn is still, today, a sleepy village of timber-framed houses and mostly unpaved streets bordered by massive, ancient oak trees. This place, even smaller than Breselenz, remained the family home until the elder Riemann died in 1855. It was the center of Bernhard's emotional world until he was almost 30 years old. He seems to have returned there at every opportunity to be amongst his family, the only surroundings in which he ever felt at ease.

In reading of Riemann's life, therefore, one must set it all against a backdrop of this environment, the environment of his home and upbringing, which he cherished, and for which, when away from it, he yearned. The flat, damp countryside; the draughty house lit only by oil lamps and candles, ill-heated in winter and ill-ventilated in summer; long spells of sickness among siblings who themselves were never quite well (they seem all to have suffered from tuberculosis); the tiny and monotonous social round of a parson's family in a remote village; the inadequate and unbalanced diet on the stodgy side of a stodgy national cuisine ("For a long time he suffered from chronic constipation," notes Neuenschwander[3]). How did they stand it? But they knew nothing else and simple affection is sufficient to sustain the human spirit amid shared hardships.

IV. The multitude of states—kingdoms, principalities, duchies, and grand duchies—that made up North Germany in Riemann's time were largely independent of each other and each made its own internal policy. This loose structure generated local pride and competition between the states.

In most respects they took their lead from Prussia. The eastern part of that kingdom was the only German state to keep some measure of independence from Napoleon after the defeats of 1806–1807. Under the stimulus of that brooding threat, the Prussians concentrated on internal reforms, overhauling their system of secondary education in 1809–1810 under the direction of the philosopher, diplomat, and linguist Wilhelm von Humboldt. Von Humboldt (whose brother Alexander was a great explorer and natural scientist) was a classicist and an ivory-tower man, who once said, "*Alles Neue ekelt mich an.*"—"All that is new disgusts me." Yet oddly, the reforms brought in by this stern reactionary eventually made the educational systems of the German states the most academically advanced in Europe.

At the heart of the system was the 10-year gymnasium school, the years in question being age 10 to 20. In its earliest form, the curriculum at these schools was divided as follows.

Latin	25 percent
Greek	16 percent
German	15 percent
Math	20 percent
History and geography	10 percent
Science	7 percent
Religion	7 percent

By contrast, it is reported (by Jonathan Gathorne-Hardy in *The Public School Phenomenon*) that the great English boys' schools of 1840 allocated 75–80 percent of teaching time—40 hours a week—to classics.

Quickborn had no gymnasium and Riemann did not begin his proper schooling until age 14, four years into the gymnasium course. This was in Hanover, the kingdom's capital city, 80 miles from Quickborn. The location was determined by the fact of his maternal grandmother's living in Hanover so that Riemann's family was spared boarding fees. Before attending this gymnasium Riemann was educated by his father with some assistance from a village schoolteacher named Schultz.

Riemann, aged 14, was terribly unhappy in Hanover, morbidly shy and homesick. His only extracurricular activity, so far as we know, was seeking out such presents as he could afford to buy for his parents and siblings, to send to them on their birthdays. The death of his grandmother in 1842 led to a slight improvement. Riemann was transferred to another gymnasium, this one in the town of Lüneburg. Dedekind has this to say about the new situation.

> The greater proximity to home, and the opportunity this offered to spend vacations with his family, made these later schooldays very happy for him. To be sure, the journeys to and fro, mostly by foot, were physically exhausting in a way he was not used to.[4] His mother, whom sad to say he was soon to lose, expressed anxious concern for his health in her letters, adding many heartfelt warnings to him to avoid excessive physical effort.

Riemann does not seem to have been a good scholar. He had the type of mind that could hold only those things it found interesting, mathematics mostly. Furthermore, he was a perfectionist to whom conscientiousness in producing flawless essay compositions was more important than timeliness in delivering them. To improve his work the school director arranged for him to board with a teacher of Hebrew called Seffer or Seyffer. Under the care of this gentleman

Riemann improved sufficiently that in 1846 he was admitted to the University of Göttingen as a student of theology. The idea was that he would follow his father into the ministry.

V. Göttingen was the only university within the sphere of the Hanover church so it was the logical choice. The name "Göttingen" will crop up all through this book, so a few words about the history of the university may not be out of order. Founded in 1734 by George II of England (who was also Elector of Hanover[5]), Göttingen quickly became one of the better German provincial universities, with more than 1,500 students registered in 1823.

The 1830s, however, were a troubled time. Political agitation by both students and faculty lowered attendance to less than 900 in 1834. Three years later matters came to a head, and Göttingen attained a moment of Europe-wide fame. King William IV of England and Hanover died in 1837 without legitimate issue and the English throne passed to his niece, Victoria. Hanover, however, subscribed to the Salic Law of the medieval Franks, according to which only a male could succeed to the throne. England and Hanover thereupon parted company. The new ruler of Hanover was Ernest Augustus, oldest surviving son of George III.

Ernest Augustus was a great reactionary. Almost his first act was to set aside the liberal constitution granted by William IV four years earlier. Seven eminent professors at Göttingen University refused to swear an oath to uphold the new constitution and were dismissed. Three of them were actually exiled from the kingdom. These dismissed scholars became known as "the Göttingen Seven" and were heroes to social and political reformers all over Europe.[6] Among them were the two brothers Grimm of fairy-tale fame, who were academic philologists.

In the changes that followed the continent-wide upheavals of 1848, Hanover got a new liberal constitution. At least one of the

Göttingen Seven, the physicist Wilhelm Weber, was reinstated. The university soon recovered its luster, eventually to become a great seat of learning, as we shall see. When Bernhard Riemann arrived in 1846, though, these upward trends were still in the future. He found Göttingen University a subdued place, attendance not yet recovered from the ructions of nine years earlier.

Göttingen did, however, have one major attraction for the young Riemann. It was the home of Carl Friedrich Gauss, the greatest mathematician of his age, and possibly of any age.[7]

Gauss was already 69 years old when Riemann arrived at Göttingen. His best work was behind him and he did little lecturing, regarding it as an annoying waste of time. Still his presence must have impressed Riemann, who had already been bitten by the math bug. We know that Riemann attended Gauss's lectures on linear algebra and those of Moritz Stern on the theory of equations. At some point during this year 1846–1847 Riemann must have confessed to his father that he was far more interested in math than in theology and his father, who seems to have been a kind parent, gave his consent to mathematics as a career. And so Bernhard Riemann became a mathematician.

VI. Of Riemann's adult personality, very little has come down to us. The primary source is the short memoir by Dedekind that I mentioned at the beginning of this chapter. The memoir was written 10 years after Riemann's death and was appended to the first edition of his *Collected Works* (but never, so far as I know, translated into English).[8] I have depended heavily on it for this book, so that many of the statements here and in Chapter 8 should really be tagged "… according to Dedekind." You must take this as understood. Though Dedekind might, of course, have been mistaken on points of fact, he was the closest thing Riemann had to a friend. He was an honest and upright man and I have never seen any suggestion that he was less than scrupulously truthful about his subject, with a single understandable exception that I shall mention in a moment. Other sources are Riemann's private letters, many of which have survived, and some random recorded comments by students and colleagues.

These accounts tell us the following.

- Riemann was an extremely shy man. He avoided human contact as far as possible and was ill at ease in company. His only close ties—and they were very close indeed—were with his family, and his only other ties of any sort were with other mathematicians. When not among his family at the vicarage in Quickborn he suffered from homesickness.
- He was very pious, in the German Protestant style. (Riemann was Lutheran.) His opinion was that the essence of religion is, to translate literally from Dedekind's German, "Daily self-examination before the face of God."
- He thought deeply about philosophy and saw all his mathematical work in a larger philosophical context.
- He was a hypochondriac, in both the old and new senses of the word. (It was formerly a synonym for "depressive.") Dedekind avoids this word, apparently out of consideration for Riemann's widow, who begged that Riemann's hypochondria not be made known. Dedekind makes it plain, though, that Riemann was subject to spells of very deep unhappiness, especially after the death of his father, whom he worshiped. Riemann dealt with these episodes by losing himself in work.
- His health was never good and was destroyed by the long years of privation to which a poor man had to resign himself if he was to get an advanced education in that time and place.

It is tempting to find Riemann a rather sad and slightly pathetic character. And yet that would be to consider only the outward appearance and manner of the man. Within that diffident, withdrawn exterior was a mind of great brilliance and staggering boldness. However timid and listless he may have appeared to casual observers, Riemann's mathematics has the fearless sweep and energy of one of Napoleon's campaigns. His mathematical friends and colleagues knew this, of course, and revered him.

Riemann brings to my mind an episode from Somerset Maugham's novel *The Moon and Sixpence*, inspired by the life of the painter Gauguin. Maugham's hero, like Gauguin an artist, dies of leprosy in a hut on a Pacific island, whither he has fled to pursue his vision of art. Hearing that the man is dying, a local doctor goes to his hut. It is a poor construction, shabby and dilapidated. When the doctor steps inside, however, he is astonished to find the interior walls all painted from floor to ceiling with brilliant, mysterious pictures. As with that hut, so it was with Riemann. Outwardly he was pitiable; inwardly, he burned brighter than the sun.

VII. In the realm of higher education, Wilhelm von Humboldt's reforms had as yet left a mark only in Berlin, the Prussian capital. The situation in other German universities was as described by Heinrich Weber in his introduction to Riemann's *Collected Works.*

> The purpose of the universities was conceived by their princely patrons as a place for the preparation of lawyers and physicians, teachers and preachers, as well as a place where the sons of the nobility and the well-to-do could pass their time conspicuously and respectably.

Indeed, the von Humboldt reforms had for a while a negative effect on German higher education. They caused a demand for an increased supply of well-trained secondary-school teachers, and the only way this demand could be met was for the universities to do the training. Even the mighty Gauss was teaching mainly elementary courses at Göttingen University in 1846–1847. In search of a meatier diet, Riemann transferred to Berlin University. Two years at that institution, under instruction from the best mathematical minds in Germany, brought Riemann to full maturity as a mathematician.

(Here and throughout these early historical chapters, you should understand that in Europe before the post-Napoleonic shift of attitudes, and in some countries for longer, there was a clear distinction between *universities*, whose purpose was to teach and train whatever of a cognitive elite the nation was thought to require, and *academies* or *societies*, which existed for the purpose of research—this being understood, to a greater or lesser degree depending on the time, the place, and the inclination of the ruler, to be for the practical advantage of the state. Institutions like Berlin University, founded in 1810, where some research was done, and the early St. Petersburg Academy, where teaching went on, were rare exceptions to this general rule. The Berlin Academy, where the Riemann Hypothesis first saw the light of day, was a pure-research establishment modeled on England's Royal Society.)

We know next to nothing about Riemann's everyday life in Berlin outside his mathematical studies. Dedekind records only one incident worth noting. In March 1848 the Berlin mob, inspired by the February revolution in Paris, took to the streets, demanding the unification of the German states into a single empire. Barricades went up, the army tried to clear them, and blood was shed. The Prussian king at the time was Friedrich Wilhelm IV, a rather dreamy and unworldly man, much under the influence of the Romantic Movement, with a sentimental view of his people and an ideal of the state as a paternalistic monarchy. He proved maladroit in the crisis, sending the army back to camp and leaving his palace unprotected before the insurrectionists had been dispersed. The university students formed a loyal guards corps to protect the king and Riemann served a spell of guard duty with this corps from 9:00 one morning until 1:00 the following afternoon, a grand total of 28 hours.

After returning to Göttingen in 1849, Riemann began work for his doctorate, which he attained two years later, at age 25, having submitted a dissertation on complex function theory. He became a lecturer at Göttingen three years after that and an associate professor in 1857—his first salaried position. (Ordinary lecturers were expected to survive on fees paid by whatever students they could attract to their lectures. The job title was *Privatdozent*—"private lecturer.")

The year 1857 was also what we should call, in the language of current celebrity biography, Riemann's "breakout year." His 1851 doctoral dissertation is nowadays regarded as a classic of nineteenth-century mathematics, but it drew little attention at the time in spite of having been enthused over by Gauss. His other written papers of the early 1850s were not widely known and were published in an accessible form only after his death. To the degree that he had become known at all, it was mainly through the content of his lectures; and much of that content was too far ahead of its time to be appreciated. In 1857, however, Riemann published a paper on analysis that was at once recognized to be a major contribution. Its title was "Theory of Abelian Functions."[9] In it, he tackled topical problems by ingenious and innovative methods. Within a year or two his name was known to mathematicians all over Europe. In 1859 he was promoted to full professor at Göttingen, at last attaining sufficient income to allow him to marry—which he did, three years later. His bride was Elise Koch, a friend of his oldest sister.

On August 11 of that same year, 1859, shortly before his 33rd birthday, Bernhard Riemann was also appointed a corresponding member of the Berlin Academy. The Academy based their decision on the only two of Riemann's papers that were well known, the 1851 doctoral dissertation and the 1857 work on Abelian functions. To be elected a member of the Berlin Academy was a great honor for a young mathematician. It was the custom to acknowledge such appointments by submitting an original paper to the Academy, describing some research one was engaged in. The paper Riemann submitted was titled "On the Number of Prime Numbers Less Than a Given Quantity" (*Über die Anzahl der Primzahlen unter einer gegebenen Grösse*).

Mathematics has not been quite the same since.

3
The Prime Number Theorem

I. Well, how many primes are there less than a given quantity? I'm going to tell you very soon, but first, the five-minute refresher course on prime numbers.

Take a positive whole number—I'll take 28 as an example. What numbers divide *exactly* into it? The answer is: 1, 2, 4, 7, 14, and 28. These are the *factors* of 28. We say: "28 has six factors."

Now, every number has 1 as a factor; and every number has itself as a factor. These are not very interesting factors. They are, to use a word mathematicians rather like, "trivial" factors. The interesting factors are the others: 2, 4, 7, and 14. These are called the *proper factors.*

The number 28, therefore, has four proper factors. The number 29, however, has no proper factors. Nothing divides into 29 exactly, except, of course, 1 and 29. It is a *prime number.* A prime number is one with no proper factors.

Here are all the prime numbers up to 1,000.

2	3	5	7	11	13	17	19	23	29	31	37	41	43
47	53	59	61	67	71	73	79	83	89	97	101	103	107
109	113	127	131	137	139	149	151	157	163	167	173	179	181
191	193	197	199	211	223	227	229	233	239	241	251	257	263
269	271	277	281	283	293	307	311	313	317	331	337	347	349
353	359	367	373	379	383	389	397	401	409	419	421	431	433
439	443	449	457	461	463	467	479	487	491	499	503	509	521
523	541	547	557	563	569	571	577	587	593	599	601	607	613
617	619	631	641	643	647	653	659	661	673	677	683	691	701
709	719	727	733	739	743	751	757	761	769	773	787	797	809
811	821	823	827	829	839	853	857	859	863	877	881	883	887
907	911	919	929	937	941	947	953	967	971	977	983	991	997

As you can see, there are 168 of them. At this point, someone usually objects that 1 is not included in this or any other list of primes. It fits the definition, doesn't it? Well, yes, strictly speaking, it does, and if you want to be a barrack-room lawyer about it, you can write in a "1" at the start of the list for your own satisfaction. Including 1 in the primes, however, is a major nuisance, and modern mathematicians just don't, by common agreement. (The last mathematician of any importance who did seems to have been Henri Lebesgue, in 1899.) Even including 2 is a nuisance, actually. Countless theorems begin with: "Let *p* be any odd prime. . . ." However, 2 pays its way on balance; 1 doesn't, so we just leave it out.

If you look closely at the list of primes, you'll see that they thin out as you go along. Between 1 and 100 there are 25 primes; between 401 and 500, 17; and between 901 and 1,000, only 14. The number of primes in any block of 100 whole numbers seems to decline. If I continued the list to show all the primes up to a million, you would see that there are only eight primes in the last hundred-block (i.e., from 999,901 to 1,000,000). If I took it to a trillion, there would be just four in the last hundred-block. (Here they are: 999,999,999,937; 999,999,999,959; 999,999,999,961; and 999,999,999,989.)

II. The question naturally arises, Do the primes eventually thin out to nothing? If I continued the list into the trillions of trillions, and trillions of trillions of trillions of trillions, would I eventually reach a point beyond which there are no more primes, so that the last prime on my list would be the last prime, the biggest prime?

The answer to that was found by Euclid around 300 B.C.E. No, the primes never thin out to nothing. There are always more. There is no biggest prime. However big a prime you find, there is always a bigger one yet to be found. The primes go on forever. Proof: Suppose N is a prime. Form this number: $(1 \times 2 \times 3 \times \ldots \times N) + 1$. This number doesn't divide exactly by any number from 1 to N—you always get remainder 1. So either it doesn't have any proper factors—and therefore is itself a prime bigger than N—or its smallest proper factor is some number bigger than N. Since any number's smallest proper factor is bound to be a prime—if it wasn't, it could be factored down into something smaller—this proves the result. If N is 5, for example, then $1 \times 2 \times 3 \times 4 \times 5 + 1$ is 121, whose smallest prime factor is 11. Whichever prime you start with, you end up with a bigger one. (I shall give another proof of the infinity of primes in Chapter 7.iv, after showing you the "Golden Key.")

Having had that point settled so early in the history of mathematics, the next thing mathematicians were naturally curious about was: Can we find a rule, a law, to describe the thinning-out? There are 25 primes up to 100. If primes were distributed perfectly evenly, there would of course be 10 times that many—250—up to 1,000. In fact there are only 168 primes up to 1,000, because of the thinning out. Why 168? Why not 158, or 178, or some other number? Is there a rule, a formula, to tell me how many primes there are less than a given number?

And there we are, back with the question that I, and Bernhard Riemann, started with: how many primes are there less than a given quantity?

III. Let's do a little reverse engineering. I actually know the answer to that last question for quite impressively large numbers. Table 3-1 shows some.

TABLE 3-1

N	How many primes less than *N*?
1,000	168
1,000,000	78,498
1,000,000,000	50,847,534
1,000,000,000,000	37,607,912,018
1,000,000,000,000,000	29,844,570,422,669
1,000,000,000,000,000,000	24,739,954,287,740,860

That's nice, but not actually terribly informative. Yes, the primes sure do thin out. If they kept up the pace set in the first 1,000, where there are 168 primes, there would be 168,000,000,000,000,000 or so in that last box. In fact there are only one-seventh that number.

In a moment I am going to perform a trick that will send a flash of light through this rather murky situation. First, though, a word about functions.

IV. A two-column table like Table 3-1 is an illustration of a function. "Function" is one of the most important concepts in all of math, the second or third most important, I should think, after "number" and possibly "set." The main idea of a function is that some number (the one in the right-hand column) depends on some other number (the one in the left-hand column) according to some fixed rule or procedure. In the case of Table 3-1, the procedure is: "Count how many primes there are up to the number in the left-hand column."

Another way to say the same thing is: a function is a way to turn (mathematicians say "map") a number into another number. The function in Table 3-1 turns, or maps, the number 1,000 into the number 168—again, by way of some definite procedure.

Terms of art: Because "the number in the left-hand column" and "the number in the right-hand column" are awfully tedious things to have to keep saying, mathematicians refer to them as the "argument" and the "value" (or the "function value") respectively. So the essence of a function is that you get a *value* by applying some rule or procedure to an *argument.*

One more key term of art. The rule that stands at the heart of a function might apply to some numbers, or some kinds of numbers, but not others. The rule, "subtract the argument from 1 and take the reciprocal," for example, defines a perfectly respectable function—the function a mathematician would call $1/(1 - x)$, which we shall look at more closely in Chapter 9.iii—but it can't be applied to the argument "1," since that would involve dividing by zero, which mathematics doesn't allow. (No use to ask: "What happens if I do?" You can't. It's against the rules. If you try, the game stops and everyone goes back to his last legal position.)

For another example, consider the function whose rule is "count the number of factors the argument has." You find that 28 has six factors (I'm including trivial factors here), while 29 has only two. So this function turns 28 into 6; it turns 29 (or any other prime number) into 2. This is another useful and respectable function, usually written "$d(N)$." However, this function has a meaning only for whole numbers—really has any point only for *positive* whole numbers. How many factors does $12\frac{7}{8}$ have? How many factors does π have? Beats me. That's not what this function is for.

The term of art here is "domain." The domain of a function is the numbers it can have as arguments. The function $1/(1 - x)$ can have any number except 1 as an argument; its domain is all numbers except 1. The function $d(N)$ can have any positive whole number as its argument; that's its domain. The domain of the function $\sqrt{x}$ is all non-negative numbers, since negative numbers don't have square roots (though I reserve the right to change my mind about this later in the book).

Some functions allow *all* numbers as their domain. The squaring function x^2, for example, works for any number. Any number can be squared (i.e., multiplied by itself). The same applies to any *polynomial* function—that is, a function whose value is got by adding and subtracting powers of the argument. Here is an example of a polynomial function: $3x^5 + 11x^3 - 35x^2 - 7x + 4$. The domain of a polynomial is all numbers. This fact will be important in Chapter 21.iii. Most interesting functions, however, have some limits on their domain. Either there are some arguments for which the rule doesn't work, usually because you would have to divide by zero, or else the rule only applies to certain kinds of numbers.

It's important to understand that a table like Table 3-1 is only a *sample* of its function. How many primes are there less than 30,000? Less than seven million? Less than 31,556,926? Well, I could tell you by putting more rows into the table; but given that I'm trying to hold this book to a reasonable number of pages, there is obviously a limit to how much of that I can do. This table is just a sample of the function, a snapshot, with arguments I have chosen for a very deliberate purpose.

In the case of most functions, there is in fact no good way to show a function in all its glory. A graph is sometimes helpful to illustrate some particular feature of a function, but in this case a graph is pretty useless. If you try plotting Table 3-1 as a graph, you will see what I mean. My efforts to provide you with a graph of the zeta function in Chapter 9.iv will drive the point home. Mathematicians generally get a feel for a particular function by working intimately with it for a long time, observing all its features and peculiarities. A table or a graph rarely encompasses the whole thing.

V. Another thing to be noted about functions is that the important ones have names; and the *really* important ones have special symbols to denote them. The function I've sampled in Table 3-1 has the name "The Prime Counting Function" and the symbol $\pi(N)$, which is pronounced "pi of *N*."

Yes, I know, this is confusing. Isn't π the ratio of a circle's circumference to its diameter, the ineffable

$$3.14159265358979323846264\ldots?$$

It is indeed, and this new use of the symbol π is nothing whatever to do with that. The Greek alphabet has only 24 letters and by the time mathematicians got round to giving this function a symbol (the person responsible in this case is Edmund Landau, in 1909—see Chapter 14.iv), all 24 had been pretty much used up and they had to start recycling them. I am sorry about this; it's not my fault; the notation is now perfectly standard; you'll just have to put up with it.

(If you have ever done any serious computer programming, you will be familiar with the concept of *overloading* a symbol. This use of π for two utterly different purposes is a sort of overloading of the π symbol.)

So $\pi(N)$ is defined to be the number of primes up to *N* (inclusive, though it rarely matters, and I shall be sloppy about saying "less than" when I should say "less than or equal to"). Back to our main question: Is there some rule, some neat formula, that will give me $\pi(N)$ without putting me to the trouble of counting?

Allow me to perform a small trick on Table 3-1. I am going to divide the first column by the second, the arguments by the values. I'm not aiming for terrific precision. In fact, I shall use the $6 pocket calculator I take to the supermarket. Here goes. 1,000 divided by 168 gives 5.9524; 1,000,000 divided by 78,498 gives 12.7392. Four more similar calculations give me Table 3-2.

TABLE 3-2

N	$N / \pi(N)$
1,000	5.9524
1,000,000	12.7392
1,000,000,000	19.6666
1,000,000,000,000	26.5901
1,000,000,000,000,000	33.5069
1,000,000,000,000,000,000	40.4204

Look closely at the values here. They go up by 7 each time. Or rather, by a number that wobbles between 6.7 and 7.0. This might not strike you as very wonderful, but a large light bulb goes on over a mathematician's head when he sees a table like that, and a particular word comes into his mind. Let me explain.

VI. There is a certain family of functions that is terrifically important in math, the *exponential* functions. Chances are you know something about them. The word "exponential" is one of those that have escaped from math into ordinary language. We all hope our mutual funds will grow exponentially—that is, faster and faster.

From the point of view I have adopted here—functions illustrated by two-column tables, like Table 3-1—I can give you a loose definition of an exponential function as follows. If you pick your arguments so that they go up by regular *addition* from row to row, and then apply the function rule to them, and if it turns out that the resulting values go up by regular *multiplication* from row to row, you are looking at an exponential function. "Regular" here means that the same number is being added, or multiplied, each time.

Here's an example, for which the rule is "Work out $5 \times 5 \times 5$ …, where there are N fives in the expression."

N	5^N
1	5
2	25
3	125
4	625

See how the arguments go up by addition of 1 each time while the values go up by a multiple of 5 each time? That's an exponential function. The arguments go up by addition while the values go up by multiplication.

I chose the arguments to go up by adding 1 each time and shall continue to do that, just for convenience. In this particular function, this makes the values multiply by 5. Of course, there is nothing special about 5. I could pick on a function with 2 as a multiplier, or 22, or 761, or 1.05 (which would give a table showing the accumulation of compound interest at five percent), or even 0.5. Each gives me an exponential function. That's why I started by saying "a family of functions."

Here's another term mathematicians are fond of: "canonical form." When you have a situation like this, in which a certain phenomenon (in this case an exponential function) can show up in many different ways, there is generally one way mathematicians prefer to represent the whole phenomenon. So it is here. There is one exponential function mathematicians prefer above all others. If you were to take a guess at it, you might suppose it is the one in which the multiplier is 2—the simplest number to multiply by, after all. Nope. The canonical form of the exponential function has multiplier 2.718281828459045235360287.... This is another of those magic numbers, like π, that turn up all over the place in math.[10] It has already turned up in this book (Chapter 1.vii). It's irrational,[11] so the decimal never repeats itself, and can't be rewritten as a fraction. The symbol for it is e, named for Leonhard Euler, of whom much more in the next chapter.

Why this number? Isn't it an awfully clumsy number to take for your canonical form? Wouldn't 2 be much simpler? Well, yes, it probably would, for these purposes. I can't explain the importance of *e* without going into calculus, though, and I have sworn a solemn oath to explain the Riemann Hypothesis with the utter minimum of calculus. I am, therefore, just going to beg you to take on faith that *e* is a really, *really* important number, and that no other exponential function can hold a candle to this one, the function e^N.

N	e^N
1	2.718281828459
2	7.389056098930
3	20.085536923187
4	54.598150033144

(To 12 places of decimals.) The main principle remains, of course. The left-hand columns—the arguments—go up by adding 1 each time. As they do so, the right-hand columns—the values—are multiplied by *e* each time.

VII. What about the contrary situation? Suppose I find myself looking at a function whose rule is: when the arguments go up by multiplication, the values go up by addition? What kind of function is that?

Here we have entered the realm of *inverse* functions. Mathematicians are very keen on inverting things—turning them inside out. If y is 8 times x, what is x in terms of y? It's $y/8$, of course. Division is the inverse of multiplication. There's a thing you like to do called squaring numbers, where you multiply a number by itself? OK, what is the inverse? If $y = x^2$, what is x equal to, in terms of y? Well, it's the square root of y. If you know a bit of calculus, you know there's a process called "differentiation," that you can use to turn a function f into another function *g* that will tell you the instantaneous rate of change of *f* at any argument. What's the inverse? It's integration. And so on. Inversion is going to be a key topic later, when I get deep into Riemann's 1859 paper.

From the point of view of my approach here, showing a function as a table, inversion just means flipping the table round, right to left, left to right. This is actually a quick way to make trouble for yourself. Take the squaring function—probably the first non-trivial function you learned in high school. To square a number, you multiply it by itself.

N	N^2
−3	9
−2	4
−1	1
0	0
1	1
2	4
3	9

(I'm assuming you remember the rule of signs[12] here, so that −3 times −3 is 9, not −9.) Now, if you flip columns, you get the inverse function.

N	$\sqrt{N}$
9	−3
4	−2
1	−1
0	0
1	1
4	2
9	3

But hold on here. What's the function value for argument 9? Is it −3 or 3? Couldn't this function be rewritten like this.

N	$\sqrt{N}$
0	0
1	1, or maybe −1
4	2, or just possibly −2
9	3, or can it also be −3?

This won't do at all—too messy. Well ... as a matter of fact, there *is* a mathematical theory of many-valued functions. Bernhard Riemann was a master of that theory and I shall offer a glimpse of his ideas about it in Chapter 13.v. This is not the time or the place, though, and I'm not going to have any truck with such things here. As far as I am concerned, the iron rule is, one argument, at most one value (no value at all, of course, when the argument isn't in the function's domain). The square root of 1 is 1; the square root of 4 is 2; the square root of 9 is 3. Does this mean I don't acknowledge that −3 times −3 is 9? Sure I acknowledge it. I just don't include it in my definition of the term "square root." Here, for the time being at any rate, is my definition of a square root. The square root of N is the single non-negative number (if any) which, when multiplied by itself, gives N.

VIII. Fortunately the exponential function doesn't give any of these problems. You can cheerfully invert it to give you a function that, when you pick arguments going up by multiplication, gives you values going up by addition. Of course, as with exponential functions, there is a whole family of inverse functions, depending on the multiplier; and as with the exponential function, mathematicians much, much prefer the one that goes up in additions of 1 when the arguments go up in multiples of e. The function you have then is called the log function, and "log" is the word that came into the mathematician's mind, under the illumination of the light bulb, when he saw Table 3-2. If $y = e^x$, then $x = \log y$. (From which it follows, by way of straightforward substitution, that for any positive number y, $y = e^{\log y}$, a fact I'll pick up on later, a lot.)

In the mathematical topics relevant to this book—relevant, that is, to the Riemann Hypothesis—the log function is everywhere. I shall have much more to say about it in Chapters 5 and 7, and it will play a starring role when I actually turn the Golden Key in Chapter 19. For the time being, just take it on faith that it is a function in the sense I have just described, a really important function, and the inverse of the exponential function: If $y = e^x$, then $x = \log y$.

I'm going to cut right to the chase at this point and show you the log function, but instead of going up in multiples of e, I shall let the arguments go up in multiples of 1,000. As I said, when showing a function as a table, I get to pick the arguments (and also the number of decimal places, in this case four). It's still the same function, I swear. To help you see what's happening, I have tacked two extra columns on at the right, the first just the right-hand column from Table 3-2, the second giving column 2's percentage difference from column 3. The result is Table 3-3.

TABLE 3-3

N	$\log N$	$N / \pi(N)$	% error
1,000	6.9078	5.9524	16.0503
1,000,000	13.8155	12.7392	8.4490
1,000,000,000	20.7233	19.6666	5.3727
1,000,000,000,000	27.6310	26.5901	3.9145
1,000,000,000,000,000	34.5388	33.5069	3.0795
1,000,000,000,000,000,000	41.4465	40.4204	2.5385

The following statement seems reasonable: $N/\pi(N)$ is close to $\log N$; and the larger N gets, the closer (proportionally) it gets.

Mathematicians have a special way to write this: $N/\pi(N) \sim \log N$. (Pronounced "*N* over pi of *N* tends asymptotically to log *N*." That wavy line is properly called a "tilde," pronounced "*til*-duh." However, in my experience, mathematicians more often refer to it as a "twiddle" sign.[13])

If you just rearrange this according to the ordinary rules of algebra, you get:

The Prime Number Theorem

$$\pi(N) \sim \frac{N}{\log N}$$

Of course, I haven't proved this, I have just shown that it's plausible. It is a very important result; so important that it is called "the Prime Number Theorem." Not "*a* prime number theorem." This is "*the* Prime Number Theorem." Note the capital letters, which I shall use when referring to the theorem. Very often, in fact, when the context is sufficiently plain, number theorists simply write "PNT," a practice I shall follow in this book.

IX. Finally, two consequences of the PNT, supposing it is true. To derive those consequences, let me point out that there is a sense—a *logarithmic* sense!—in which, when dealing with all the numbers up to some large N, most of those numbers resemble N in size. Of all the numbers from 1 to 1 trillion, for example, over 90 percent have 12 or more digits, and in that respect resemble 1 trillion (which has 13 digits) more than they resemble, say, 1,000 (which has only 4 digits).

If there are $N/\log N$ primes from 1 to N, the average density of primes in that range is $1/\log N$; and since most numbers in that range are like N in size—in the very rough sense that I just described—it is fair to conclude that around N, the density of primes is $1/\log N$. So it is. At the end of the first section in this chapter I counted primes in the last block of 100 numbers before 100, 500, 1,000, 1 million and 1 trillion. The counts were: 25, 17, 14, 8, and 4. The corresponding values of $100/\log N$ (i.e., for $N = 100, 500$, etc.) are, to the nearest whole numbers: 22, 16, 14, 7, and 4. Another way to say this is that in the neighborhood of a big number N, the probability of a number being prime is $\sim 1/\log N$.

By the same rough logic, we can estimate the size of the N-th prime. Consider a range of numbers from 1 to K, for some big number K. If the count of primes in that range is C, then on average we should expect to find the first of those numbers at $K \div C$, the second at $2K \div C$, the third at $3K \div C$, and so on. The N-th will be around $NK \div C$, and the C-th, which is to say the last in this range, will be around $CK \div C$, which means, of course, K. Now, if the PNT is true, then the count C is actually $K/\log K$, so that the N-th prime is actually around $NK \div (K/\log K)$, which is to say, around $N \log K$. Since most numbers in this range resemble K in size, I can take K and N to be interchangeable, and the N-th prime is $\sim N \log N$. I know it looks fishy, but in fact this is not a bad estimate, and gets proportionately better and better on the twiddle principle. It predicts, for example, that the trillionth prime will be 27,631,021,115,929; in fact, the trillionth prime is 30,019,171,804,121, an 8 percent error. Percent errors at a thousand, a million, and a billion are 13, 10, and 9.

Consequences of the PNT

$$\text{The probability that } N \text{ is prime is} \sim \frac{1}{\log N}.$$

$$\text{The } N\text{-th prime number is} \sim N \log N.$$

Not only are these consequences of the PNT; *it* is also a consequence of *them.* If you could mathematically prove the truth of either, the PNT would follow. Each of these results is equiponderant with (i.e., has the same weight as) the PNT, and can be considered just an alternative way of stating it. In Chapter 7.viii I shall show another, more important way to express the PNT.

X. In this chapter I have broached the topic of *the distribution of the prime numbers.* This topic is vast and multifaceted. The Riemann Hypothesis is central to it, as we shall see.

The distribution of the primes has two especially outstanding characteristics. One is the *thinning-out* that I have been discussing, for which the PNT gives a good approximate formula. The other is a certain quality of *randomness.*

Look back at my list of primes up to 1,000, at the beginning of this chapter. The four consecutive primes from 821 to 829 are separated, first to last, by only 8 (because 829 minus 821 is 8). That is an average spacing—remember that there are three spaces separating four numbers—of less than 3. The four consecutive primes from 773 to 809, by contrast, are separated by 36, an average spacing of 12. These patterns of clumping—David Hilbert (see Chapter 12.ii) used the term "condensation"—and spreading continue for as far as the eye can see, at all scales.

The function $\pi(N)$ gives the number of primes up to N. That was the topic of Riemann's 1859 paper, and that is what I have concentrated on in this book. The other outstanding feature of the distribution of the primes—the element of *randomness,* their clumping and spreading—is closely bound up with the properties of $\pi(N)$, though, and must always be kept in mind when contemplating the prime numbers at large.

4

On the Shoulders of Giants

I. The first person to whom the truth contained in the Prime Number Theorem (PNT) occurred was Carl Friedrich Gauss, whose dates were 1777 to 1855. Gauss has, as I mentioned in Chapter 2.v, a good claim to having been the greatest mathematician who ever lived. In his lifetime he was known as *Princeps Mathematicorum*—the Prince of Mathematics—and at his death the King of Hanover, George V, ordered a commemorative medal in his honor, with that title on it.[14]

Gauss came from extremely humble origins. His grandfather was a landless peasant; his father was a jobbing gardener and bricklayer. Gauss attended the poorest kind of local school. A famous incident, reported from that school, is much more likely to be true than most such stories are. One day the schoolmaster, to give himself a half-hour break, set the class to adding up the first 100 numbers. Almost instantly, Gauss threw his slate onto the master's table, saying, "*Ligget se!*" which in the peasant dialect of that place and time meant, "There it is!" Gauss had mentally listed the numbers horizontally in order (1, 2, 3, ..., 100), then in reverse order (100, 99, 98, ..., 1) then added the two lists vertically (101, 101, 101, ..., 101). That is 100 occurrences of 101, and since all the numbers were listed twice, the required answer is half this sum: 50 times 101, which is 5,050. Easy when you have been told it, but not a method that would occur to the average 10-year-old; nor even the average 30-year-old, for that matter.

It was Gauss's good luck that his schoolmasters recognized his ability and were willing to go to some pains to promote it. It was his even greater luck to live in the small German duchy of Brunswick—the blob that separates the two parts of Hanover on the map in Chapter 2.ii. Brunswick was ruled at this time by Carl Wilhelm Ferdinand, who rejoiced in the title *Herzog zu* [that is to say, "Duke of"] *Braunschweig-Wolfenbüttel-Bevern.* We have met this Duke already without knowing it at the time. A keen soldier all his life, he held the rank of field marshal in the Prussian army and was in charge of the joint Prussian-Austrian force that the French stopped at Valmy on September 20, 1792.

Carl Wilhelm truly was a gentleman. If there is a mathematicians' Heaven, some sumptuous apartments must be set aside in it for him, for his use whenever he feels inclined to visit. Hearing of the boy Gauss's talent, the Duke asked to see him. Young Gauss cannot have possessed much in the way of social polish at this point. Later in life, after much acquaintance with courts and universities, he is described as mild and affable; but he always had the rough-cut features and stocky physique of his peasant origins. However, the Duke was sufficiently discerning that he took to the boy at once, remained his friend until death parted them, and provided the steady financial support that enabled young Gauss to embark on a long brilliant career as a mathematician, physicist, and astronomer.[15]

The Duke's ability to support Gauss ended very tragically. In 1806 Napoleon was at the height of his career. In the previous year's campaigning, he had defeated the combined armies of Russia and Austria at the battle of Austerlitz, having temporarily bought off the Prussians by offering them Hanover. He had then established the Confederation of the Rhine, bringing all the western part of what is now Germany under French rule, and reneged on the Hanover deal, offering it now to Britain. Only Prussia and Saxony held out against him; and their only ally was Russia, gun-shy from the defeat at Austerlitz.

To prevent Saxony from becoming a French satellite, the Prussians occupied it, calling the Duke of Brunswick out of retirement—he was 71 years old at this point—to lead their forces. Napoleon declared war and his army struck northwest through Saxony toward Berlin. The Prussians tried to concentrate forces, but the French were too fast for them, and crushed the main Prussian units at Jena. The Duke was with a detachment at Auerstädt a few miles to the north; one of Napoleon's flanking corps caught him and routed his troops.

Defeated and mortally wounded, the Duke asked Napoleon, via an emissary, for leave to return to his home to die. The Emperor, a thoroughly modern dictator who was not much given to chivalry, laughed in the messenger's face. The unfortunate Duke, blinded and dying, had to be hurried away in a cart to the free territories beyond the Elbe. Napoleon's secretary, Louis de Bourienne, tells the melancholy end of the tale in his *Memoirs.*

> The Duke of Brunswick, grievously wounded at the battle of Auerstädt, arrived at Altona [across the Elbe, just west of Hamburg] on October 29. His entry into this city was a new and striking example of the vicissitudes of fortune. People beheld a sovereign prince, enjoying, whether rightly or wrongly, a great military reputation, and but lately powerful and tranquil in his capital, and now wounded to death, making his entry into Altona on a miserable stretcher borne by ten men, without officers, without servants, escorted by a crowd of children. While the Duke continued to live, he saw nobody but his wife, who reached him November 1. He persisted in refusing all visits and died November 10.

He had passed through Brunswick on the way, and it is said that Gauss saw the cart from the window of his room opposite the castle gate. The Duchy of Brunswick was then wound up, incorporated into

Napoleon's puppet "Kingdom of Westphalia." The Duke's heir, Friedrich Wilhelm, was dispossessed and had to flee to England. He, too, died fighting Napoleon, at the battle of Quatre Bras in 1815, a few days before Waterloo, but not before his duchy had been restored to him.

(In strict fairness to Napoleon, I should add that on a later razzia through western Germany, when Gauss was installed at Göttingen, the Emperor spared the city because "the greatest mathematician of all time is living there.")

II. Having lost his patron, Gauss had to find a job. He was offered, and took, the position of director of the observatory at Göttingen University, arriving there in late 1807.[16] Göttingen was already known as one of the better-equipped provincial German universities. Gauss had studied there himself in 1795–1798, apparently attracted by its splendid library, where he had spent most of his time. Now he became head of astronomy at the university and stayed at Göttingen until his death in February 1855, a few weeks short of his 78th birthday. In the last 27 years of his life, he slept away from his beloved observatory only once, to attend a conference in Berlin.

To tell of Gauss's connection with the PNT, I must explain his chief peculiarity as a mathematician. Gauss published much less than he wrote. We know—from his correspondence, his surviving unpublished papers, and circumstantial evidence in his published works—that what he presented to the world was only part of what he discovered. Theorems and proofs that would have made another man's reputation, Gauss left languishing in his personal diaries.

There seem to have been two reasons for this apparent carelessness. One was a lack of ambition. A serene, self-contained, and frugal man, who grew up without material possessions and seems never to have acquired the taste for them, Gauss had little need of anyone's approval and did not seek social advancement. The other factor, much more common among mathematicians in all ages, was perfectionism. Gauss could not bring himself to present any result to the world until it was polished smooth, all in faultless logical order. His personal seal showed a tree with only sparse fruit, and the motto, *Pauca sed matura*—"Few, but ripe."

This is, as I said, a common failing among mathematicians and often makes the reading of published mathematical papers a very tedious business. In one of the minor classics of modern psychological literature, *The Presentation of Self in Everyday Life*, Erving Goffman develops a theory of "performances," in which a product or activity created in conditions of disorder and opportunity in some "back" environment is presented as a smooth, finished creation at the "front." Restaurants illustrate the point. Dishes prepared in the clatter, breakage, and yelling of an overheated kitchen appear in the public area as flawless arrangements on spotless plates, delivered by dapper murmuring waiters. A great deal of intellectual work is like this. Says Goffman:

> [I]n those interactions where the individual presents a product to others, he will tend to show them only the end product, and they will be led into judging him on the basis of something that has been finished, polished and packaged. In some cases, if very little effort was actually required to complete the object, this fact will be concealed. In other cases it will be the long, tedious hours of lonely labor that will be hidden....

Published mathematical papers often have irritating assertions of the type: "It now follows that...," or: "It is now obvious that...," when it doesn't follow, and isn't obvious at all, unless you put in the six hours the author did to supply the missing steps and checking them. There is a story about the English mathematician G.H. Hardy, whom we shall meet later. In the middle of delivering a lecture, Hardy arrived at a point in his argument where he said: "It is now obvious that...." Here he stopped, fell silent, and stood motionless with furrowed brow for a few seconds. Then he walked out of the lecture hall. Twenty minutes later he returned, smiling, and began: "Yes, it *is* obvious that…."

If he lacked ambition, however, Gauss also lacked tact. He made a great deal of trouble for himself with his fellow mathematicians by referring to discoveries he had made, but not published, years before someone else discovered and published them. This was not vanity—Gauss was free of vanity—but what Dr. Johnson called "stark insensibility." In a book published in 1809, for example, Gauss referred to his discovery in 1794 of the method of least squares (a way of finding the best "fit" for a number of experimental observations). He had, of course, not published the discovery at the time he made it. The older French mathematician Adrien-Marie Legendre had discovered, and published, the method in 1806 and was furious at Gauss's claim to prior discovery. There is no doubt of the truth of Gauss's claim—we have documentary evidence—but if he wanted the credit, he really should have published. He did not care about the credit, though; and would not publish a paper if he hadn't enough time to polish it to perfection.

III. In December 1849 Gauss exchanged letters with the astronomer Johann Franz Encke (after whom a famous comet is named). Encke had made some remarks about the frequency of primes. Gauss's letter opened:

> The kind communication of your remarks about the frequency of primes was of interest to me as more than just a reference. It recalled to me my own work in the same subject, whose beginnings were in the distant past, in 1792 or 1793…. One of the first things I did was direct my attention to the decreasing frequency of primes, to which purpose I counted the same in several chiliads and jotted down the results on the attached white pages. *I soon perceived that*

> *beneath all of its fluctuations, this frequency is, on average, close to inversely proportional with the logarithm*.... I have very often (since I have no patience for a continuous count of the range) spent an idle quarter of an hour to count another chiliad here and there; but I gave it up at last without quite getting through a million.
>
> [My italics]

By "chiliads" Gauss meant blocks of 1,000 numbers. So beginning in 1792—when he was fifteen years old!—Gauss had amused himself by tallying all the primes in blocks of 1,000 numbers at a time, continuing up into the high hundreds of thousands ("without quite getting through a million").

To get a feeling for the effort involved here, I set myself the task of extracting the primes from the chiliad 700,001 to 701,000, using just the aids that would have been available to Gauss: a pencil, some sheets of paper, and a list of the primes up to 829, which is as many as you need in order to apply the basic prime-finding process to numbers up to 701,000.[17] I confess I gave up after an hour, when I had worked through prime divisors up to 47...which means I had 130 prime divisors still to go. You are welcome to try the same exercise yourself. This was Gauss's "idle quarter of an hour" (*unbeschäftigte Viertelstunde*).

The sentence I italicized in the extract from Gauss's letter to Encke is the first of the two PNT-related results I showed in Chapter 3.ix. It is, as I remarked there, equivalent to the PNT. There is no doubt that Gauss was indeed working on this in the early 1790s. His claim is well documented, just as other claims of the same kind were. He just never bothered to publish.

IV. Oddly, the first *published* work touching on the PNT came from that same Adrien-Marie Legendre who had been so vexed by Gauss's claim to have discovered the method of least squares. In 1798—that is, five or six years after Gauss had unearthed the PNT, without making his results known to the world—Legendre published a book titled

Essay on the Theory of Numbers, in which he conjectured, on the basis of some prime counts of his own, that

$$\pi(x) \sim \frac{x}{A \log x + B}$$

for some numbers A and B, "to be determined." In a later edition of the book he refined this conjecture (which he could not prove) to

$$\pi(x) \sim \frac{x}{\log x - A}$$

where A, for large values of x, tended to some number near 1.08366. Gauss discusses Legendre's conjectures in his 1849 letter to Encke. He demolishes the 1.08366 value but comes to no other very definite conclusions.

No doubt the Encke letter, if he had read it, would have caused poor Legendre to throw another conniption. Fortunately he had died some years before it was written.[18]

V. Because I am surveying here relevant discoveries and conjectures before 1800, and because he was the author of the "Golden Key," of which I am going to make so much in later chapters, this is the right place to introduce the other first-rank mathematical genius born in the eighteenth century, Leonhard Euler (pronounced "oiler"). Euler (1707–1783) was, says E.T. Bell in *Men of Mathematics*, "probably the greatest man of science that Switzerland has produced" and he is, so far as I know, the only mathematician to have *two* numbers named after him: *e*, which I have already mentioned, equal to 2.71828..., and the Euler-Mascheroni number, which I have not had enough space to describe properly in this book,[19] equal to 0.57721.... In order to introduce Euler, I must first open up a new geographical region in the history of this topic, Russia.

Russia, as I think is well known, entered the modern age somewhat behind the rest of Europe, and her entry was accomplished mainly by the energy and imagination of Peter the Great, who as a boy of 10 was crowned Tsar in 1682. Peter's regnal dates are commonly given as 1682–1725. In fact, for the first seven of those years he reigned jointly with his blind, lame, and speech-impaired half-brother Ivan, and the government was actually controlled by Ivan's sister, Sophia. Peter attained sole power only in 1689, at age 17. Even then he displayed no interest in statecraft and spent the next five years amusing himself. Fortunately he was a man of keen intelligence and great curiosity, and most of his amusements were of an improving sort. He was especially fond of the company of foreigners, of whom at that time there was a large settlement near Moscow, in the so-called "German suburb." Here, among Scottish mercenaries, Dutch merchants, and German and Swiss engineers, Peter took in European science and culture and indulged his passion for fireworks and boats (in between riotous banquets and all-night drinking bouts). In 1692–1693, at Lake Pleschev near Moscow, Peter actually built a warship himself, from the keel up. The following year, 1694, his mother died and Peter took power in earnest.

In 1695–1696 this extraordinary, extraordinary-*appearing* man—he stood 6 feet 7 inches and suffered from occasional, but terrifying, facial twitches—attacked the Black Sea port of Azov and wrested it from the Ottoman Turks. In 1697–1698 he traveled incognito in France, Britain, and Holland, the first Russian sovereign to go abroad, learning as he traveled. (From his British trip the following story is well known, though it is almost certainly apocryphal. Staying at John Evelyn's country house outside London, Peter marched into the drawing-room one day with a shotgun over his arm and announced, in thick English: "I haff shot a peasant." "No, no, my dear fellow," replied his host, laughing. "You mean a *pheasant*." "*Nyet*," said Peter, shaking his head. "It voss a peasant. He voss insolent, unt so I shot him.") Returning to Russia, he began his great campaign of reform, ordering the nobility to shave their beards, humbling the Church and crushing the old Muscovite imperial guard, the Streltsy, which had terrorized his childhood. In 1700 he began his 20-year war with Charles of Sweden; in 1703 he broke through Swedish territory and occupied the length of the river Neva, from Lake Ladoga to the shores of the Baltic. There, on land still in the legal possession of a powerful, undefeated enemy, on the boggy estuary of the Neva, he founded his new capital, St. Petersburg.

One of those astonishing personalities that put the lie to any notion of history as a mere mechanical shadow-play of impersonal forces, Peter went on to reform the government, the nobility, trade, education, and even the customary dress of his people. Not all of it worked; that is, not all of it "stuck"; and not all of it penetrated very far into the gloomy wooded depths of that vast old country; but there is no doubt that Peter left Russia a very different place from the one he found.

Most to the point so far as this book is concerned, he made her a nation hospitable to mathematics and mathematicians.[20]

VI. In January 1724, Peter issued a decree establishing an Academy at St. Petersburg. The decree explained that in the normal way of things an academy, where learned scholars carried out research and produced inventions for the use of the state, was different from a university, which existed to teach young people. Because of the dearth of learning in Russia, however, the St. Petersburg Academy would include a university and a gymnasium (that is, a secondary school) under its authority. It would also have its own observatories, laboratories, workshops, publishing house, print shop, and library. Peter did not do things by halves.

The dearth of learning in Russia was indeed so great that there were no Russians capable of acting as academicians. In fact, since Russia lacked any significant number of elementary or secondary schools, there were not even any Russian youngsters qualified to attend as students at the attached university. These problems were solved by simply importing the required personnel. This was well-established practice in Europe. The first director of the Paris Academy of Sciences, founded 60 years earlier, had been the Dutch physicist Christiaan Huygens. St. Petersburg was a long way from the great European centers of culture, though, and Western Europeans still thought of Russia as a dark and barbarous land, so generous terms had to be offered. Eventually, though, it all got off the ground, the shortage of students for the university being solved by importing eight German youngsters. The St. Petersburg Academy opened its doors in August, 1725—too late for Tsar Peter to preside over the ceremony; he had died six months earlier.

Among the foreign scholars who showed up at the first session of the St. Petersburg Academy were two brothers, Nicholas and Daniel Bernoulli. Aged 30 and 25 respectively, they were sons of Johann Bernoulli of Basel in Switzerland—the gentleman we met in Chapter 1.iii in connection with the harmonic series. (There was a whole dynasty of mathematical Bernoullis; in this generation, in fact, there was a third brother, who followed his father into the chair of mathematics at Basel University, and who "personified the mathematical genius of his native city in the second half of the eighteenth century," according to the *Dictionary of Scientific Biography*.)

Unfortunately, after less than a year in St. Petersburg, Nicholas Bernoulli died ("of a hectic fever"—*D.S.B.*), creating a vacancy at the Academy. Daniel Bernoulli had known Leonhard Euler in Basel and recommended him. Euler, glad of the chance of an academician's post at such a young age, arrived in St. Petersburg on May 17, 1727, a month after his 20th birthday.

That date was also, unfortunately, 10 days after the death of Empress Catherine, Peter's wife, who had succeeded Peter on the throne and followed through on his plans for the Academy. It was a bad time to come to Russia. The 15-year period between Peter's death and the reign of Elizabeth, his daughter, was one of feeble leadership, clique politics, and occasional outbreaks of xenophobia. The warring cliques all maintained networks of spies and informers, and the atmosphere in the capital (which St. Petersburg now was) went from bad to worse.

Under the cruel and brutish Empress Anna, who reigned 1730–1740, Russia descended into one of those spells of state terrorism that she seems particularly prone to, with endless treason trials, mass executions, and other atrocities. This was the notorious *Bironovschina*, named for Anna's favorite, the German Ernst Johann Biron, on whom ordinary Russians put the blame.

Euler stuck it out for 13 years, burying himself in work, staying well clear of the court and its intrigues. "Common prudence forced him into an unbreakable habit of industry," writes E.T. Bell, and this seems as good an explanation as any for Euler's astonishing productivity. Even now the full edition of his collected works is not complete. To date it comprises 29 volumes on mathematics, 31 on mechanics and astronomy, 13 on physics, and 8 volumes of correspondence.

For Euler's friend Daniel Bernoulli, with whom he lodged during the early years in St. Petersburg, the stifling political atmosphere of Russia after Peter was all too much. In 1733 Daniel left to return to Basel, and Euler took over the chair of mathematics at the Academy. This brought him sufficient income to get married. He chose a Swiss girl, Catherine Gsell, whose father was a painter living in St. Petersburg.

It was in these circumstances that Euler solved the Basel problem in 1735; I'll describe that problem in the next chapter. Two years later, in a small memorandum on infinite series, Euler discovered the result that I have called "the Golden Key," and to which I have devoted the first half of Chapter 7. He was, in short, a principal player in the story I am telling—but this will emerge more clearly later, as the mathematical side of the story unfolds.

VII. By 1741 Euler had had enough of secret-police spies and the public impaling of "traitors." Frederick the Great was now on the throne of Prussia and had already embarked on his plan to make the

Kingdom of Prussia—a mere duchy until 1700—one of the great powers of Europe. He planned an Academy of Science in Berlin to replace, or re-vivify, that city's moribund Society of Sciences and invited Euler—by now famous throughout Europe—to be the Academy's Director of Mathematics. Euler arrived in Berlin on July 25, 1741, after a one-month sea and land journey from St. Petersburg. Frederick's mother, Sophia Dorothea of England—she was George II's sister—took a shine to young Euler (he was still only 34) but could not get him to say much. "Why won't you talk to me?" she asked him. Euler replied, "Because, Madame, I have come from a country where every person who speaks is hanged."

In fact, part of Frederick's aim in bringing Euler to Berlin was precisely that he *should* speak. Frederick wanted his court to be a sort of salon, full of brilliant people saying brilliant things to each other. Euler was a very brilliant man indeed, but unfortunately only in mathematics. His opinions on matters of philosophy, literature, religion, and worldly affairs, while well-informed and sensible, were commonplace and uninspired. Further, Frederick was a manipulative egotist who, while in principle wishing to surround himself with geniuses, in practice preferred second-raters who would flatter him. Setting aside a few luminaries like Voltaire and Euler, the general intellectual level at Frederick's court was probably less than scintillating. In 1745–1747 Frederick built the Sans Souci summer palace for himself at Potsdam, 20 miles outside Berlin. (Euler helped design a system of water pumps for the place.) A visitor to Sans Souci asked one of the royal princes: "What do you do here?" The prince replied: "We conjugate the verb *s'ennuyer*." *S'ennuyer* means "to be bored." The language of Frederick's court was French, the language of high society all over Europe.[21]

Euler stuck *that* out for 25 years, through all the horrors of the Seven Years War, when foreign armies twice occupied Berlin, and one in ten of Frederick's subjects died of hunger, disease, or by the sword. By then a second Catherine, Catherine the Great, was on the throne of Russia. (It is interesting that for two-thirds of the eighteenth century—67 years out of 100—Russia, one of the most difficult nations to govern, was ruled by women, for the most part very successfully.) Catherine showed every sign of being an enlightened monarch, firmly in control of her throne. She was, furthermore, a German princess, and it is possible Euler had some acquaintance with her at Frederick's court before she was shipped off to St. Petersburg to marry Peter the Great's grandson. Be that as it may, he left the genteel intrigues of Sans Souci to resume his position in St. Petersburg—which, incredibly, had been held open for him. He spent his last 17 years in Russia, productive to the end, and died in an instant, in full possession of all his powers but sight, at age 76, with a grandchild on his knee.

VIII. I have had to restrain myself considerably in this sketch of Leonhard Euler, because he is one of my favorite people in the history of mathematics for a number of reasons. One is that his work is a pleasure to read. Euler always expresses himself briefly and clearly, without any fuss, and without much of that polishing that Gauss went in for. Euler wrote mainly in Latin, but this is not much of an obstacle to appreciating him, as he had a spare and utilitarian style.[22]

Euler's crystal-clear Latin makes one realize what western civilization lost when scholars ceased writing in that language. Gauss was the last important mathematician to do so; this was one of those changes that came upon us after the Napoleonic wars. It is a curious thing that while the Congress of Vienna, which marked the end of those wars, was a gathering of reactionaries intent on restoring the *status quo ante* to Europe, in fact the wars had changed everything, and nothing could be the same after them. The historian Paul Johnson has written a good book about this, *Birth of the Modern.*

Another reason I find Euler so attractive is that, without being striking or eccentric or interesting in any particular way, he was a very admirable human being. When you read about his life you get a strong impression of serenity and inner strength. Euler lost the sight in his right eye when he was barely 30 (the heartless Frederick called him "My Cyclops") and went completely blind in his early 60s. Neither the partial nor the full disability seems to have slowed him down a bit. Of his thirteen children, only five survived into adolescence, and only three outlived him. His wife Catherine died when Euler was 69; a year later he remarried—to another Gsell, Catherine's half-sister.

He loved children, and it is reported that he could do serious mathematics with infants playing at his feet. (As a writer working at home, with two small children running around, this is very impressive indeed to me.) He seems to have been incapable of intrigue, seems never to have lost a friend other than by death, and was frank in all his dealings—though, if Strachey is to be believed, willing to bend his principles a little for the sake of a quiet life.[23] He wrote one of the first pop-science bestsellers, *Letters to a German Princess*, explaining to ordinary readers why the sky is blue, why the moon looks larger when it rises, and similar points of common bafflement.[24]

Underneath it all was a rock-solid religious faith. Euler had been raised a Calvinist and never wavered in his belief. His father, like Riemann's, had been the pastor of a village church, and Euler, like Riemann, had originally been intended for a clerical career. We are told that while living in Berlin, "He assembled the whole of his family every evening, and read a chapter of the Bible, which he accompanied with an exhortation." This, while attending a court at which, according to Macaulay, "the absurdity of all the religions known among men was the chief topic of conversation." Hardworking, pious, stoical, devoted to his family, plain-living and plain-spoken—no wonder Frederick didn't like him. But it is time to turn from the life to the work, and to look at Euler's first great triumph, the Basel problem.

5 RIEMANN'S ZETA FUNCTION

I. *—The Basel Problem—*

Find a closed form for the infinite series

$$1+\frac{1}{2^2}+\frac{1}{3^2}+\frac{1}{4^2}+\frac{1}{5^2}+\frac{1}{6^2}+\frac{1}{7^2}+\cdots$$

The Basel problem[25] is named from the Swiss city in whose university two of the Bernoulli brothers successively served as professor of mathematics (Jakob, 1687–1705, Johann, 1705–1748). I mentioned in Chapter 1.iii that both Bernoullis found proofs for the divergence of the harmonic series. In the book where he published his brother's proof, and then his own, Jakob Bernoulli stated the above problem and asked anyone who could figure it out to tell him the answer. (I shall explain the term "closed form" in just a moment.)

Notice that the series the Basel problem is concerned with—I shall call it "the Basel series"—is not far removed from the harmonic series. Each term is, in fact, the square of the corresponding term in the harmonic series. Now, if you square a number smaller than 1, you get a still smaller number; the square of one-half is one-quarter, which is smaller. The smaller the number you start with, the stronger is this effect; one-quarter is only modestly smaller than one-half, but the square of one-tenth is one-hundredth, which is a *lot* smaller than one-tenth.

Every term in the Basel series is, therefore, smaller than the corresponding term in the harmonic series, and as you go along they get *much* smaller. Since the harmonic series only barely diverges, it is not too much to hope that the Basel series, made up of smaller, and then much smaller, terms, *con*verges. Calculation suggests that this is indeed so. The sum of the series to 10 terms is 1.5497677…, the sum to 100 terms is 1.6349839…, the sum to 1,000 terms is 1.6439345…, and the sum to 10,000 terms is 1.6448340…. It really does seem to be converging to some number in the neighborhood of 1.644 or 1.645. But what number?

In situations like this, mathematicians are not satisfied just to get an approximation, especially when the series under investigation converges rather slowly, as this one does. (That sum to 10,000 terms is still 0.006 percent short of the true, final, infinite sum, which is 1.6449340668….) Is the answer a fraction, $\frac{9108}{5537}$ perhaps, or $\frac{560837199}{340948133}$? Or something more complicated, perhaps involving roots, $\sqrt{\frac{46}{17}}$, or the fifth root of $\frac{11983}{995}$, or the eighteenth root of 7766? What *is* it? A lay person might think that it would be satisfying enough to know the number to half a dozen places of decimals. No: mathematicians want to know it *exactly*, if they can. Not just because they are weird obsessives, but because they know from experience that getting that *exact* value often opens unexpected doors and throws light on the underlying math. The mathematical term of art for this exact representation of a number is "closed form." A mere decimal approximation, however good, is an "open form." The number 1.6449340668… is an open form. Look—those three dots tell you that it is open at the right-hand end, open for you to compute a few more digits, if you feel like it.

That was the Basel problem: to find a closed form for the series of reciprocal squares. The problem was finally cracked in 1735, 46 years after being posed, by the young Leonhard Euler, toiling away in St. Petersburg. The astonishing answer was $\pi^2/6$. This is the familiar π, the magic number 3.14159265…, the ratio of a circle's circumference to its diameter. What is it doing in a question that has nothing to do with circles, or with geometry at all? This is not very astounding to modern mathematicians, who are used to seeing π turn up all over the place, but it was very striking in 1735.

The Basel problem opens the door to the zeta function, which is the mathematical object the Riemann Hypothesis is concerned with. Before we can pass through that door, though, I must recapitulate some essential math: powers, roots, and logs.

II. Powers arise in the first place from repeated multiplication. The number 12^3 is $12 \times 12 \times 12$, with three multiplicands; 12^5 is $12 \times 12 \times 12 \times 12 \times 12$, with five. What happens if I multiply 12^3 by 12^5? That would be $(12 \times 12 \times 12) \times (12 \times 12 \times 12 \times 12 \times 12)$, which of course is 12^8. I just add the powers, $3 + 5 = 8$. This is the first great rule of powers.

$$\textit{Power Rule 1:} \quad x^m \times x^n = x^{m+n}$$

(Let me just add here that the whole of this section is only concerned with positive values of x. Raising zero to powers is mostly a waste of time, and raising negative numbers to powers brings up tricky problems I shall deal with later.)

What happens if I divide 12^5 by 12^3? That is $(12 \times 12 \times 12 \times 12 \times 12) / (12 \times 12 \times 12)$. I can cancel out three of the twelves top and bottom, leaving 12×12, which, of course, is 12^2. You can see that this is equivalent to just subtracting the powers.

$$\textit{Power Rule 2:} \quad x^m \div x^n = x^{m-n}$$

Suppose I cube 12^5: $(12 \times 12 \times 12 \times 12 \times 12) \times (12 \times 12 \times 12 \times 12 \times 12) \times (12 \times 12 \times 12 \times 12 \times 12)$ is 12^{15}. Here the powers are being multiplied.

$$\textit{Power Rule 3:} \quad (x^m)^n = x^{m \times n}$$

These are the most fundamental rules for powers. I shall refer to them as "Power Rule 1" and so on throughout this book, without further explanation. I am not quite through with power rules, though. I need to add a few more, because so far I have used only powers that are positive whole numbers. What about negative powers and fractional powers? What about a zero power?

To take the last first, if a^0 is going to mean anything at all, it might as well be consistent with the power rules I already have, since they are so commonsensical. Suppose I put m equal to n in Power Rule 2; then the right-hand side will indeed be a^0. The left-hand side will be $a^m \div a^m$. Now, if I divide anything by itself, the answer is 1.

$$\textit{Power Rule 4:} \quad x^0 = 1 \text{ for any positive number } x$$

Power Rule 2 can also be used to give meaning to negative powers. Divide 12^3 by 12^5; by Power Rule 2, the answer should be 12^{-2}. The answer is in fact $(12 \times 12 \times 12)/(12 \times 12 \times 12 \times 12 \times 12)$, which, canceling out three 12s top and bottom, is $\frac{1}{12^2}$.

$$\textit{Power Rule 5:} \quad x^{-n} = \frac{1}{x^n} \text{ (and in particular, } x^{-1} = \frac{1}{x}\text{)}$$

Power Rule 3 gives a clue to what fractional powers ought to mean. What could I do with $x^{\frac{1}{3}}$? Well, I could cube it; and if I did, by Power Rule 3, I ought to get x^1, which is just x. Therefore, $x^{\frac{1}{3}}$ is just the cube root of x. (Definition of "cube root of x": That number which, if cubed, gives x.) Power Rule 3 then tells us the meaning of any fractional power; $x^{\frac{2}{3}}$ is the cube root of x, squared—or the cube root of x^2, which works out to the same thing.

Power Rule 6: $x^{\frac{m}{n}}$ is the n-th root of x^m

Since 12 is 3×4, it follows that 12^5 is $(3 \times 4) \times (3 \times 4) \times (3 \times 4) \times (3 \times 4) \times (3 \times 4)$. This can be rearranged as $(3 \times 3 \times 3 \times 3 \times 3) \times (4 \times 4 \times 4 \times 4 \times 4)$. In a nutshell, $12^5 = 3^5 \times 4^5$. This is generally true.

Power Rule 7: $(x \times y)^n = x^n \times y^n$

What about raising x to an irrational power? What would $12^{\sqrt{2}}$ mean, or 12^{π}, or 12^e? Here we are back in the realm of analysis. Recall that sequence from Chapter 1.vii, the one that converges to $\sqrt{2}$. It looked like this: $\frac{1}{1}$, $\frac{3}{2}$, $\frac{7}{5}$, $\frac{17}{12}$, $\frac{41}{29}$, $\frac{99}{70}$, $\frac{239}{169}$, $\frac{577}{408}$, $\frac{1393}{985}$, $\frac{3363}{2378}$, ... By taking the sequence far enough, you can get as close as you please to $\sqrt{2}$. Now, since Power Rule 6 tells me the meaning of any fractional power, I can work out 12 to the power of any of those fractions. Of course, 12^1 is 12. And $12^{\frac{3}{2}}$ is the square root of 12, cubed: 41.569219381.... And $12^{\frac{7}{5}}$ is the fifth root of 12, raised to the seventh power, which comes out to 32.423040924.... Similarly, $12^{\frac{17}{12}}$ is 33.794038815..., $12^{\frac{41}{29}}$ is 33.553590738..., $12^{\frac{99}{70}}$ is 33.594688567..., and so on. As you can see, these fractional powers of 12 are closing in on a number—actually, the number 33.588665890.... Since the fractions themselves close in on $\sqrt{2}$, I am highly justified in saying that $12^{\sqrt{2}} = 33.588665890...$

Given a positive number x, I can, therefore, raise x to any power at all—positive, negative, fractional, or irrational; and doing so always obeys the Power Rules I have stated, because I rigged my definitions to make sure of that! Figure 5-1 shows graphs of x^a for various numbers a, ranging from -2 to 8. Notice particularly the zero-th power of x, which is just a horizontal line at height 1 above the x-axis—what mathematicians call "a constant function" (and Intensive Care Unit nurses call "a flat trace"). For every argument x, the function value is 1. Notice also how fast the whole-number powers of x

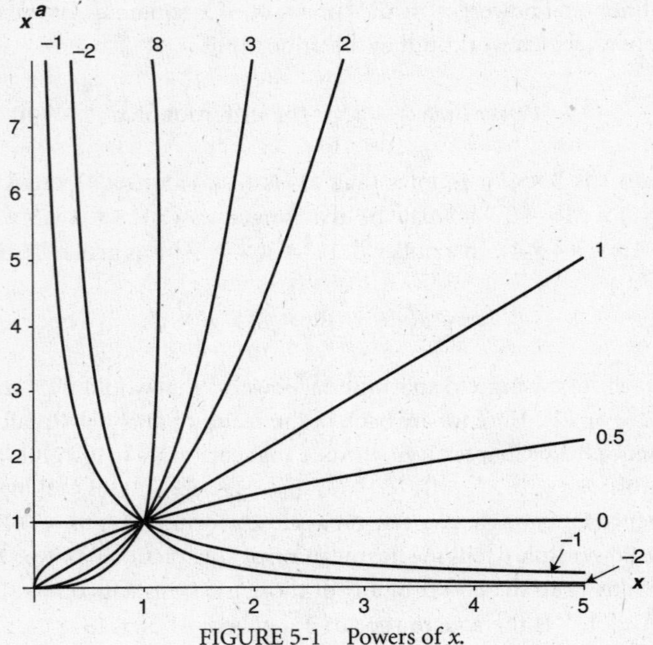

FIGURE 5-1 Powers of *x*.

(x^2, x^3, x^8) increase; and, much more to the point of this book, how *slowly* positive fractional powers like $x^{0.5}$ do so.

III. Raising numbers to powers—the proper term is "exponentiation"—is, in the beginning, analogous to multiplication. Multiplication is first presented as repeated addition: $12 \times 5 = 12 + 12 + 12 + 12 + 12$. Then you move on to a higher level and learn how to do $12 \times 5\frac{1}{2}$, which is a bit more than just repeated addition. So it is with powers. We can define 12^5 very easily. It's repeated multiplication, $12 \times 12 \times 12 \times 12 \times 12$. To get to grips with $12^{5\frac{1}{2}}$ needs more explanation—the explanation I attempted to provide in the previous section.

As I said before, mathematicians love to invert expressions. I have an expression for P in terms of Q? All right, let's see if I can get Q in terms of P. This is where the analogy between exponentiation and multiplication breaks down. Inverting multiplication is easy. If $x = a \times b$, then $a = x \div b$ and $b = x \div a$. Division provides a complete solution to the problem of inverting multiplication.

The analogy breaks down there because, while $a \times b$ is always, invariably and infallibly, equal to $b \times a$, it is unfortunately not true, except occasionally and accidentally, that $a^b = b^a$. (The only whole-number case with different a and b is $2^4 = 4^2$.) For instance, 10^2 is 100, but 2^{10} is 1,024. If I seek to invert $x = a^b$, therefore, I am going to need two different methods: one to get a in terms of x and b, another to get b in terms of x and a. The first is a breeze. Raising both sides to the power $\frac{1}{b}$, Power Rule 3 gives me $a = x^{\frac{1}{b}}$, which, by Power Rule 6, means that a is the b-th root of x. But what is b in terms of x and a? The Power Rules offer no clue.

This is where logs make their appearance. The answer is, b is the log of x to base a. That is just the definition of log. The log of x to base a (generally written "$\log_a x$") is defined to be the number b that makes $x = a^b$ true. From this flows the whole family of log functions: log of x to base 2, log of x to base 10 (which older readers will remember as an aid to calculation taught in high schools up to about 1980), and so on. I could present them all in graphs, as I did the graphs of x^a in Figure 5-1.

I am not going to do this because I am deeply indifferent to all members of the log family except one, log to base e, where e is the extremely important, though unfortunately irrational, number 2.71828182845.... Log to the base e is the only kind of log I care about, and the only kind I shall use in this book. In fact, I shall not say "log to base e" any more, just "log." So what is the log of x? By the above definition, it is the number b that makes $x = e^b$ a true statement.

Since $\log x$ is the b that makes $x = e^b$ a true statement, it is obvious that $x = e^{\log x}$. This is just the definition of "log x" written mathematically; but it is so important in what follows that I am going to make a rule out of it.

$$\textit{Power Rule 8:} \quad x = e^{\log x}$$

That is true for every positive number x. The log of 7, for example, is 1.945910..., because $7 = 2.718281^{1.945910}$, to six decimal places. Negative numbers don't have logs (though this is another thing I reserve the right to change my mind about later); zero doesn't have a log, either. There is no power you can raise e to with a negative or zero result. The domain of the log function is all positive numbers.

The log function is everywhere in this region of math. We have already seen it in Chapter 3.viii-ix, in the Prime Number Theorem and its equivalents. It will show up again and again in this book in everything to do with prime numbers and the zeta function.

With the log function all over like this, I should give some more detailed coverage of it. Figure 5-2 is a graph[26] of $\log x$, for arguments out to 55. I've particularly marked the function values for arguments 2, 6, 18, and 54. These arguments go up in multiples of 3; and you can see from the graph that the corresponding function values go up in equal steps—that is, by addition. That's the point I made about the log function in Chapter 3.viii.

It's worth enlarging on a little. The great thing about the log function is that it turns multiplication into addition. Look at those lines I marked on the graph. The arguments are 2, 6, 18, 54—I start with 2, multiply by 3, multiply by 3 again, then multiply by 3 again. The function values, holding myself to four places of decimals here, and putting up with a small rounding error, are 0.6931, 1.7918, 2.8904, 3.9890—which start with 0.6931, add 1.0987, add 1.0986, then add 1.0986 again. The log function turned multiplication (by 3) into addition (of log 3, which is 1.09861228866810...).

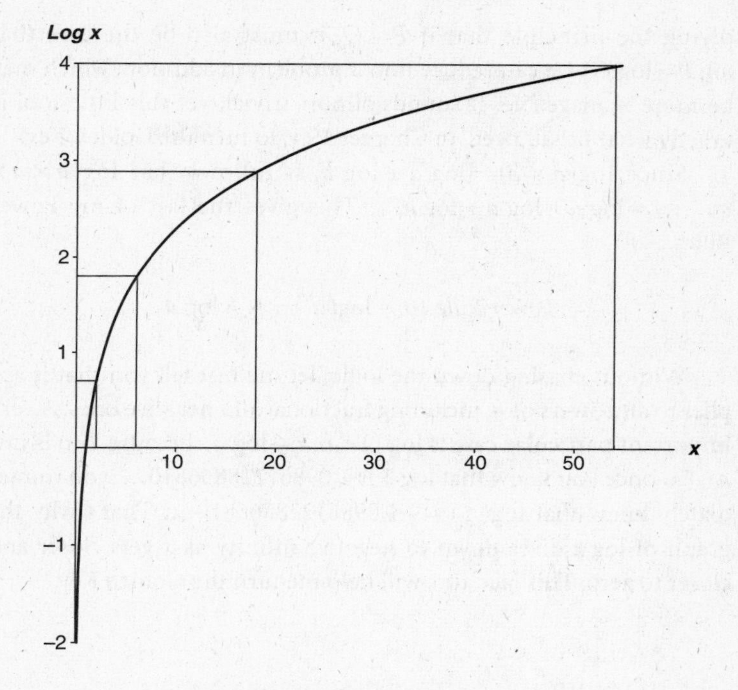

FIGURE 5-2 The log function.

This follows from the definition of log x and from the Power Rules. From Power Rule 8, if a and b are any two positive numbers, $a \times b = e^{\log a} \times e^{\log b}$. But from Power Rule 1, I can replace the right-hand side like this, $a \times b = e^{\log a + \log b}$. However, $a \times b$ is just a number itself, and so, from Power Rule 8 again, $a \times b = e^{\log (a \times b)}$. Equating the two different expressions I just got for $a \times b$ gives a new Power Rule.

$$\textit{Power Rule 9:} \quad \log(a \times b) = \log a + \log b$$

This is a wonderful thing. It means that, when faced with a difficult problem involving multiplication, by "taking logs" (i.e., by applying the principle that if $P = Q$, it must also be the case that $\log P = \log Q$) we can reduce it to a problem in addition, which may be more manageable. It sounds almost trivial; yet this little tool is exactly what I shall need, in Chapter 19.v, to turn the Golden Key.

Since $\log(a \times b) = \log a + \log b$, it follows that $\log(a \times a \times a \times \ldots) = \log a + \log a + \log a \ldots$. This gives the last of my Power Rules.

$$\textit{Power Rule 10:} \quad \log(a^N) = N \times \log a$$

Without chasing down the logic, let me just tell you that it applies to all powers of a, including fractional and negative ones. A very important particular case is $\log(1/a) = -\log a$, because $1/a$ is just a^{-1}. So once you know that log 3 is 1.09861228866810..., you immediately know that $\log(\frac{1}{3}) = -1.09861228866810\ldots$. That's why the graph of $\log x$ dives down to negative infinity as x gets closer and closer to zero. This fact, too, will help me turn the Golden Key.

IV. Log x increases slowly, as you can see. The slowness with which $\log x$ increases is a very fascinating and important thing all by itself. The main point is that $\log x$ increases slower than any power of x. At first thought, that might seem to be very obvious. When I say "power of x," you probably think of squares and cubes; and you know that a graph of the squaring function or the cubing function zooms up out of sight as the argument increases, way beyond the feeble inching-up of the log function. True, but that's not the point. What I have in mind here is not a power like this, x^2, or like this, x^3, but rather a power like this, $x^{0.1}$.

Figure 5-3 shows some graphs of x^a for small numbers a. I've chosen $a = 0.5$, 0.4, 0.3, 0.2, and 0.1 (with the log function—dotted line—for comparison). You can see that the smaller a is, the flatter the graph of x^a is. You can also see that for values of a below a certain

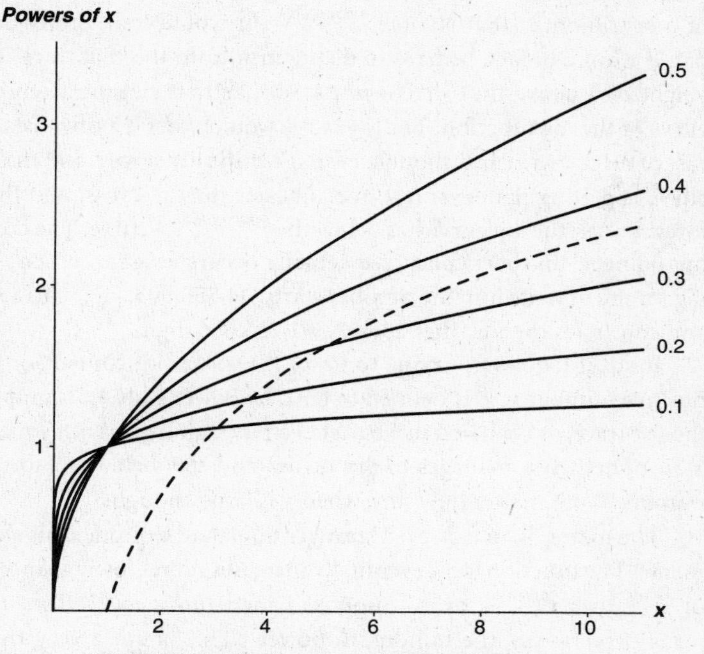

FIGURE 5-3: The functions x^a, for small positive a.

point (actually, for a less than $1/e$, which is 0.3678794...) the log curve cuts the curve x^a not very far east (never further than e^e, which is 15.1542...).

Well, no matter how small you make a, the graph of log x is eventually flatter than the graph of x^a. If a is bigger than $1/e$, this is true already, even in this diagram. If a is less than $1/e$, then by going far enough east—by taking a big enough argument x—the log x curve eventually cuts the x^a curve *again*, and then, forever after, lies below it.

Of course, you might have to go some way out. The log curve recrosses the $x^{0.3}$ curve slightly east of $x = 379$; it re-crosses the $x^{0.2}$ curve around $x = 332{,}105$; it does not re-cross the $x^{0.1}$ curve until past $x = 3{,}430{,}631{,}121{,}407{,}801$. If I were to plot the graph of x to the power of one trillionth (that is, of $x^{0.000000000001}$), it would look pretty darn flat. It would, in fact, be hard to distinguish from the "flat trace" at a height of 1 above the x-axis—not at all like that elegant ascending curve of the log function. The log curve would cross it a tiny distance east of e. It's increasing, though, even if terrifically slowly, and the log curve is getting flatter; and sooner or later they re-cross, and then, forever after, the log curve lies below the $x^{0.000000000001}$ curve. The crossing point in this particular case actually occurs at an argument too big for me to write out; the number starts: 44,556,503,846,304,183 ... and continues for a further 13,492,301,733,606 digits.

It's as if log x were trying to be x^0. It is *not* x^0, of course; for any positive number x, x^0 is defined to be 1, by Power Rule 4. Its graph is the flat trace, as I showed earlier. Yet even though log x is not equal to x^0, it nonetheless manages to dip below, and stay below, x^{ε}, for any number ε, no matter how tiny, when x is large enough.[27]

The matter is, in fact, even stranger than that. Consider this statement: "The function log x eventually increases more slowly than $x^{0.001}$, or $x^{0.00001}$, or $x^{0.0000001}$, or...." Suppose I raise *this whole statement* to some power—say, the hundredth power. (This is not a very rigorously mathematical procedure, I admit, but it gives a true result.) Applying Power Rule 3, the statement will then read: "The function $(\log x)^{100}$ eventually increases more slowly than $x^{0.1}$, or $x^{0.001}$, or $x^{0.00001}$, or...." In other words, since log x increases more slowly than any power of x, *the same is true of any power of* log x. Each one of the functions $(\log x)^2$, $(\log x)^3$, $(\log x)^4$, ..., $(\log x)^{100}$, ..., increases more slowly than any power of x. *Any power of* log x *eventually increases more slowly than any power of* x. The graph of $(\log x)^N$ will eventually drop below, and for ever after stay below, the graph of x^{ε}, no matter how big N is or how small ε is.

This is hard to visualize. Those functions $(\log x)^N$ increase fast, and then *very* fast. Still, if you go far enough out to the east in Figure 5-3, every one of them will eventually, at some argument of stupendous size, drop below the $x^{0.3}$ curve, the $x^{0.2}$ curve, the $x^{0.1}$ curve, and any other curve of this family you might care to draw. You need to go out east to the neighborhood of $x = 7.9414 \times 10^{3959}$ before $(\log x)^{100}$ drops below the $x^{0.1}$ curve; but eventually it does.

V. Some of this I am going to use right away; some I shall leave lying here for future reference. All of it is important to the understanding of the Riemann Hypothesis, and I urge you to try out some of the main points, to check your understanding of them, before proceeding. A pocket calculator is good for this. You might, for example, find log 2 (which is 0.693147...) and log 3 (which is 1.098612...) and confirm that by adding them together you do indeed get log 6 (which is 1.791759...). Please notice, however, that because of the older use for base-10 logs that I mentioned, the "log" key on many pocket calculators delivers log to base 10. For the only log I care about, log to base *e*, such calculators generally provide an alternative key labeled "ln." That's the key you need. (The "n" stands for "natural"; log to base *e* is properly called "the natural log.")

Now, let's return to the Basel problem.

VI. As an illustration of what I said in section I about the search for closed-form solutions yielding important insights, Euler's solution of the Basel problem not only gave a closed form for the reciprocal-squares series; as a by-product, it also gave closed forms for $1+\frac{1}{2^4}+\frac{1}{3^4}+\frac{1}{4^4}+\frac{1}{5^4}+\ldots$, $1+\frac{1}{2^6}+\frac{1}{3^6}+\frac{1}{4^6}+\frac{1}{5^6}\ldots$, and so on. So long as N is an even number, Euler's result tells you the precise value, as a closed form, of the infinite series shown in Expression 5-1.

$$1+\frac{1}{2^N}+\frac{1}{3^N}+\frac{1}{4^N}+\frac{1}{5^N}+\frac{1}{6^N}+\frac{1}{7^N}+\frac{1}{8^N}+\frac{1}{9^N}+\frac{1}{10^N}+\frac{1}{11^N}+\cdots$$

Expression 5-1

When N is 2, the series converges on $\pi^2/6$, as I said. When N is 4, it converges on $\pi^4/90$; when N is 6 it converges on $\pi^6/945$, and so on. Euler's argument provided an answer for every even number N. He himself, in a later publication, took the argument all the way out to $N = 26$, when the series converges to $1315862\,\pi^{26}/11094481976030578125$.

But what if N is odd? Euler's result has nothing to say about that. Neither has any other result in the 260-odd years since. We have no clue about the closed form for $1+\frac{1}{2^3}+\frac{1}{3^3}+\frac{1}{4^3}+\frac{1}{5^3}+\ldots$, if there even is one, nor the equivalent for any other odd number. Nobody has been able to find closed forms for these series. We know that they converge, and we can of course, by brute calculation, get their values to any required degree of accuracy. We just don't know what they mean. They are, in fact, very difficult numbers. It was not until 1978 that $1+\frac{1}{2^3}+\frac{1}{3^3}+\frac{1}{4^3}+\frac{1}{5^3}+\ldots$ was even proved irrational.[28]

So by the middle of the eighteenth century, quite a lot of mathematicians were thinking about the infinite series in Expression 5-1. Precise values—closed forms—were known for all even numbers N, while for the odd numbers, approximate values could be got by just adding up enough terms. Remember that when N is 1, the series is just the harmonic series, which diverges. Table 5-1 shows values for Expression 5-1—which, just to remind you, is $1+\frac{1}{2^N}+\frac{1}{3^N}+\frac{1}{4^N}+\frac{1}{5^N}+\frac{1}{6^N}+\ldots$

TABLE 5-1

N	Value of Expression 5-1
1	(No value)
2	1.644934066848
3	1.202056903159
4	1.082323233711
5	1.036927755143
6	1.017343061984

(to 12 decimal places). This looks like one of those snapshots of a function that I spoke about in Chapter 3.iv. Sure enough, it is. Recall the statement of the Riemann Hypothesis that I gave in my Prologue.

The Riemann Hypothesis
All non-trivial zeros of the zeta function
have real part one-half.

Table 5-1 is your first glimpse of Riemann's zeta function, and therefore a first step toward understanding the Riemann Hypothesis.

VII. Since, in the earlier sections of this chapter, I went to the trouble of defining the meaning of "x^a" for any number *a*, not just whole numbers, I am under no obligation to restrict the number *N* in Expression 5-1 to whole numbers. I can, in my imagination, let it roam freely over fractions, negative numbers, and irrational numbers. There is no guarantee that the infinite series will converge for all numbers—we already know from Chapter 1.iii that it doesn't when $N = 1$. But we can at least entertain the possibility.

In honor of this new realization, I am going to change the "*N*" to a different letter, one that has less traditional association with whole numbers. The obvious choice is, of course, "*x*." Riemann himself, however, did not use "*x*" in his 1859 paper. These matters were not so settled in his day. He used "*s*" instead; and so momentous was that 1859 paper that every succeeding mathematician has followed him. In studies of the zeta function, the argument is always given as "*s*."

Here then, at last, is the Riemann zeta function (zeta, written "ζ," being the sixth letter of the Greek alphabet).

$$\zeta(s) = 1 + \frac{1}{2^s} + \frac{1}{3^s} + \frac{1}{4^s} + \frac{1}{5^s} + \frac{1}{6^s} + \frac{1}{7^s} + \frac{1}{8^s} + \frac{1}{9^s} + \frac{1}{10^s} + \frac{1}{11^s} + \cdots$$

Expression 5-2

VIII. Before going any further, let me introduce a handy mathematical notation that cuts down on typing. (Do you think it's easy, getting stuff like Expression 5-2 into Microsoft Word?)

If mathematicians want to add up a lot of terms that all have the same pattern, they use the Σ sign. That's a capital sigma, the eighteenth letter of the Greek alphabet, the Greek "s" (for "sum"). The way it works is, you stick the pattern "under" (which actually means to the right, though we illogically say "under") the sigma sign. Then at bottom and top of the sigma, you declare where your sum will start and end. This expression, for example,

$$\sum_{n=12}^{n=15} \sqrt{n}$$

is mathematicians' shorthand for $\sqrt{12}+\sqrt{13}+\sqrt{14}+\sqrt{15}$. The sigma says "add 'em up"; the expressions at the top and bottom of the sigma tell us when to start and when to finish adding; and the expression "under" (to the right of) the sigma tells us what, exactly, is being added—in this case, $\sqrt{n}$.

Mathematicians are not especially strict about the style of these expressions. That one, for example, would probably be written

$$\sum_{12}^{15} \sqrt{n}$$

since it is obvious that it must be "n" that's going from 12 to 15. Now, using the sigma sign, I can save myself a lot of fiddling around with symbols by rewriting Expression 5-2 as

$$\zeta(s)=\sum_{n=1}^{\infty}\frac{1}{n^s}$$

Or equivalently, bearing in mind Power Rule 5,

$$\zeta(s)=\sum_{n=1}^{\infty}n^{-s}$$

In fact, since "*n*" is so commonly and obviously used to stand for the positive whole numbers 1, 2, 3, 4, ..., mathematicians are generally even more terse, just writing

$$\zeta(s) = \sum_n n^{-s}$$

which, again, is Riemann's zeta function. This is pronounced "Zeta of *s* is defined to be the sum over all *n* of *n* to the power of minus *s*." Here, "all *n*" is understood to mean "all positive whole numbers *n*."

IX. Having got the zeta function set up as a neat expression, let's turn our attention to that argument "*s*." We know, from Chapter 1.iii, that when *s* is 1, the series diverges, so that the zeta function has no value. When *s* is 2, 3, 4, ... it always *con*verges, though, and we get values for the zeta function (see Table 5-1). In fact, you can show that the series converges for *any* number bigger than 1. When *s* is 1.5, it converges to 2.612375.... When *s* is 1.1, it converges to 10.584448.... When *s* is 1.0001 it converges to 10,000.577222.... It might seem odd that the series diverges when $s = 1$ but yet manages to converge for $s = 1.0001$. This is a common situation in math, though. In fact, when *s* gets very close to 1, the zeta function behaves remarkably like $1/(s-1)$. This, too, has a value for any number *s* except when *s* is precisely equal to 1, because the denominator is then zero, and you can't divide by zero.

Perhaps a graph will make things clearer. Figure 5-4 is a graph of the zeta function. You can see that as *s* approaches the number 1 from the right, the function value shoots up to infinity; and as *s* itself goes off to infinity at the far right, the function value gets closer and closer to 1. (I've drawn in the line $s = 1$ and the constant function 1, both dashed.)

The graph doesn't show any part of the function to the left of the line $s = 1$. That's because so far I've been assuming that *s* is greater

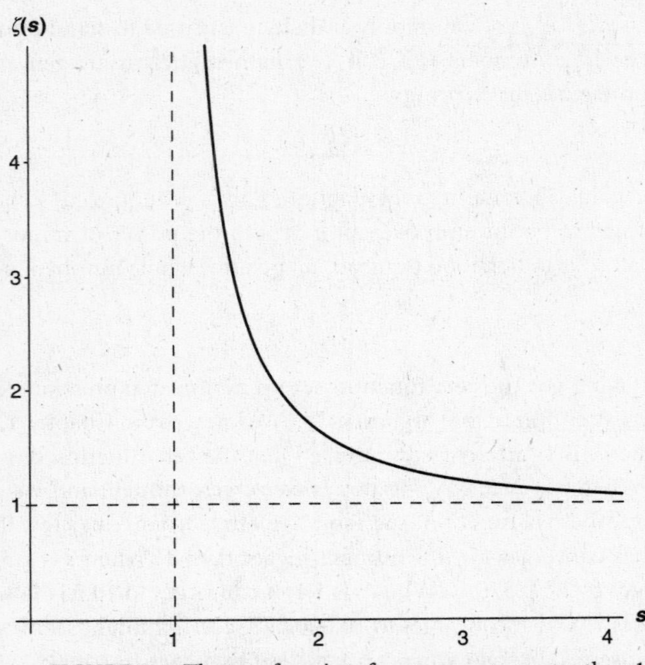

FIGURE 5-4 The zeta function, for arguments greater than 1.

than 1. What if it isn't? What if, for example, *s* is zero? Well, then Expression 5-2 would look like this:

$$\zeta(0) = 1 + \frac{1}{2^0} + \frac{1}{3^0} + \frac{1}{4^0} + \frac{1}{5^0} + \frac{1}{6^0} + \frac{1}{7^0} + \frac{1}{8^0} + \frac{1}{9^0} + \frac{1}{10^0} + \frac{1}{11^0} + \cdots$$

By Power Rule 4, this sum is 1 + 1 + 1 + 1 + 1 + 1 +..., which pretty obviously diverges. Add a hundred terms, the sum is a hundred; add a thousand, the sum is a thousand; add a million, the sum is a million. Yep, it diverges.

For negative numbers, things are even worse. What value does Expression 5-2 have if *s* is −1? From Power Rule 5, 2^{-1} is just $\frac{1}{2}$, 3^{-1} is just $\frac{1}{3}$, and so on. Since $1/\frac{1}{2}$ is just 2, $1/\frac{1}{3}$ is just 3, etc., the series looks like this, 1 + 2 + 3 + 4 + 5 +.... Definitely divergent. How about $x = \frac{1}{2}$? Since $2^{\frac{1}{2}}$ is just $\sqrt{2}$ etc., the series is

$$\zeta(\tfrac{1}{2}) = 1 + \frac{1}{\sqrt{2}} + \frac{1}{\sqrt{3}} + \frac{1}{\sqrt{4}} + \frac{1}{\sqrt{5}} + \frac{1}{\sqrt{6}} + \frac{1}{\sqrt{7}} + \frac{1}{\sqrt{8}} + \cdots$$

Since the square root of any whole number is smaller than the number, each term in this series is bigger than the corresponding term in $1 + \frac{1}{2} + \frac{1}{3} + \frac{1}{4} + \frac{1}{5} + \frac{1}{6} + \frac{1}{7} + \ldots$. (Basic algebra: if a is smaller than b, then $1/a$ is bigger than $1/b$. For example, 2 is smaller than 4, but $\frac{1}{2}$ is bigger than $\frac{1}{4}$.) That series diverges, so this one must, too. Sure enough, if you take the trouble to actually work out the sums and add them up, you see that the first ten terms add up to 5.020997899..., the first hundred add up to 18.589603824..., the first thousand add up to 61.801008765..., the first ten thousand to 198.544645449...and so on.

It seems that the graph shows all that can be shown of the Riemann zeta function. There's no more. The function has values only when s is greater than 1. Or, as we now know to say, using the proper term of art, the domain of the zeta function is all numbers greater than 1. Right? Wrong!

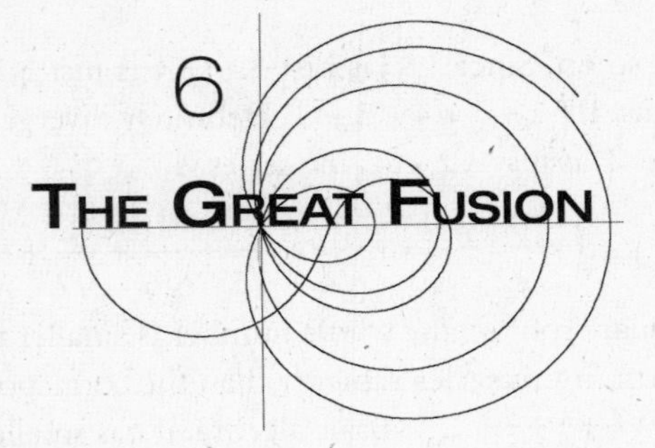

6 The Great Fusion

I. The Chinese word *Taiye* (pronounced "tie-yeah") translates literally as "ultimate grandpa." It is the title given in my wife's family to her paternal grandfather. Visiting China in the summer of 2001, our first duty was to call on *Taiye.* The family is immensely proud of him, for he has lived to age 97 in good health and with a clear head. "Ninety-seven years old now!" they all told me. "You should see him!" Well, I did see him—a fine cheerful Buddha of a man, his face glowing ruddy and his mind still sharp. Whether he was actually 97 at the time is, however, an interesting point.

Taiye was born on the third day of the twelfth lunar month of the lunar year named *yi si* in the traditional "Heaven-Earth" year-numbering system. This day was December 28, 1905, on the Western calendar. Since my visit occurred early in July 2001, *Taiye*'s age at the time was, in the modern western reckoning, 95½ years and a few days. So why was everyone telling me that he was 97? Because in the old Chinese style, which *Taiye* clove to, he was one year old at birth, and another one year old when each Lunar New Year rolled round—which one did, on January 24, 1906, by our calendar, 27 days after his birth. Not yet one month in the world, and he was already two years old! Thus, when the Lunar New Year arrived in 2001 (also on January 24, as it happened, though Lunar New Year can fall on any date between January 21 and February 20), Ultimate Grandpa hailed it as his 97th year.

There is nothing wrong with the logic behind this traditional Chinese system of reckoning age. You come into the world on a certain day. That day belongs to a certain year. Obviously, that is your first year. If, 28 days later, a new year dawns—well, that will be your second year. It all makes perfect sense. The only reason it seems odd is that in the matter of computing our ages, modern people (in China as well as in the West) have got accustomed to dealing with time as something to be *measured*. In *Taiye*'s young days, the Chinese thought of a person's age as something to be *counted*.

II. This distinction between numbers for counting and numbers for measuring reaches deep into human habits of thought and speech. It is as if with one part of our minds we perceive the world as made up of distinct, solid objects that can be tallied; while with another, we see it as a collection of fabrics, grains, or fluids, to be divided up and measured. Keeping the two notions straight does not come easily. My son, six years old, still confuses "many" with "much." To a friend, after the Christmas festivities: "How much presents did you get?"

Our perceptions of the world are mirrored in our languages. The English language takes the world to be mainly a countable place: one cow, two fishes, three mountains, four doors, five stars. Somewhat less frequently, our language takes the world to be measurable: one blade of grass, two sheets of paper, three head of cattle, four grains of rice, five gallons of gasoline. The words "blade," "sheet," "head," "grain," and "gallon," though of course some of them have lives of their own, here are acting as units of measurement. The Chinese language, by contrast with English, takes very nearly the whole of creation to be measurable. One of the minor chores of learning Chinese is memorizing the right "measure word" (that is a precise translation of the Chinese grammatical term *liang ci*) for each noun: one head of cow, two sticks of fish, three plinths of mountain, four fans of door, five grains of star. In the entire Chinese language there are only two words that can *always* be let loose grammatically without a measure word: "day" and "year." Everything else—cows, fishes, mountains, doors, stars—is a kind of *stuff* that must be divided up and measured out before we can talk about it.

The much/many confusion has occasioned much argument and many inconveniences. At the time of the millennium, for example, which most of us celebrated when the year 1999 turned into the year 2000, there was an irritating minority of dissidents who said we had it all wrong. The source of their complaint was the true fact that our common calendar was set up without a year zero. The first day of the year 1 C.E. was preceded by the last day of the year 1 B.C.E. This was because Dionysus Exiguus, the sixth-century monk who imposed a Christian year-numbering system on the months and days of Julius Caesar's calendar, regarded years as countable things, just as our *Taiye* does. The first year of the Christian era was, therefore, to be the year 1, the second was to be year 2, and so on.

The error is easily understood. Look at a common desk ruler. (Not for the first time in this book. It is amazing how much math—even higher math—can be referred back to the marks on a $1.89 ruler.) Yes, there are 12 inches marked on it. Yes, you can count them: 1, 2, 3, 4, …, 12. Ah, but if you are an ant, and you begin walking from the left-hand end of the ruler to the right, and you have just covered the first half-inch, where are you? In the middle of the first inch? Yes. In the middle of inch 1, then? Sure, if you like. But what is the precise *measure* of the distance you have walked? Well, it is 0.5 inches. Since walking is a continuous process—since the ant will eventually traverse *every point* of the ruler—this is a much more interesting and important number for the mathematician. He therefore prefers to say that you are halfway (that is, .5 of the way) through the zero-th inch, giving a position 0.5.

Modern people are sufficiently sophisticated about mathematics that they think like this quite naturally most of the time. That, in fact, was the source of confusion for those millennium complainers—or for the revelers late on the night of December 31, 1999, depending on which point of view you want to take. The complainers were saying: "If you *measure* the time from the starting instant of the common era to the very end of the year 1999, you only have 1,999 complete years. You should wait until 2,000 complete years have elapsed." They were imposing measuring logic on a system created according to counting logic. The revelers, on the other hand, were saying: "Here comes year number 2,000! Whoopee!"—pure counting logic. Yet these same revelers might fall back on measuring logic if asked the age of their new baby: "Oh, he's just half a year old." Which is to say, his age is 0.5 years—measuring logic, at least by contrast with the traditional Chinese approach. (They might, of course, confuse the issue further by saying: "Six months. . . .")

I once got into a mild controversy with the writer and word-lover William F. Buckley, Jr., about the word "data." Is this a singular word or a plural word? The word originated with the Latin verb *dare*, "to give." From this, by the ordinary processes of Latin grammar, a gerund (that is, a verbal noun) can be formed: *datum*, meaning "that which is given." From this, in turn, you can make a plural: *data*—"those things that are given." However, we are speaking English, not Latin. Plenty of Latin plurals are used as English singulars—*agenda*, for example. Nobody says "The agenda are prepared." English is *our* language; if we borrow a word from another tongue, we may do with it as we please.

Having worked with data all my adult life, I know very well what it is. It is a *stuff*, made up of innumerable tiny particles, indistinguishable one from another—like rice, sand, or grass. This kind of stuff needs to be referred to, in English, with singular verb forms ("The rice is cooked") or measure words. If you want to pluck out one particle and address it, you use a measure word: "A grain of rice," "An item of data." This is, in fact, how people who make a living handling data *do* speak, by instinct. Among people whose business is data, nobody ever says "One datum, two data." If people did say this, nobody would understand them. The grammarians, however, still want us to say: "The data are. . . ." I predict they will lose the battle eventually.

As a final example, one that used to puzzle me in my Church of England schooldays, consider the three days that Jesus Christ lay in his tomb before being resurrected, according to his own prophecy: "After three days I will rise again." Three days? The Crucifixion occurred on a Friday—Good Friday. The Resurrection occurred on a Sunday. That's 48 hours, measure-wise, but of course 3 days (Friday, Saturday, Sunday) counting-wise, which is how the Hellenized intellectuals who compiled the New Testament reckoned it.

III.

The Riemann Hypothesis
All non-trivial zeros of the zeta function
have real part one-half.

The Riemann Hypothesis was born out of an encounter, what my chapter heading calls a great fusion, between counting logic and measuring logic. To put it in precise mathematical terms; it arose when some ideas from *arithmetic* were combined with some from *analysis* to form a new thing, a new branch of the mathematical tree, analytic number theory.

To summarize the traditional categories of mathematics that I gave in Chapter 1.viii.

- *Arithmetic*—The study of whole numbers and fractions.
- *Geometry*—The study of figures in space.
- *Algebra*—The use of abstract symbols to represent mathemati-

cal objects (numbers, lines, matrices, transformations), and the study of the rules for combining those symbols.

- *Analysis*—The study of limits.

This fourfold scheme was well established in people's minds around 1800, and the great fusion I am going to describe in this chapter was a fusion of ideas which, until 1837, had lived separate lives under two of the above headings, arithmetic and analysis. This fusion created the discipline of *analytic number theory.*

We are quite blasé about these leaps of imagination nowadays, and perhaps a little better at them. Today, in fact, as well as analytic number theory, there is an algebraic number theory and a geometric number theory. (I shall introduce some algebraic number theory in Chapter 20.v.) In the 1830s, however, it was a very striking thing, to yoke together concepts from two areas previously thought to be unconnected. Before I can introduce you to the principal player in this phase of the story, though, I need to say a little more about those two disciplines he brought together.

IV. At the time I am speaking of—the early nineteenth century—analysis was still the newest and sexiest branch of math, where the greatest advances were being made and the keenest minds were working. We knew more about arithmetic, geometry, and algebra at the end of the nineteenth century than we did at the beginning, but we knew *way* more about analysis. At the opening of that century, in fact, the most fundamental concept of analysis, the concept of a limit, was not clearly understood even by the best minds. If you had asked Euler, or even the young Gauss,what analysis was all about, he would have said: "It is about the infinite and the infinitesimal." If you had then asked Euler what, precisely, the infinite *is,* he would have had a coughing fit and left the room, or else opened a discussion about the meaning of "is."

Analysis really dates from the invention of calculus by Newton and Leibnitz in the 1670s. Certainly the idea of *limit*, the idea that separates analysis from the rest of math, is fundamental to calculus. If you ever sat through a calculus class at school, you probably have some dim memory of a graph showing a curve with a straight line intersecting it at two points. "Now," says the instructor, "if you let the two points come closer and closer together, *in the limit*..." and you forget the rest.

Calculus is not the whole of analysis—the divergence of the harmonic series is a theorem in analysis, but it does not belong to calculus, which did not exist in Nicole d'Oresme's time. There are other quite large areas of analysis that do not strictly belong in calculus. Measure theory, for example, developed by Henri Lebesgue in 1901, and a chunk of set theory. I think it's fair to say, though, that even these newer non-calculus areas of analysis were opened up with the idea of improving calculus—in Lebesgue's case, of getting a better definition of "integral."

The concepts that analysis deals with—"the infinite and the infinitesimal," as Euler would have said; "limits and continuity," his modern counterpart would insist—are among the most difficult for the human mind to grasp. This is why calculus is so fearsome to so many intelligent people. The causes of all the bafflement were stated very early on in the history of math—in about 450 B.C.E., by a Greek philosopher named Zeno. How (asked Zeno) is motion possible? How can we say that an arrow moves, if, at any given instant, it must be *somewhere*? If all time is composed of instants, and motion is not possible in any given instant, then how is motion possible at all?

In the early eighteenth century, when calculus first became known to the general educated public, the notion of infinitesimals came in for much scorn. The Irish philosopher George Berkeley (1685–1753—the California town is named after him) was a notable skeptic: "And what are these evanescent increments? They are neither finite quantities, nor quantities infinitely small, nor yet nothing. May we not call them the ghosts of departed quantities?"

The difficulty people have in grasping these ideas is a reminder that mathematical thinking is, at some level, deeply unnatural. It goes against all the grain of human thought and language. Never mind analysis, this is true even of basic arithmetic. In the preface to *Principia Mathematica*, Whitehead and Russell note that

> [T]he very abstract simplicity of the ideas of this work defeats language. Language can represent complex ideas more easily. The proposition "a whale is big" represents language at its best, giving terse expression to a complicated fact; while the true analysis of "one is a number" leads, in language, to an intolerable prolixity.

(They weren't kidding. *Principia Mathematica* takes 345 pages to define the number "1.")

This is surely right. A whale is, by any standard of complexity that makes sense, a vastly more complicated thing than "five," yet it is a much easier thing for the human mind to apprehend. Any tribe of human beings that was acquainted with whales would certainly have a word for them in their language; yet there are peoples whose language has no word for "five" even though five-ness is there, quite literally, at their fingertips! I repeat, mathematical thinking is a deeply unnatural way of thinking, and this is probably why it repels so many people. And yet, if that repulsion can just be overcome, what benefits flow! Consider the 2,000-year struggle to domesticate the concept of "zero"—a number widely accepted as mathematically legitimate only about 400 years ago. Where should we be nowadays without it?

Arithmetic, by contrast with analysis, is widely taken to be the easiest, most accessible branch of math. Whole numbers? Obviously useful for counting. Negative numbers? Indispensable if you want to know the temperature on a cold day. Fractions? Well, of course I know that a $\frac{3}{8}$ nut won't fit onto a $\frac{13}{32}$ bolt. If you gave me a little time with paper and pencil, I could probably tell you whether a $\frac{15}{23}$ nut could fit on a $\frac{29}{44}$ bolt. What's to be afraid of?

In fact, arithmetic has the peculiar characteristic that it is rather easy to state problems in it that are ferociously difficult to prove. It was in 1742 that Christian Goldbach put forward his famous conjecture that every even number greater than 2 can be expressed as the sum of two primes. Twenty-six decades of effort by some of the best minds on the planet have failed to prove or disprove this simple assertion (which has inspired at least one novel, Apostolos Doxiadis's *Uncle Petros and Goldbach's Conjecture*[29]). There are a thousand conjectures like this in arithmetic[30]; some proved, most still open.

This is undoubtedly what Gauss had in mind when he declined to enter into a prize competition for the solution of Fermat's Last Theorem. To Heinrich Olbers, who had urged him to compete, Gauss replied "I confess that Fermat's Theorem ... has very little interest for me, because I could easily lay down a multitude of such propositions, which one could neither prove nor dispose of."

Gauss's indifference is in this case a minority viewpoint, it must be said. A problem that can be stated in a few plain words, yet which defies proof by the best mathematical talents for decades or—in the case of Goldbach's Conjecture or Fermat's Last Theorem—for centuries, has an irresistible attraction for most mathematicians. They know that they can achieve great fame by solving it, as Andrew Wiles did when he proved Fermat's Last Theorem. They know, too, from the history of their subject, that even failed attempts can generate powerful new results and techniques. And there is, of course, the Mallory factor. When the *New York Times* asked George Mallory why he wanted to climb Mount Everest, Mallory replied: "Because it's there."

V. The connection between measuring and continuity is this. Since there is no theoretical limit to the accuracy with which a quantity can be measured, the list of all possible measurements is infinite, and infinitely fine. Between a measurement of 2.3 inches and one of 2.4 inches there are intermediate, more precise, measurements of 2.31, 2.32, 2.33..., 2.39 inches; and these in turn can be subdivided *ad infinitum.* We can, therefore, in imagination, travel connectedly from any measuring number to any other, passing over the infinitude of other measuring numbers that lie between them, without ever finding ourselves without (so to speak) a number to stand on. This idea of connectedness—of traversing some space or some interval without ever having to leap over a void—lies behind the vitally important mathematical concepts of *continuity* and *limit.* In other words, it lies behind all of analysis.

When counting, by contrast, there is nothing between seven and eight; we must leap from one to the other, with no stepping-stones in between. You can measure something at seven and a half units, but you can't count seven and a half objects. (You might object to this, saying "What if I say I have seven and a half apples? Isn't that a counting statement?" To which the answer is "I might allow you to say so ... but only if you're sure that is *precisely* one-half of an apple, as precisely as Larry, Curly, and Moe are precisely three people. Could it not be 0.501 of an apple, or 0.497...?" And at once, if we want to resolve the issue, we must pass into the realm of measuring. "Seven and a half string quartets" is just *cheating.*)

The great fusion between arithmetic and analysis—between counting and measuring, between numbers *staccato* and numbers *legato*—came about as the result of an inquiry into prime numbers, conducted by Lejeune Dirichlet in the 1830s. Dirichlet (1805–1859) was, names notwithstanding, German, from a small town near Cologne, where he got most of his education.[31] The fact that he was a German deserves a brief detour by itself; for the fusion of ideas from arithmetic and analysis, carried out by Dirichlet and Riemann, happened within a broader social change in mathematics at large, the rise of the Germans.

VI. If you draw up a list of the dozen or so greatest mathematicians at work in 1800, it looks something like this: Argand, Bolyai, Bolzano, Cauchy, Fourier, Gauss, Germain, Lagrange (just), Laplace, Legendre, Monge, Poisson, Wallace. A different writer, or this writer in a different mood, might of course add a name here or subtract one there, but without making any difference to the most striking feature of the list, which is the near-total absence of Germans. Gauss is the only one. There is one Scot, one Czech, one Hungarian, and one "disputed" (Lagrange, baptized Giuseppe Lagrangia, is claimed by both Italy and France). The rest are all French.

There were a great many more mathematicians at work in 1900, so a list made up for that year would be correspondingly more likely to start a fistfight. However, I believe that the following attains some local minimum of controversiality: Borel, Cantor, Carathéodory, Dedekind, Hadamard, Hardy, Hilbert, Klein, Lebesgue, Mittag-Leffler, Poincaré, Volterra. Four Frenchmen, an Italian, an Englishman, a Swede, and *five Germans.*[32]

The rise of the Germans to prominence in mathematics is intimately related to some of the historical events I sketched in Chapters 2 and 4. For all of Frederick the Great's reforms, the defeat at Jena in 1806 showed the Prussians that they still had some way to go in modernizing and strengthening their state. The rising nationalist passions stimulated by the long wars against Napoleon, and by the Romantic Movement, were an added spur to reform, in spite of having been thwarted (as the nationalists saw it) by the failure of the Congress of Vienna to unify the German-speaking peoples. In the years after Jena, the Prussian army was reorganized on a basis of universal conscription, serfdom was abolished, restrictions on industry were lifted, taxation and the whole financial system were overhauled, and the educational reforms of Wilhelm von Humboldt, already mentioned in Chapter 2.iv, were instituted. The lesser German states took their lead from Prussia, and Germany at large soon became a place hospitable to science, industry, progress, education—and, of course, mathematics.

It should perhaps be added that there was another, lesser, reason for the rise of nineteenth-century German mathematics. There was Gauss. His is the only German name in that list I drew up for 1800; but there go ten dimes to the dollar, and one Gauss was worth at least ten ordinary mathematicians. The fact that Gauss was in his observatory at Göttingen and teaching (though he disliked teaching and did as little as he could get away with) was sufficient to put Germany, and Göttingen, on the mental map of anyone interested in mathematics.

VII. That is the world in which Lejeune Dirichlet grew up. Born in 1805, he was of the generation before Riemann. The son of a postmaster in a small town 20 miles southwest of Cologne, in Prussia's Rhine province, Dirichlet was also among the first generation to benefit from von Humboldt's reformed gymnasium system of secondary education. He must have been an exceptionally quick study, for by age 16 he had acquired all the qualifications necessary for university entrance. Already hooked on mathematics, he set off for what was still the world capital of mathematical knowledge, Paris, carrying with him the book he treasured above all others, Gauss's *Disquisitiones Arithmeticae.* In Paris, 1822–1825, Dirichlet attended lectures given by many of the great French stars of that time, including at least four from the list I presented earlier: Fourier, Laplace, Legendre, and Poisson.

In 1827, now 22 years old, Dirichlet returned to Germany to teach at the University of Breslau in Silesia. (Breslau is now in Poland, and appears on modern maps as the city of Wrocław.) He gained this position with the assistance and encouragement of Alexander von Humboldt, the explorer, and brother of Wilhelm. Both von Humboldts were key players in these early nineteenth-century German cultural developments.

Outside Berlin, however, German universities were in the condition I have described in Chapter 2.vii, given over mainly to the training of teachers, lawyers, and so on. Dissatisfied with Breslau, Dirichlet got a position in Berlin and spent most of his professional career—1828–1855—teaching there. Among those he taught was a brilliant but shy young scholar from the Wendland region of north Germany, Bernhard Riemann, who had transferred from the University of Göttingen in search of the finest mathematical instruction. I shall have much more to say about Dirichlet's influence on Riemann in Chapter 8; here I note only the connection, and the fact that through it, Riemann came to revere Dirichlet, considering him to be the second greatest mathematician alive, after Gauss.

Dirichlet married Rebecca Mendelssohn, one of the sisters of the composer Felix Mendelssohn, thereby forming one of the many Mendelssohn-mathematics connections.[33]

We have some sketches of Dirichlet and his teaching style during his Berlin years from Thomas Hirst, an English mathematician and diarist who spent much of the 1850s traveling in Europe, taking in mathematics wherever he could find it. During the fall and winter of 1852–1853 he was in Berlin, where he befriended Dirichlet and attended his lectures. From Hirst's diary:

> *31st October 1852*: Dirichlet cannot be surpassed for richness of material and clear insight into it: as a speaker he has no advantages—there is nothing like fluency about him, and yet a clear eye and understanding make it dispensable: without an effort you would not notice his hesitating speech. What is peculiar in him, he never sees his audience—when he does not use the black-board at which time his back is turned to us, he sits at the high desk facing us, puts his spectacles up on his forehead, leans his head on both hands, and keeps his eyes, when not covered with his hands, mostly shut. He uses no notes, inside his hands he sees an imaginary calculation, and reads it out to us—that we understand it as well as if we too saw it. I like that kind of lecturing.

> *14th November 1852*: ... Wednesday evening I spent with Dirichlet: saw Mrs. Dirichlet again, found she was sister to Mendelssohn—she played me several of her brother's pieces, to which I listened with great willingness.
>
> *20th February 1853*: ... Dirichlet has also his peculiarities—one is of forgetting time; he pulls his watch out, finds it past three, and runs out *without even finishing the sentence.*

VIII. For the purposes of this story, Dirichlet's principal significance is as follows. Inspired by a result Euler had proved precisely 100 years before, a result I hereby name "the Golden Key," Dirichlet in 1837 brought together ideas from analysis and arithmetic to prove an important theorem about prime numbers. This is generally considered to be the beginning of analytic number theory; of arithmetic with limits. The title of Dirichlet's groundbreaking paper was, I am sorry to say, *Beweis des Satzes, dass jede unbegrenzte arithmetische Progression, deren erstes Glied und Differenz ganze Zahlen ohne gemeinschaftlichen Factor sind, unendlich viele Primzahlen enthält*—"Proof of the theorem that each unlimited arithmetic progression, whose first member and difference are whole numbers without common factor, contains infinitely many prime numbers."

Take any two positive whole numbers and repetitively add one to the other. If the two numbers have a common factor, every resulting number has that factor, too; repetitively adding 6 to 15 gives you 15, 21, 27, 33, 39, 45, ... all of which have 3 as a factor. If the two numbers have no common factor, however, there is the possibility of getting some primes in the list. If, for example, I repetitively add 6 to 35, I get 35, 41, 47, 53, 59, 65, 71, 77, 83, ... which has lots of primes—along, of course, with many non-primes like 65 and 77. How many primes? Could this sequence contain an infinity of primes? In other words, could it be that, for any number *N*, no matter how big, I could, by repetitively adding 6 to 35 enough times, turn up more than N primes? Could any sequence like this, made from any two numbers with no common factor, contain an infinity of primes?

Yes, it could. This is, in fact, precisely the case. Take any two numbers with no common factor and repetitively add one to the other. You will generate an infinity of primes (mixed with an infinity of non-primes). Gauss had conjectured that this was the case—knowing Gauss's powers, one is tempted to say that he intuited it—but it was decisively proved by Dirichlet in that 1837 paper. It was in Dirichlet's proof that the first part of the great fusion was accomplished.

The truth is even more interesting. Take any positive whole number, say, 9. How many of the numbers less than 9 have no factor in common with it, not counting 1 as a factor? Well, there are six such numbers, and here they are: 1, 2, 4, 5, 7, 8. Take each one of these in turn, and repetitively add 9 to it.

1: 10, 19, 28, 37, 46, 55, 64, 73, 82, 91, 100, 109, 118, 127, ...
2: 11, 20, 29, 38, 47, 56, 65, 74, 83, 92, 101, 110, 119, 128, ...
4: 13, 22, 31, 40, 49, 58, 67, 76, 85, 94, 103, 112, 121, 130, ...
5: 14, 23, 32, 41, 50, 59, 68, 77, 86, 85, 104, 113, 122, 131, ...
7: 16, 25, 34, 43, 52, 61, 70, 79, 88, 97, 106, 115, 124, 133, ...
8: 17, 26, 35, 44, 53, 62, 71, 80, 89, 98, 107, 116, 125, 134, ...

Not only does every one of those sequences contain an infinity of primes (I have underlined them), but each of the six sequences contains the *same proportion* of primes. In other words, if you imagine each sequence stretching out to the neighborhood of some very large number N, instead of merely to the neighborhood of 134, then each contains about the same number of primes, about $\frac{1}{6}(N/\log N)$, if the Prime Number Theorem is true (which had not yet been proved in Dirichlet's time). If N is 134, $\frac{1}{6}(N/\log N)$ is about 4.55983336.... The six sequences I've shown turn up 5, 5, 4, 5, 4, and 5 primes, for an average of 4.6666...; high by 2.3 percent, which is pretty good for such a small sample size.

To prove his result, Dirichlet began with a form of arithmetic developed at great length by Gauss in *Disquisitiones Arithmeticae.* Mathematicians call it "the arithmetic of congruences." You can think of it as clock arithmetic. Temporarily replace the 12 on a clock face with 0. The 12 hours of the clock now read 0, 1, 2, 3, ... up to 11. If the time is eight o'clock, and you add 9 hours, what do you get? Well, you get five o'clock. So in this arithmetic, $8 + 9 = 5$; or, as mathematicians say, $8 + 9 \equiv 5 \pmod{12}$, pronounced "eight plus nine is congruent to five, modulo twelve." The phrase "modulo twelve" means "I am working from a clock-face with twelve hours marked, 0 to 11." This may seem trivial, but in fact the arithmetic of congruences goes very deep and is full of strange and difficult results. Gauss was a great grand master of it; not one of the seven sections of *Disquisitiones Arithmeticae* is free from that "≡" sign.

The *Disquisitiones*, remember, was the constant companion of Dirichlet's younger years. When he came to this problem, in 1836 or 1837, he was in his early 30s and must have completely internalized Gauss's work on congruences. Then somehow, Euler's 1737 result—"the Golden Key"—came to his attention. It gave him an idea; he put the two things together, applied some elementary techniques of analysis, and got his proof.

IX. Dirichlet was thus the first to pick up the Golden Key, the link between arithmetic and analysis, and make serious use of it. In terms of the analogy I am using, it would be a bit too much to say that he turned the key. I would rather say that he picked it up, sensed its beauty and potential power, set it down again, then used it as a model for a similar key—a silver key, you might say—to unlock the particular problem he had in front of him. The great fusion, analytic number theory, did not appear in its full glory until 22 years later, in Riemann's paper of 1859.

Recall, though, that Riemann was one of Dirichlet's students and certainly knew of the older man's work. In the opening paragraph of the 1859 paper, in fact, he mentioned Dirichlet's name in conjunction with that of Gauss. They were his two mathematical idols. If it was Riemann who turned the key, it was Dirichlet who first showed it to him and demonstrated that it *was* a key to something or other; and it is to Dirichlet that the immortal glory of inventing analytic number theory properly belongs.

But what, exactly, is this Golden Key? What was it that Leonhard Euler, working away by candlelight in his room, the secret police of the *Bironovschina* prowling the streets of St. Petersburg outside, left lying around for Dirichlet to find a hundred years later?

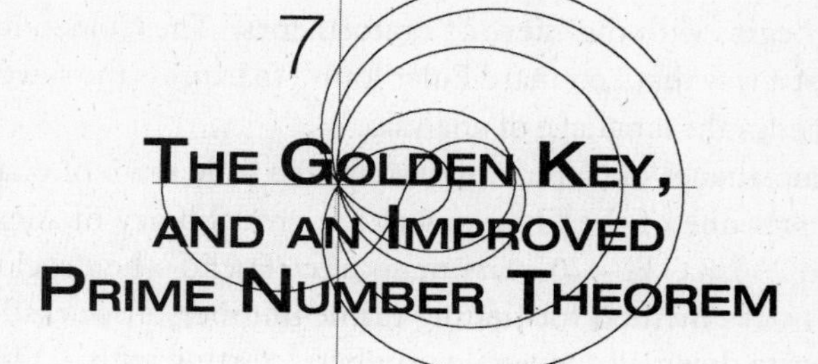

7 The Golden Key, and an Improved Prime Number Theorem

I. The patient reader will have noticed that the mathematical chapters of this book have been moving on two tracks. Chapters 1 and 5 were all about those infinite sums, leading up to a mathematical object named, by Riemann, "the zeta function"; Chapter 3 was concerned with primes, taking its lead from the title of Riemann's 1859 paper and proceeding from there to the Prime Number Theorem (PNT). Obviously, both issues—the zeta function and the primes—are connected through Riemann's interest in them. In fact, by yoking the two concepts together in a certain way, by turning the Golden Key, Riemann opened up the whole field of analytic number theory. But how did he do that? What's the connection? What *is* the Golden Key? In this chapter I aim to answer that question—to show you the Golden Key. Then I shall begin preparations for turning the Golden Key by offering an improved version of the PNT.

II. It begins with the "sieve of Eratosthenes." The Golden Key is, in fact, just a way that Leonhard Euler found to express the sieve of Eratosthenes in the language of analysis.[34]

Eratosthenes of Cyrene (nowadays the little town of Shahhat in Libya) was one of the librarians at the great library of Alexandria. Around 230 B.C.E.—70 years or so after Euclid—he developed his famous sieve method for finding prime numbers. It works like this. First, write down all the whole numbers, starting with 2. Of course, you can't write them all, so let's make do with 100 or so.

2	3	4	5	6	7	8	9	10	11	12	13	14	15
16	17	18	19	20	21	22	23	24	25	26	27	28	29
30	31	32	33	34	35	36	37	38	39	40	41	42	43
44	45	46	47	48	49	50	51	52	53	54	55	56	57
58	59	60	61	62	63	64	65	66	67	68	69	70	71
72	73	74	75	76	77	78	79	80	81	82	83	84	85
86	87	88	89	90	91	92	93	94	95	96	97	98	99
100	101	102	103	104	105	106	107	108	109	110	111	112	113

Now, starting from 2, and leaving 2 untouched, remove every second number from 2 on. The result is

2	3	.	5	.	7	.	9	.	11	.	13	.	15
.	17	.	19	.	21	.	23	.	25	.	27	.	29
.	31	.	33	.	35	.	37	.	39	.	41	.	43
.	45	.	47	.	49	.	51	.	53	.	55	.	57
.	59	.	61	.	63	.	65	.	67	.	69	.	71
.	73	.	75	.	77	.	79	.	81	.	83	.	85
.	87	.	89	.	91	.	93	.	95	.	97	.	99
.	101	.	103	.	105	.	107	.	109	.	111	.	113

The first number left unscathed after 2 is 3. Leave 3 untouched, but remove every third number from 3 on, if it hasn't already been removed. The result is

2	3	.	5	.	7	.	.	.	11	.	13	.	.
.	17	.	19	.	.	.	23	.	25	.	.	.	29
.	31	.	.	.	35	.	37	.	.	.	41	.	43
.	.	.	47	.	49	.	.	.	53	.	55	.	.
.	59	.	61	.	.	.	65	.	67	.	.	.	71
.	73	.	.	.	77	.	79	.	.	.	83	.	85
.	.	.	89	.	91	.	.	.	95	.	97	.	.
.	101	.	103	.	.	.	107	.	109	.	.	.	113

The first number left unscathed after 3 is 5. Leave 5 untouched, but remove every fifth number from 5 on, if it hasn't already been removed. The result is

2	3	.	5	.	7	.	.	.	11	.	13	.	.
.	17	.	19	.	.	.	23	.	.	.	.	.	29
.	31	.	.	.	.	.	37	.	.	.	41	.	43
.	.	.	47	.	49	.	.	.	53	.	.	.	.
.	59	.	61	.	.	.	.	.	67	.	.	.	71
.	73	.	.	.	77	.	79	.	.	.	83	.	.
.	.	.	89	.	91	.	.	.	.	.	97	.	.
.	101	.	103	.	.	.	107	.	109	.	.	.	113

The first number left unscathed after 5 is 7. The next step would be to leave 7 untouched, but remove every seventh number from 7 on, if it hasn't already been removed. The first number left unscathed after 7 would then be 11, and so on.

If you keep doing this for ever, the numbers you are left with are all the primes. That is the sieve of Eratosthenes. If you stop just before processing prime *p*—that is, just before removing every *p*th number that wasn't already removed—you have all the primes less than p^2. Since I stopped before processing 7, I have all the primes up to 7^2, which is 49. After that you see some numbers, like 77, that are not prime.

III. The sieve of Eratosthenes is pretty straightforward, and 2,230 years old. How does it get us into the middle of the nineteenth century, and deep results in function theory? Here's how.

I am going to repeat the process I went through above. (That's why I went through it so painstakingly.) This time, however, I'm going to apply it to Riemann's zeta function, which I defined at the end of Chapter 5. Here is the zeta function for some number *s* bigger than 1.

$$\zeta(s)=1+\frac{1}{2^s}+\frac{1}{3^s}+\frac{1}{4^s}+\frac{1}{5^s}+\frac{1}{6^s}+\frac{1}{7^s}+\frac{1}{8^s}+\frac{1}{9^s}+\frac{1}{10^s}+\frac{1}{11^s}+\cdots$$

Note that writing it in this way involves writing out all the positive whole numbers—which is how we started off the sieve of Eratosthenes (except that this time I included 1).

What I'm going to do is multiply both sides of the equals sign by $\frac{1}{2^s}$. This gives me

$$\frac{1}{2^s}\zeta(s)=\frac{1}{2^s}+\frac{1}{4^s}+\frac{1}{6^s}+\frac{1}{8^s}+\frac{1}{10^s}+\frac{1}{12^s}+\frac{1}{14^s}+\frac{1}{16^s}+\frac{1}{18^s}+\cdots$$

because of Power Rule 7 (which, for example, makes 2^s times 7^s equal to 14^s). Now I'll subtract the second of these expressions from the first. On the left-hand sides I have one of $\zeta(s)$, and I have $\frac{1}{2^s}$ of it. Subtracting

$$\left(1-\frac{1}{2^s}\right)\zeta(s)=1+\frac{1}{3^s}+\frac{1}{5^s}+\frac{1}{7^s}+\frac{1}{9^s}+\frac{1}{11^s}+\frac{1}{13^s}+\frac{1}{15^s}+\frac{1}{17^s}+\frac{1}{19^s}+\cdots$$

The subtraction eliminated all the even-numbered terms from the infinite sum. I'm left with just the odd-numbered terms.

Remembering the sieve of Eratosthenes, I'll now multiply both sides of this equals sign by $\frac{1}{3^s}$, 3 being the first unscathed number on the right-hand side.

$$\frac{1}{3^s}\left(1-\frac{1}{2^s}\right)\zeta(s)=$$

$$\frac{1}{3^s}+\frac{1}{9^s}+\frac{1}{15^s}+\frac{1}{21^s}+\frac{1}{27^s}+\frac{1}{33^s}+\frac{1}{39^s}+\frac{1}{45^s}+\frac{1}{51^s}+\cdots$$

Now subtract this expression from the one before. When subtracting the left-hand sides, treat $\left(1-\frac{1}{2^s}\right)\zeta(s)$ as a blob, a single number (which of course it is, for any given s). I have one of this blob, and I have $\frac{1}{3^s}$ of it. Subtracting, I get $\left(1-\frac{1}{3^s}\right)$ of it.

$$\left(1-\frac{1}{3^s}\right)\left(1-\frac{1}{2^s}\right)\zeta(s)=$$

$$1+\frac{1}{5^s}+\frac{1}{7^s}+\frac{1}{11^s}+\frac{1}{13^s}+\frac{1}{17^s}+\frac{1}{19^s}+\frac{1}{23^s}+\frac{1}{25^s}+\frac{1}{29^s}+\cdots$$

All the multiples of 3 have vanished from the infinite sum. The first unscathed number on the right is now 5.

If I multiply both sides by $\frac{1}{5^s}$, the result is

$$\frac{1}{5^s}\left(1-\frac{1}{3^s}\right)\left(1-\frac{1}{2^s}\right)\zeta(s)=$$

$$\frac{1}{5^s}+\frac{1}{25^s}+\frac{1}{35^s}+\frac{1}{55^s}+\frac{1}{65^s}+\frac{1}{85^s}+\frac{1}{95^s}+\frac{1}{115^s}+\cdots$$

And now, subtracting this equation from the previous one, and this time treating $\left(1-\frac{1}{3^s}\right)\left(1-\frac{1}{2^s}\right)\zeta(s)$ as a single blob, I have one of it, and I have $\frac{1}{5^s}$ of it. Subtracting

$$\left(1-\frac{1}{5^s}\right)\left(1-\frac{1}{3^s}\right)\left(1-\frac{1}{2^s}\right)\zeta(s)=$$

$$1+\frac{1}{7^s}+\frac{1}{11^s}+\frac{1}{13^s}+\frac{1}{17^s}+\frac{1}{19^s}+\frac{1}{23^s}+\frac{1}{29^s}+\frac{1}{31^s}+\cdots$$

All the multiples of 5 vanished in the subtraction, and the first number left unscathed on the right is 7.

Notice the resemblance to the sieve of Eratosthenes? Actually, you should first notice the difference. When doing the original sieve, I chose to leave each original prime standing, deleting only its multiples by 2, 3, 4, ... Here, I eliminate the original prime from the right-hand side in the subtraction, along with all its multiples.

If I keep doing this up to some decently large prime, let's say 997, I have this.

$$\left(1-\frac{1}{997^s}\right)\left(1-\frac{1}{991^s}\right)\cdots\left(1-\frac{1}{5^s}\right)\left(1-\frac{1}{3^s}\right)\left(1-\frac{1}{2^s}\right)\zeta(s)=$$

$$1+\frac{1}{1009^s}+\frac{1}{1013^s}+\frac{1}{1019^s}+\frac{1}{1021^s}+\cdots$$

Now, that right-hand side, if s is any number bigger than 1, is just a tiny bit bigger than 1 itself. If s is 3, for example, it works out to 1.0000000673103608 1534.... So it is not too improbable to say that if you repeated the process forever, you'd get the result shown in Expression 7-1.

$$\cdots\left(1-\frac{1}{13^s}\right)\left(1-\frac{1}{11^s}\right)\left(1-\frac{1}{7^s}\right)\left(1-\frac{1}{5^s}\right)\left(1-\frac{1}{3^s}\right)\left(1-\frac{1}{2^s}\right)\zeta(s)=1$$

Expression 7-1

for any number s bigger than 1, with the left-hand side having one bracketed expression for *every* prime number, stretching away forever to the left. Dividing each side of the expression repeatedly by each of the parentheses in turn, I get the result shown in Expression 7-2.

$$\zeta(s)=\frac{1}{1-\frac{1}{2^s}}\times\frac{1}{1-\frac{1}{3^s}}\times\frac{1}{1-\frac{1}{5^s}}\times\frac{1}{1-\frac{1}{7^s}}\times\frac{1}{1-\frac{1}{11^s}}\times\frac{1}{1-\frac{1}{13^s}}\times\cdots$$

Expression 7-2

IV. That is the Golden Key. To show it to you in all its elegance, let me clean it up a little. I don't like fractions with fractional denominators any more than you do, and there is a useful bit of mathematical notation I can introduce here to save on typing.

First, remember from Power Rule 5 that a^{-N} means $1/a^N$, and a^{-1} means $1/a$. I can therefore write Expression 7-2 somewhat more succinctly as

$$\zeta(s) = \left(1-2^{-s}\right)^{-1}\left(1-3^{-s}\right)^{-1}\left(1-5^{-s}\right)^{-1}\left(1-7^{-s}\right)^{-1}\left(1-11^{-s}\right)^{-1}\ldots$$

There is an even neater way to write this. Recall the Σ notation from Chapter 5.viii. When I am adding up a bunch of terms with the same pattern, I can write the sum in shorthand using the Σ sign. Well, there is an equivalent thing for when I am *multiplying* terms that all conform to a pattern, the Π sign. That's a capital Greek letter "pi," for "product." Here is Expression 7-2 written using the Π sign.

$$\zeta(s) = \prod_p \left(1-p^{-s}\right)^{-1}$$

This is pronounced "Zeta of *s* equals the product over all primes of one minus *p* to the minus *s*, to the minus one." The little "*p*" beneath the Π sign is understood to mean "over all primes."[35] Remembering the definition of $\zeta(s)$ as an infinite sum, I can rewrite the left-hand side and get Expression 7-3.

The Golden Key

$$\sum_n n^{-s} = \prod_p \left(1-p^{-s}\right)^{-1}$$

Expression 7-3

Both the sum on the left and the product on the right go all the way to infinity. This, in fact, offers another proof that the primes never end. If they did end, the right-hand side product would end, therefore working out to some finite number, no matter what the value of *s*. When $s = 1$, however, the left-hand side is the harmonic series from Chapter 1, which "adds up to infinity." Since it cannot be the case that an infinity on the left equals a finite number on the right, the number of primes must be infinite.

V. What, you may be wondering, is so all-fired special about Expression 7-3, that I have given it such a grandiloquent name?

The answer to that won't become entirely clear until a later chapter, when I actually turn the Golden Key. At this point, the main thing to be impressed by—mathematicians, at any rate, find it extremely impressive—is the fact that on the left-hand side of Expression 7-3 we have an infinite sum running through all the positive whole numbers 1, 2, 3, 4, 5, 6, . . . , while on the right-hand side we have an infinite product running through all the prime numbers 2, 3, 5, 7, 11, 13, ...

Expression 7-3—the Golden Key—is actually named "the Euler product formula."[36] It first saw the light of day, though arranged slightly differently, in a paper with the title *Variae observationes circa series infinitas*, written by Leonhard Euler and published in 1737 by the St. Petersburg Academy. (The title translates as "Various Observations about Infinite Series"—compare the Latin with the translation, and you can see what I meant back in Chapter 4.viii when I spoke of the ease of reading Euler's Latin.) The actual statement of the Golden Key in that paper is as follows:

THEOREMA 8

Si ex serie numerorum primorum sequens formetur expressio

$$\frac{2^n \cdot 3^n \cdot 5^n \cdot 7^n \cdot 11^n \cdot \textit{etc.}}{(2^n-1)(3^n-1)(5^n-1)(7^n-1)(11^n-1)\ \textit{etc.}}$$

erit eius valor aequalis summae huius seriei

$$1+\frac{1}{2^n}+\frac{1}{3^n}+\frac{1}{4^n}+\frac{1}{5^n}+\frac{1}{6^n}+\frac{1}{7^n}+\textit{etc.}$$

The Latin means "If from the series of prime numbers the following expression be formed . . . its value will be equal to the sum of this series. . . ." Again, once you know a basic few dozen word endings ("-orum," a genitive; "-etur," a present subjunctive passive, etc.), Euler's Latin holds no terrors.

When jotting down the ideas that make up this book, I first looked through some of the math texts on my shelves to find a proof of the Golden Key suitable for non-specialist readers. I settled on one that seemed to me acceptable and incorporated it. At a later stage of the book's development, I thought I had better carry out authorial due diligence, so I went to a research library (in this case the excellent new Science, Industry and Business branch of the New York Public Library in midtown Manhattan) and pulled out the original paper from Euler's collected works. His proof of the Golden Key covers ten lines and is far easier and more elegant than the one I had selected from my textbooks. I thereupon threw out my first choice of proof and replaced it with Euler's. The proof in part III of this chapter is essentially Euler's. It's a professorial cliché, I know, but it's true nonetheless: you can't beat going to the original sources.

VI. Having shown you the Golden Key, I now have to begin preparations for turning it. This involves recapitulating a fair amount of math, including a very small quantity of calculus. In the rest of this chapter, I am going to present all the calculus you need to understand the Hypothesis and its significance. Then, making a virtue of necessity, I shall employ that calculus to present an improved version of the PNT—a version much more relevant to Riemann's work.

Calculus instruction traditionally begins with a graph. The graph I am going to start with is the one in Chapter 5.iii, showing the log function—I have reproduced it here as Figure 7-1. Imagine you are a very small—infinitesimal, if you can manage it—homunculus, climbing up the graph of the log function from left to right. At first, if you start somewhere close to zero, the ascent is very steep, and you need rock-climbing gear. As you go on, however, it gets less and less steep. By the time you get to arguments around 10, you can get upright and actually walk it.

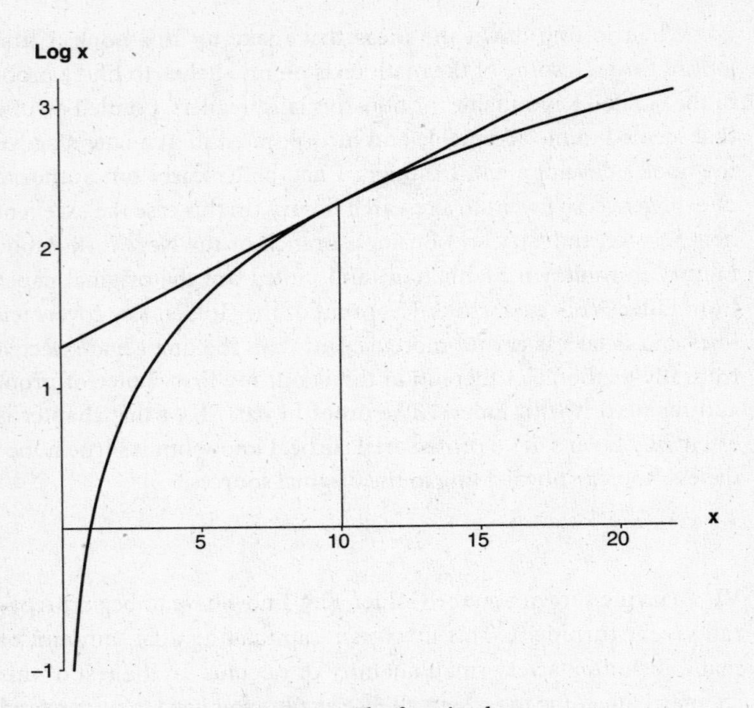

FIGURE 7-1 The function log x.

The steepness of the curve varies from point to point. At every point it has a definite numerical value, though, just as your automobile has a definite speed at any point while you are accelerating—namely, the speed you see if you glance at the speedometer. If you glance again an instant later, you see a slightly different speed; but at every point in time there is some definite speed. Just so, for any argument in its domain (which is all numbers greater than zero), the log function has some definite gradient.

How do we measure that gradient, and what is it? First, let me define "gradient" for a sloping straight line. It is the vertical rise divided by the horizontal span. If, in covering a horizontal distance of 5 units, I rise a vertical distance of 2 units, the gradient is 2 in 5, or 0.4. See Figure 7-2.

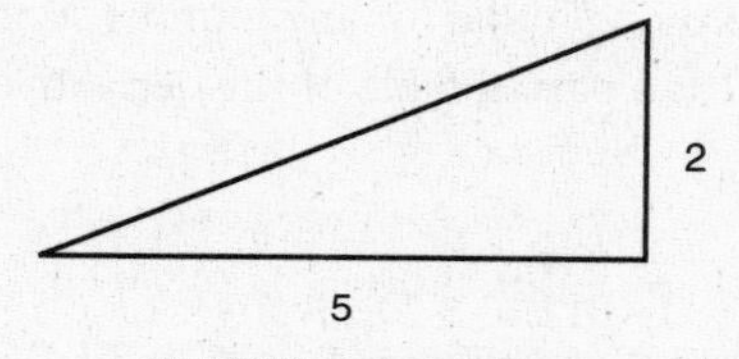

FIGURE 7-2 Gradient.

To get the gradient of a curve at any point, I construct the one straight line that touches the curve at that point. Plainly there is only one such line; if I "roll" the line a little (imagine it's a steel rod, and the curve a steel band), it touches the curve at a slightly different point. The gradient of the curve at the point is the gradient of that unique touching straight line. The gradient of log x at the argument $x = 10$ turns out, if you measure it, to be $\frac{1}{10}$. The gradient at argument 20 is less, of course; it measures $\frac{1}{20}$. The gradient at argument 5 is steeper; it measures $\frac{1}{5}$. It is, in fact, yet another amazing property of the log function that the gradient at any argument x is $1/x$, the reciprocal of x, also known as x^{-1}.

If you ever did a calculus course, this all sounds pretty familiar. The starting point of calculus is, in fact, this: From any function f I can derive another function g, which measures the gradient of f at any argument. If f is log x, then g is $1/x$. This derived function is called, believe it or not, "the derivative" of f. For example, $1/x$ is the derivative of log x. If you are presented with some function f, the process of finding the derivative is called "differentiation."

Differentiation follows some easy rules. It is, for example, transparent to several basic arithmetic operations. If the derivative of f is g, then the derivative of $7f$ is $7g$. (So the derivative of 7 times log x is 7 times $1/x$.) The derivative of f-plus-g is the derivative of f plus the derivative of g. This breaks down for multiplication, though; the derivative of f-times-g is *not* the derivative of f times the derivative of g.

In this book, the only functions besides log x whose derivatives I am concerned with are the simple power functions x^N. I am going to tell you without proving it that for any number N, the derivative of x^N is Nx^{N-1}. Table 7-1 is a partial table of the derivatives of all the power functions.

TABLE 7-1 Derivatives of x^N.

Function	…	x^{-3}	x^{-2}	x^{-1}	x^0	x^1	x^2	x^3	…
Derivative	…	$-3x^{-4}$	$-2x^{-3}$	$-x^{-2}$	0	1	$2x$	$3x^2$	…

Of course x^0 is just 1, its graph a flat horizontal line. It has no gradient, zero gradient. If you differentiate any fixed number, you get zero. And x^1 is just x; the graph is a straight line going diagonally upward, exiting the graph paper at the top right corner; the gradient is a steady 1. Note that there is no power whose derivative is x^{-1}, though x^0 looks to be in the right position. That is not surprising, since we already know that the derivative of log x is x^{-1}. Once again, log x looks as though it is trying to pass itself off as x^0.

VII. You no doubt recall my saying more than once that mathematicians love to be able to invert things. Here is P in terms of Q; what is Q in terms of P? That is how I brought up the log function in the first place—as the inverse of the exponential function. If $a = e^b$, then what is b in terms of a? It's log a.

Suppose then, that I differentiate function f and get function g. Then g is the derivative of f. And f is the … what? of g? What is the inverse of differentiation? The derivative of log x is $1/x$, so log x is the … what? of $1/x$? Answer: It's the *integral*, that's what. The inverse of a derivative is an integral, and the inverse of differentiation is *integration.* Since the whole business is transparent to multiplication by a fixed number, turning Table 7-1 upside down and fiddling a little gives the inverse table shown in Table 7-2.

TABLE 7-2 Integrals of x^N.

Function	...	x^{-3}	x^{-2}	x^{-1}	x^0	x^1	x^2	x^3	...
Integral	...	$-\frac{1}{2}x^{-2}$	$-x^{-1}$	$\log x$	x	$\frac{1}{2}x^2$	$\frac{1}{3}x^3$	$\frac{1}{4}x^4$	...

And in fact, so long as N is not equal to -1, the integral of x^N is $x^{N+1}/(N+1)$. (And looking at that table, you see again how the function $\log x$ strives to behave as if it were x^0, which of course it is not.)

If derivatives are good for telling us the gradient of a function—that is, the rate at which it is changing at any point—what are integrals good for? Answer: For finding the areas under graphs.

The function I've shown in Figure 7-3—it is actually the function $1/x^4$, which is to say of x^{-4}—embraces a certain area between the arguments $x = 2$ and $x = 3$. To calculate that area, you first figure out the integral function of x^{-4}. That, by the general rule above, is $-\frac{1}{3}x^{-3}$, that is, $-1/(3x^3)$. Like any other function, this has a value for every argument in its domain. To find the area from argument 2 to argument 3, you calculate the value of the integral at argument 3, then calculate the value at argument 2, then subtract the second value from the first.

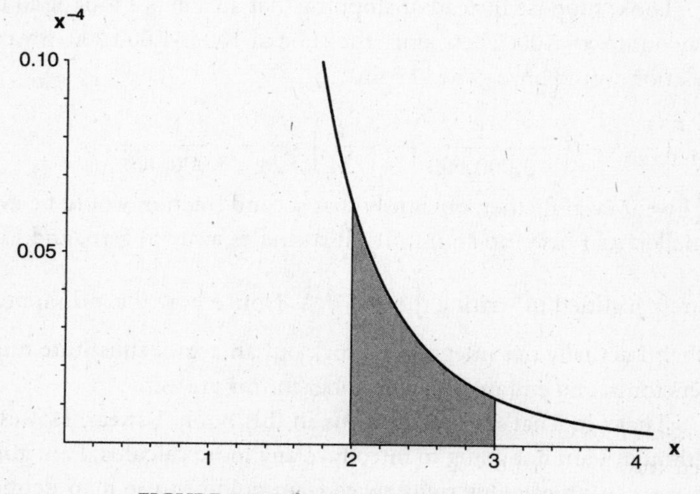

FIGURE 7-3 What integration is good for.

When $x = 3$, the value of $-1/(3x^3)$ is $-\frac{1}{81}$; when $x = 2$ it is $-\frac{1}{24}$. Subtract, remembering that subtracting a negative number is the same as adding the corresponding positive number, $\left(-\frac{1}{81}\right) - \left(-\frac{1}{24}\right) = \frac{1}{24} - \frac{1}{81}$, which is $\frac{19}{648}$, about 0.029321.

Mathematicians have a way to write this, $\int_2^3 x^{-4}\,dx$, read as "the integral of x to the minus fourth power, with respect to x, from 2 to 3." (Don't worry too much about that "with respect to x." Its purpose is to declare x as the main variable we are working with, whose integral has to be figured out. If there happen to be any other variables under the integral sign, they are just hanging out there; they are not being integrated. Chapter 19 has an illustration.)

Now, sometimes you can let the right-hand end of the integration go off to infinity and get a finite area. It's like infinite sums. If the values are right, they can converge to a finite value. Same here. If the function is right, the area under it can be finite even though infinitely long. Integrals are connected to sums at a deep level. The integral sign—first used by Leibniz in 1675—is just an elongated "S" for "sum."

Look, suppose instead of stopping that area at 3, I took it all the way out to $x = 100$. Then, since the cube of 100 is 1,000,000, my calculation would have gone like this:

$$\left(-\frac{1}{3{,}000{,}000}\right) - \left(-\frac{1}{24}\right) = \frac{1}{24} - \frac{1}{3{,}000{,}000}$$

If I went even further, obviously that second fraction would be even smaller. As I head off to infinity, it dwindles away to zero, and I am surely justified in writing $\int_2^{\infty} x^{-4}\,dx = \frac{1}{24}$. Notice how the x disappears when I actually use integrals to work out an area. I substitute numbers for it, and end up with a number for my answer.

That's it. That's all the calculus in this book, I swear. However, although I am not going to introduce any more calculus, I am going to start using calculus right away. I am going to use it to define a completely new function, one that is terrifically important in the theory of prime numbers and the zeta function.

VIII. First, consider the function $1/\log t$. Figure 7-4 shows a graph of it. I have changed my symbol for argument from *x* to *t*, because I have a use for *x* other than as a dummy variable.

I have also shaded an area under the graph, because I am going to do a spot of integration. Integration, as I presented it just now, is a way to calculate the area under a function. First you figure out the integral of the function, then you hit the calculator. So, what is the integral of $1/\log t$?

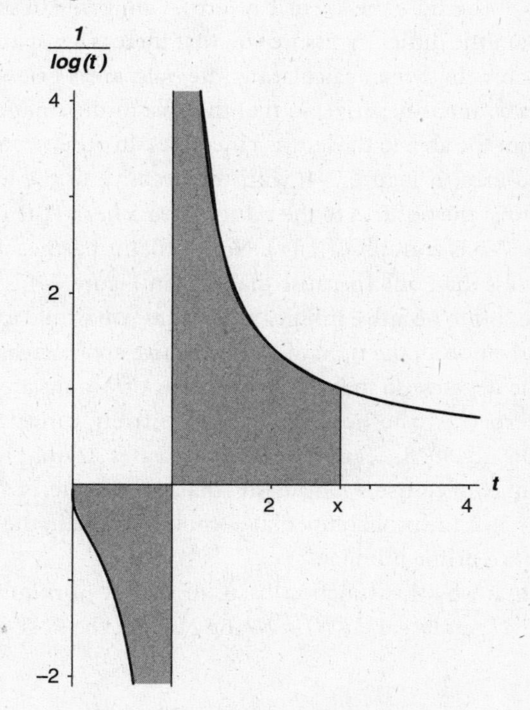

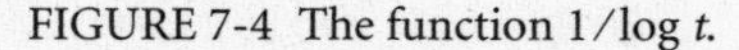
FIGURE 7-4 The function $1/\log t$.

Unfortunately, there is no ordinary household function that can be used to express the integral of $1/\log t$. This integral is, however, very important. It turns up a lot in our researches into the Riemann Hypothesis. Since we don't want to keep having to write $\int_0^x (1/\log t)\, dt$ every time we refer to the darn thing, we simply define it to be a new function, and issue it a certificate declaring it a sound and respectable function in good standing with its peers.

This new function has the name "the log integral function." The usual symbol for it is $Li(x)$. (Sometimes "$li(x)$.") It is defined to be[37] the area under that graph—the graph of $1/\log t$—from zero to x.

This involves a certain sleight of hand, because $1/\log t$ has no value at $t = 1$ (because the log of 1 is zero). I am going to skate breezily over that little difficulty, assure you that there is a way to finesse it, and note only that when calculating integrals, areas below the horizontal axis count as negative, so that the area to the right of 1 works to cancel out the area to the left, as t increases. In other words, $Li(x)$ is the shaded area in Figure 7-4, with the negative to the left of $t = 1$ netted against the positive to the right (when x lies to the right).

Figure 7-5 is a graph of $Li(x)$. Notice that it has negative values when x is less than one (because that area in Figure 7-4 is negative), that it dives off to negative infinity at $x = 1$ (as you would expect), but that as x advances to the right of 1, the positive area increasingly cancels out the negative so that $Li(x)$ comes back from negative infinity, reaches zero (i.e., the negative area is entirely canceled out) at $x = 1.4513692348828\ldots$, and thereafter increases steadily. Its gradient at any point is, of course, $1/\log x$. And that, please note, is, as I showed in Chapter 3.ix, the probability that a whole number in the neighborhood of x is a prime number.[38]

Which is why this function is so important in number theory. You see, as N gets larger, $Li(N) \sim N/\log N$. Now, the PNT asserts that

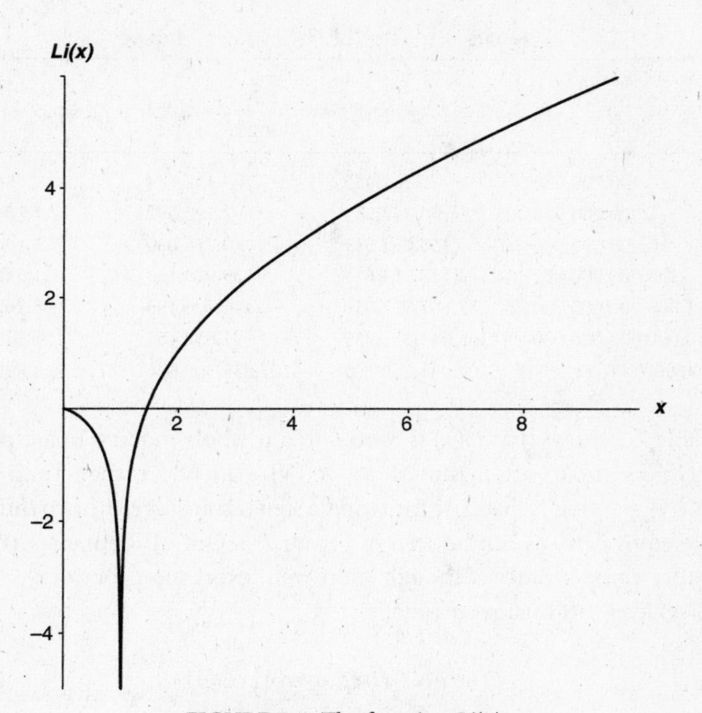

FIGURE 7-5 The function $Li(x)$.

$\pi(N) \sim N/\log N$. A moment's thought will convince you that the twiddle sign is transitive—that is, if $P \sim Q$ and $Q \sim R$, then it must be the case that $P \sim R$. So if the PNT is true—which we know it is, it was proved in 1896—then it must also be true that $\pi(N) \sim Li(N)$.

This is not merely true; it is, in a manner of speaking, *truer.* I mean, $Li(N)$ is actually a better estimate of $\pi(N)$ than $N/\log N$ is. A *much* better estimate.

TABLE 7-3

N	$\pi(N)$	$\frac{N}{\log N} - \pi(N)$	$Li(N) - \pi(N)$
100,000,000	5,761,455	−332,774	754
1,000,000,000	50,847,534	−2,592,592	1,701
10,000,000,000	455,052,511	−20,758,030	3,104
100,000,000,000	4,118,054,813	−169,923,160	11,588
1,000,000,000,000	37,607,912,018	−1,416,706,193	38,263
10,000,000,000,000	346,065,536,839	−11,992,858,452	108,971
100,000,000,000,000	3,204,941,750,802	−102,838,308,636	314,890

Table 7-3 shows that $Li(x)$ is central to our whole inquiry. In fact, the PNT is most often stated as $\pi(N) \sim Li(N)$, rather than as $\pi(N) \sim N/\log N$. Because the twiddle sign is transitive, the two things are equivalent, as can be seen in Figure 7-6. Out of Riemann's 1859 paper came a precise, though unproven, expression for $\pi(x)$, and $Li(x)$ leads off that expression.

The PNT (Improved Version)

$$\pi(N) \sim Li(N)$$

Note just one more thing about Table 7-3. For all the values of N shown in the table, $N/\log N$ gives a low estimate for $\pi(N)$, while $Li(x)$ gives a high one. I am just going to leave that lying there as a comment, for future reference.

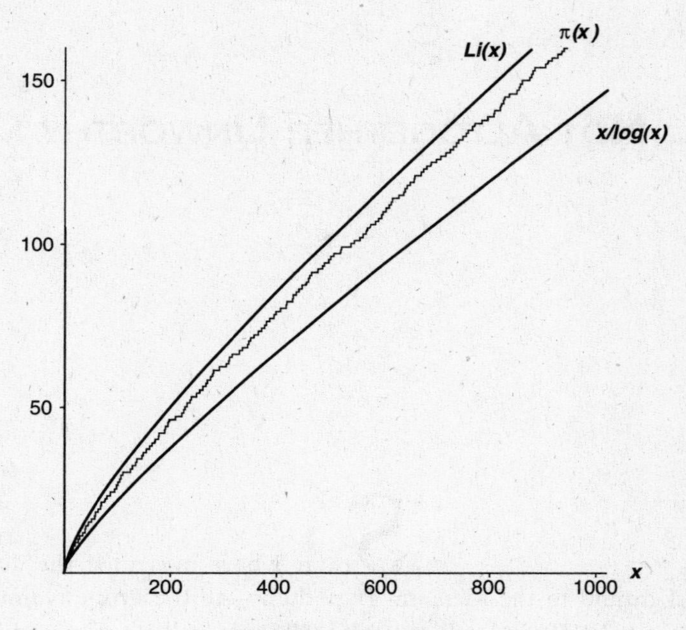

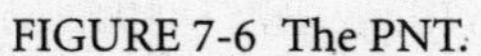
FIGURE 7-6 The PNT.

8

NOT ALTOGETHER UNWORTHY

I. So far I have presented the deep background to the Riemann Hypothesis—to the Prime Number Theorem (PNT) and to Riemann's 1859 paper on that topic, in which the Hypothesis was first stated. In this chapter I shall describe the immediate background to that paper. This is really two stories intertwined: the story of Bernhard Riemann and the story of Göttingen University in the 1850s, with brief side trips to Russia and New Jersey for some local color.

You should keep in mind a broad general picture of European intellectual life in the 1830s, 1840s, and 1850s. It was, of course, a time of great change. The upheavals of the Napoleonic wars had let loose new forces of nationalism and reform. The Industrial Revolution was on the march. The shifts in thought and feeling we customarily collect under the heading "the Romantic Movement" had seeped down to the general population everywhere. The 1830s, when spirits had revived after the exhaustion of the long wars, were an unsettled time, marked by the July revolution in France, a nationalist uprising in Poland (at that time part of the Russian empire[39]), agitation among the Germans for national unity, and the great Reform Bills in Britain. Alexis de Tocqueville visited the United States and wrote a penetrating analysis of that curious new experiment in popular government. In the following decade darker forces stirred, culminating in 1848, "the year of revolutions," whose disturbances, as we saw in Chapter 2, penetrated for a moment even the deep reserve of Bernhard Riemann.

Göttingen was for all this period a provincial backwater illuminated mainly by the presence of Gauss. The university's one moment of political prominence occurred in 1837 with the dismissal of the "Göttingen Seven" that I already mentioned, the main effect of which was to lower the prestige of the university. Paris remained the great center of mathematical research, with Berlin rising fast. In Paris Cauchy and Fourier had overhauled analysis, laying the foundations of the modern treatment of limits, continuity, and the calculus. In Berlin new advances were being made by Dirichlet in arithmetic, by Jacobi in algebra, by Steiner in geometry, and by Eisenstein in analysis. Anyone who wanted to do serious mathematics in the 1840s needed to be in Paris or Berlin. That is why young Bernhard Riemann, 20 years old in the spring of 1847, disappointed with the standard of instruction at Göttingen and very keen indeed to do serious mathematics, went to Berlin. He studied there for two years, during which the greatest influence on him was Lejeune Dirichlet, the man who had picked up the Golden Key in 1837. Dirichlet took a personal liking to the shy, poverty-stricken young Riemann, an attitude which Riemann, in the words of Heinrich Weber, "reciprocated with respectful gratitude."

Returning to Göttingen after the Easter break in 1849, Riemann embarked upon his doctorate course, under the supervision of Gauss himself. Plainly, his hope was to become a lecturer at the university. That was a long road to travel, though. To lecture at Göttingen required not only a doctorate, but also a further qualification, the "habilitation," a sort of second doctorate, with a thesis to be prepared and a trial lecture to be given. The whole thing, doctorate and habilitation, took Riemann more than five years—from age 22½ to nearly age 28—during which he had no income at all.

Right away, Riemann enrolled for some courses in physics and philosophy along with math. These were required subjects for those wishing to teach in the gymnasium high-school system, pretty much Riemann's only career choice if he could not get a lecturing position. He might have been hedging his bets by taking these courses. He had, however, a deep interest in both subjects, so it is probable that pure personal inclination was at least as much of a factor in his enrolling for them. Standards at Göttingen had improved, too. The physicist Wilhelm Weber, one of the Göttingen Seven cashiered in 1837, had returned to the university to teach, the political climate having thawed considerably. An old friend and colleague of Gauss's—the two of them had together invented the electric telegraph—Weber taught a course in experimental physics, which Riemann attended.[40]

II. Those five years of unpaid research work must have been hard ones for Bernhard Riemann. He was far from home; it was 120 miles from Göttingen to Quickborn, a two-day journey in great discomfort, and expensive. He did, though, have some company. In 1850 Richard Dedekind arrived at the university. Dedekind was 19, five years younger than Riemann, and was also aiming for a doctorate. It is plain from Dedekind's biographical note on Riemann in the *Collected Works* that he felt affection and sympathy for his older colleague, and great admiration for his mathematical abilities; it is more difficult to judge Riemann's feelings in the matter.

The two men got their doctorates within a few months of each other, Riemann in December 1851, Dedekind the following year. Both were examined by Gauss, now in his mid-70s but keenly alert to exceptional mathematical talent. On the thesis submitted by young Dedekind, still not mathematically mature, Gauss's report is little better than boilerplate approval. On Riemann's, he gushed—and Gauss was a man who rarely gushed. "A substantial and valuable work, which does not merely meet the standards required for a doctoral dissertation, but far exceeds them."

Gauss was not mistaken. (About mathematics, I doubt he ever was.) Riemann's doctoral dissertation is a key work in the history of complex function theory. I shall attempt an explanation of complex function theory in Chapter 13. For the time being, suffice it to say that it is a very deep, powerful, and beautiful branch of analysis. To this day, almost the first things you learn in a course on complex function theory are the Cauchy-Riemann equations for a function to be well behaved and worthy of further investigation. These equations first appear in their modern form in Riemann's doctoral dissertation. The paper also contains the first sketches of the theory of Riemann surfaces, a fusion of function theory with topology—the latter topic so new at the time there was really no coherent body of knowledge about it, only some scattered results going back to Euler's time.[41] Riemann's doctoral thesis is, in short, a masterpiece.

Both Riemann and Dedekind then embarked on the second leg of the academic marathon to which they had committed themselves, the habilitation thesis and trial lecture required for a teaching position at the university.

III. Let us leave Bernhard Riemann for a while, toiling away at that habilitation thesis in his room at Göttingen, and step a year or two back in time, and a thousand miles away in space, to St. Petersburg. Considerable water has flowed under the bridges of that city since last we were there, watching Leonhard Euler living contentedly and working productively, even though old and blind, under the rule of Catherine the Great. Euler died in 1783, the Empress herself in 1796. Catherine was succeeded by her eccentric and irresponsible son Paul. Four and a half years of Paul proved enough for the nobility, who staged a coup, garroted Paul, and replaced him with his son Alexander.

The nation was then soon absorbed in the conflict with Napoleon, and her French-speaking aristocracy in the glittering social scenes drawn by Tolstoy in *War and Peace.* After a postwar spell of managerial despotism under Alexander, and in spite of the failed revolt of the liberalizing faction known as Decembrists, the throne passed in 1825 to the more old-fashioned absolutism of Nicholas I.

However, the reassertion, and re-reassertion, of the absolutist principle had not prevented great social changes, most memorably the first great flowering of modern Russian literature under Pushkin, Lermontov, and Gogol.[42] The university at St. Petersburg, now a separate institution from the Academy, had grown and flourished, and new universities had been established in Moscow, Kharkov, and Kazan. In Kazan, the university boasted the presence of the great mathematician Nikolai Lobachevsky, who served as Rector until his dismissal in 1846. Lobachevsky was the inventor of non-Euclidean geometry, of which I shall have more to say shortly.[43]

And now, in 1849–1850, 25 years into the reign of Nicholas I, intellectual life in Russia was enduring another spell of repression, as Nicholas reacted to the 1848 revolutions in Europe. University enrolments were slashed and Russians studying abroad were ordered home. This was the environment in which a young lecturer at the university of St. Petersburg produced two remarkable papers on the PNT.

The first thing to be said about Pafnuty Lvovich Chebyshev is that his last name is a data-retrieval nightmare. Researching for this book, I turned up 32 different transcriptions of the name: Cebysev, Cebyshev, Chebichev, Chebycheff, Chebychev, etc., etc.

And if that unusual first name, Pafnuty, caught your eye, you are not alone. It caught the eye of mathematician Philip J. Davis around 1971. Davis embarked on a quest to find the origins of "Pafnuty," and wrote an extremely funny book about his researches, *The Thread* (1983). In very brief, the name "Pafnuty" is Coptic in origin (*Papnute* = "the man of God"), entered Europe via Egyptian Christianity, and was the name of a minor Church Father in the fourth century. Present at the Synod of Nicea, Bishop Paphnutius (as he is usually spelled) argued against priestly celibacy. A later Pafnuty noted by Davis *en passant* was St. Pafnuty of Borovsk, the son of a Tartar noble, who entered a monastery at age 20 and stayed there until he died, aged 94, in 1478. Says the hagiographer of this Pafnuty: "He was a virgin and an ascetic, and, because of this, a great wonderworker and seer." (In the middle of writing this chapter I got an e-mail from a reader of my web column asking me to suggest a name for her new dog. There is now a Pafnuty chasing squirrels somewhere in the Midwest.)

Our own Pafnuty was something of a wonder-worker himself. To him belongs the honor of having accomplished the only real advances toward a proof of the PNT in between Dirichlet's picking up the Golden Key in 1837 and Riemann's turning it in 1859. The curious thing is that his most original work did not flow into the mainstream of researches on the PNT, but started a lesser branch of the stream, which went underground, to emerge only 100 years later.

Chebyshev actually wrote two papers on the PNT. The first, dated 1849, is titled "On the Function that Determines the Totality of Prime Numbers Less Than a Given Limit"; notice the similarity to the title of Riemann's paper of 10 years later. In this paper Chebyshev picked up Euler's Golden Key, fiddled with it a little in much the way Dirichlet had 12 years before, and produced the following interesting result.

Chebyshev's First Result

If $\pi(N) \sim \frac{CN}{\log N}$ for some fixed number C,
then C must be equal to 1.

The problem, of course, was with that "if." Chebyshev could not get past it, and neither, for half a century, could anyone else.

Chebyshev's second paper, dated 1850, is much more curious. Instead of using the Golden Key, it began from a formula proved by the Scottish mathematician James Stirling in 1730 to get approximate values of the factorial function for large numbers. (The factorial of N is $1 \times 2 \times 3 \times 4 \times \ldots \times N$. The factorial of 5, for example, is 120: $1 \times 2 \times 3 \times 4 \times 5 = 120$. The usual symbol for the factorial of N is "$N!$" Stirling's formula says that for large values of N, the factorial of N is about $N^N e^{-N} \sqrt{2\pi N}$.) Chebyshev converted this into a different formula involving a step function—that is, a function that has the same value across a range of arguments, then jumps to another value.

With just these tools, and some very elementary calculus, Chebyshev got two important results. The first was a proof of "Bertrand's postulate," suggested in 1845 by the French mathematician Joseph Bertrand. The postulate states that between any number and its double (for example, between 42 and 84) there is always a prime to be found. The second was the one shown here.

Chebyshev's Second Result

$\pi(N)$ cannot differ from $\dfrac{N}{\log N}$ by more than

about ten percent up or down.

This second paper was important in two ways. First, its use of a step function might have inspired Riemann's use of a similar function in his 1859 paper, which I shall show in detail later. It is certain that Riemann knew of Chebyshev's work; the Russian mathematician's name appears in Riemann's notes (spelled "Tschebyschev").

It is Chebyshev's line of approach in that second paper that is more noteworthy, though. He got his results without using any complex function theory. Mathematicians have a shorthand way of expressing this fact. They say that Chebyshev's methods were "elementary." Riemann, in his 1859 paper, did not use elementary methods. He brought the full power of complex function theory to bear on the issue he was investigating. The results he got were so striking that other mathematicians followed him, and the PNT was proved at last using Riemann's non-elementary methods.

That it might be possible to prove the PNT by elementary methods remained an open issue, but by the time several decades had passed, the general opinion was that no such proof was possible. Thus, in Albert Ingham's 1932 text *The Distribution of Prime Numbers*, the author says in a footnote "[A] 'real variable' proof of the prime number theorem, that is to say a proof not involving explicitly or implicitly the notion of an analytic function of a complex variable, has never been discovered, and we can now understand why this should be so...."

Then, to everybody's astonishment, such a proof *was* discovered in 1949 by Atle Selberg, a Norwegian mathematician working at the Institute for Advanced Study in Princeton, New Jersey.[44] There was much controversy over the result, because Selberg had communicated some of his preliminary ideas to the eccentric Hungarian mathematician Paul Erdős, who used them to create a proof of his own at the same time. Two popular biographies of Erdős were produced after his death in 1996, and the curious reader can find a full account of the controversy in either. The proof is called the Erdős-Selberg proof in Hungary, and the Selberg proof elsewhere.

In addition to his research, Chebyshev was a great teacher and proselytizer for his subject. His disciples took his ideas and methods to other Russian universities, inspiring interest and raising standards everywhere. Active into his 70s, Chebyshev was also a keen inventor, who built a series of calculating machines still preserved at museums in Moscow and Paris. A lunar crater is named after him; it is at about 135°W 30°S.[45]

IV. I cannot leave Chebyshev without at least a passing mention of his famous bias—famous among number theorists, I mean.

If you divide a prime number (other than 2) by 4, the remainder must be either 1 or 3. Do the primes show any preference? Yes, they do: up to $p = 101$, there are 12 remainder-1 primes and 13 remainder-3's.

Up to $p = 1{,}009$, the tallies are 81 and 87. Up to $p = 10{,}007$, they are 609 and 620. Clearly, the remainder-3's have a small but persistent edge over the remainder-1's. This is an example of a Chebyshev bias, first remarked on by Chebyshev in a letter dated 1853. This particular bias is eventually violated at $p = 26{,}861$, when remainder-1's snatch a momentary lead. Even that is only a one-time aberration, though: the first real *zone* of violation is the 11 primes from $p = 616{,}877$ to 617,011. Remainder-1's hold the lead at only 1,939 of the first 5.8 million primes, which is as far as I checked. They don't hold it *once* in the last 4,988,472 of those primes.

With divisor 3, the bias is even more dramatic. Here, the remainder (once you get past $p = 3$) can be either 1 or 2, and the bias is to 2. This bias is not violated until $p = 608{,}981{,}813{,}029$. Now *that* is a bias! This violation was tracked down in 1978, by Carter Bays and Richard Hudson. I shall have occasion to mention the Chebyshev bias again, in Chapter 14.

V. In the fall of 1852, the first year of work on his habilitation thesis, Riemann met Dirichlet again. The whole episode is rather touching, and I transcribe it here from the biography by Dedekind.

> In the fall vacation of 1852, Lejeune Dirichlet stayed a while in Göttingen. Riemann, who had just returned from Quickborn, had the good fortune to see him almost daily. Both on his first visit to Dirichlet's lodgings and on the following day ... he consulted Dirichlet, who was recognized as the greatest living mathematician of the time after Gauss, for advice on his work. Riemann wrote to his father about the meeting: "The other morning, Dirichlet spent about two hours with me. He gave me notes I need for my Habilitation thesis—they are so comprehensive that my work has been substantially lessened. I would otherwise have had to spend a long time looking in the library for some of those things. He also went through my thesis with me, and all in all was very friendly towards me, which

> I could hardly have expected, considering the great disparity of rank between us. I hope he will not forget me in the future." Some days after this ... a large group of them went out on an excursion together—a very valuable trip, as after so many hours in company Riemann's reserve was much diminished. The following day, Dirichlet and Riemann met again in Weber's house. The stimulus provided by these personal contacts did Riemann a world of good. Still, he wrote about it to his father thus: "You see that I am not altogether housebound here; but the next morning I worked all the harder, and advanced as much as if I had sat with my books all day long."

That last remark shows the demands Riemann placed on himself, his powerful sense of duty, and his determination to justify every minute of his time at Göttingen to himself, to his father (who, after all, was supporting him), and to God.

The procedure for habilitating was that Riemann should first submit a written thesis, then prepare a trial lecture to be delivered before the faculty. The thesis itself—it is titled "On the representability of a function by a trigonometric series"—is a landmark paper, giving the world the Riemann integral, now taught as a fundamental concept in higher calculus courses. The habilitation lecture, however, far surpassed the thesis.

Riemann was supposed to offer three lecture titles from which Gauss, his supervisor, would pick one to be delivered. Riemann's three offerings were of two topics in mathematical physics and one in geometry. Gauss picked the lecture titled "On the Hypotheses that Lie at the Foundations of Geometry," and Riemann delivered it to the assembled faculty on June 10, 1854.

This is one of the top 10 mathematical papers ever delivered anywhere, a sensational achievement. Its reading was, declares Hans Freudenthal in the *Dictionary of Scientific Biography*, "one of the highlights in the history of mathematics." The ideas contained in this paper were so advanced that it was decades before they became fully accepted, and 60 years before they found their natural physical application, as the mathematical framework for Einstein's General Theory of Relativity. James R. Newman, in *The World of Mathematics*, refers to the paper as "epoch-making" and "imperishable" (but fails to include it in his huge anthology of classic mathematical texts). And the astonishing thing is that the paper contains almost no mathematical symbolism. Leafing through it, I see five equals signs, three square root signs and four Σ signs—an average of fewer than one symbol per page! There is just one real formula. The whole thing was written to be understood—or perhaps (see below) *mis*understood—by the average faculty member of a middling provincial university.

Riemann's starting point was some ideas Gauss had put forward in an 1827 paper titled "A General Investigation into Curved Surfaces." Gauss had been employed for the previous few years in carrying out a detailed topographical survey of the Kingdom of Bavaria (during which, by the way, he invented the heliotrope, a device for making long-distance observations by reflecting flashes of sunlight from an arrangement of mirrors). Gauss's stupendous mind had abstracted from the material he was dealing with some ideas about the properties of two-dimensional surfaces, and the way those properties might be described mathematically. Gauss's paper is generally regarded as the starting point for the subject named "differential geometry."

Riemann, in his habilitation lecture, took up these ideas and generalized them to spaces of any number of dimensions. More significantly, he brought in quite a new way of looking at the topic. Gauss saw it all, in his imagination, in terms of curved two-dimensional sheets embedded in ordinary three-dimensional space from which they could be viewed—the natural abstraction from his experiences as a land surveyor. Riemann shifted the point of view to one that was *interior* to the space under consideration.

I imagine you are familiar with the idea contained in Einstein's General Theory of Relativity, that the three dimensions of space and one of time can be dealt with mathematically as a four-dimensional space-time, and that this four-dimensional continuum is warped and puckered by the presence of mass and energy. From the Gaussian point of view, the geometry of this space-time would have been developed by imagining it imbedded in a five-dimensional continuum, in the way that Gauss thought of his two-dimensional surfaces as embedded in ordinary three-dimensional space. That modern physicists do *not* think of space-time in this way is due to Riemann. In fact, if you were to go down to your local university and sign up for a course in the General Theory of Relativity, these would be the topic headings you might cover, in order:

- The metric tensor
- The Riemann tensor
- The Ricci tensor
- The Einstein tensor
- The stress-energy tensor
- Einstein's equation $\mathbf{G} = 8\pi\mathbf{T}$

You would then have mastered the essentials of the General Theory.

Though I am concerned in this book to describe Riemann's discoveries in arithmetic and the great Hypothesis that sprang from them, these geometrical researches of his are not entirely off the topic. Riemann's general cast of mind, and all his best mathematical work, arose from a tension between two contrary sets of ideas. On the one hand he was a great globalist, whose tendency was always to see things in the large. A function was not, for Riemann, a mere set of points; still less was it any of its pictorial representations as a graph or a table; and still less a collection of expressions involving algebraic formulas. (In one of his few recorded negative comments about anyone at all, Riemann noted that the Berlin mathematician Gotthold Eisenstein "stopped at formal computation.") What, then, was a function? It was an *object*, from which none of its attributes could properly be detached. Riemann saw a function the way chess grandmasters are said to see a game, all at once, as a unified whole, a *Gestalt*.

Yet in tension with this was an opposing tendency, also very marked in Riemann's work—the tendency to reduce every mathematical topic to analysis. "Riemann ... always thought in analytic terms," says Laugwitz. The writer is thinking here of the infinitesimal aspect of analysis; of limits, continuity, smoothness—of the *local* properties of numbers, functions, and spaces. It is, when you think about it, very odd that inquiries about the infinitesimal neighborhoods of points and numbers should give us the power to explain the large global properties of functions and spaces. This is especially apparent in General Relativity theory, where you start off by analyzing microscopic regions of space-time and end by contemplating the shape of the universe and the death throes of galaxies. That we are able to think in this extraordinary way, in both pure and applied mathematics, is mainly due to the mathematicians of the early nineteenth century, and most of all to Bernhard Riemann.

That great habilitation lecture is, in fact, as much a philosophical document as a mathematical one. In this respect the much-remarked obscurity of some of its passages might have been deliberate on Riemann's part. (Though see Freudenthal's remark below.) What he was speaking about at its most fundamental was the nature of space. Now, to the average complacent elderly academic of the time—the kind of person who would have been among the Göttingen faculty listening to Riemann's lecture that June day—the nature of space was a settled matter. It had been settled 70 years earlier by Immanuel Kant in *The Critique of Pure Reason*. Space is a pre-existing part of our mental equipment, with which we organize our sense impressions, and it is necessarily Euclidean—that is, flat, with a straight line being the shortest distance between two points, and the angles of a triangle adding up to 180 degrees.

The non-Euclidean geometry described by Lobachevsky in the 1830s was, seen from this point of view, a philosophical heresy. Riemann's paper was an enlargement of that heresy; and this might be why he presented his ideas at such a very general level that their connection with non-Euclidean geometry would have escaped all but the most mathematically adept in his audience. (But not, of course, Gauss. Gauss had in fact invented non-Euclidean geometry for himself, but had not published his findings, "for fear," as he wrote in a letter to a friend, "of the hue and cry of the blockheads." Nineteenth-century Germans took their philosophy seriously.)

Hans Freudenthal, in the *Dictionary of Scientific Biography* note mentioned above, has the following to say about Riemann's philosophical abilities.

> One of the most profound and imaginative mathematicians of all time, he had a strong inclination to philosophy, indeed was a great philosopher. Had he lived and worked longer, philosophers would acknowledge him as one of them.

I am not qualified to judge whether this is true. I can, however, give wholehearted assent to another remark of Freudenthal's: "Riemann's style, influenced by philosophical reading, exhibits the worst aspects of German syntax; it must be a mystery to anyone who has not mastered German." I confess that, though I possess a copy of Riemann's collected works in the original German—it is a single volume of 690 pages—and have done my best with his actual words, where he departs from straightforward mathematical exposition—as, for example, in the habilitation lecture—I have approached his tremendous thoughts mainly through translations and secondary sources.[46]

VI. Dedekind habilitated shortly after Riemann, and both mathematicians began lecturing in the fall-winter term of 1854, Riemann now 28, Dedekind 23. For the first time in his life, Riemann had a salary. It can't have been much of a salary, though. Ordinary lecturers were paid by the students who attended their lectures (technically by the university, which forwarded the students' fees to the lecturers).

There were few students of mathematics at Göttingen at this time—Riemann's first lecture drew eight—and lectures were frequently canceled because nobody enrolled for them. Riemann and Dedekind seem to have attended each other's lectures, though whether they paid each other the requisite fees, I have not been able to discover.

There was further the problem that Riemann seems to have been a poor lecturer. Dedekind is frank about this.

> There is no doubt that lecturing caused large difficulties for Riemann in the first years of his academic career. His brilliant intellect and prescient imagination were usually not apparent. What appeared rather were large steps in the logic of his arguments, steps that were difficult for lesser intellects to follow. If he was asked to elaborate the missing links, he became flustered and could not adjust himself to the slower train of thought of the inquirer.... His attempts to judge from his students' expression whether he was going too fast or not, also disturbed him when, against his expectations, they caused him to feel that he should prove a point that seemed perfectly natural to him....

Dedekind, ever sympathetic to his subject, goes on to claim that Riemann's lecturing style improved over the years. This might be true; but surviving letters by Riemann's students suggest that as late as 1861 "His thoughts frequently failed him and he was unable to explain the simplest things." Riemann's own take on the matter is, as usual, rather touching. Writing to his father after his first lecture, which was on October 5, 1854, he says "I hope that in half a year I shall feel easier about my lectures, and the thought of them will not spoil my stay in Quickborn and my being together with you, as last time." This was a desperately shy man.

VII. The great event of that fall-winter term was the death of Gauss on February 23, 1855, at the age of 77. Though not in good health toward the end, he died quickly, of a heart attack, while sitting in his favorite chair in his beloved observatory.[47]

Gauss's professorship was immediately offered to Dirichlet, who accepted, arriving in Göttingen a few weeks later. Recalling how generously Dirichlet had treated him in Berlin, and their bonding during the older man's visit to Göttingen in 1852, Riemann must have been pleased. Gauss's brain, meanwhile, was pickled and stored in the university's physiology department, where it remains to this day.

Dirichlet was pleased, too; he had been seriously overworked in Berlin. Whether his wife was pleased is not so certain. Accustomed to the high society of Berlin, Rebecca Dirichlet, *née* Mendelssohn, must have thought Göttingen very dull and provincial. She did her best with the place, organizing balls—Dedekind mentions one attended by 60 or 70 people—and musical soirées in the Berlin style. Dedekind himself thrived in this environment, being sociable and musical. Riemann was, of course, a different case, and if his friend ever persuaded him to attend one of these functions, poor Riemann must have endured it in an agony of self-consciousness.

He experienced much deeper agony in October of that year, 1855, when his father died, followed very shortly by his younger sister Clara. Now the cherished link with Quickborn was broken. Riemann's brother had a position as postal clerk in Bremen and Riemann's three remaining sisters, having no other means of support, nor even accommodation (since the vicarage at Quickborn was taken over by the new pastor), went to live with him there.

Poor Riemann must have been devastated. He threw himself into work, and in 1857 produced the landmark paper on function theory that I mentioned in Chapter 1, the paper that made his name known. The effort, however, combined with grief, precipitated a nervous breakdown. Dedekind's family had a summer home in the Harz mountains a few miles west of Göttingen. He persuaded Riemann to spend a few weeks there and joined him briefly, going for walks with him.

After Riemann's return to Göttingen in November, he was appointed Assistant Professor at the university, with a modest salary of 300 thalers a year. But now calamities came thick and fast. His brother Wilhelm died in Bremen that same month, then, early the following year, his sister Marie. The family that Riemann adored, and that was the entire focus of his emotional life, was disappearing before his eyes. He brought the two surviving sisters to stay with him at Göttingen.

In the summer of 1858 Dirichlet suffered a heart attack while lecturing in Switzerland and was brought back to Göttingen only with much difficulty. While he was lying gravely ill, his wife died suddenly of a stroke. Dirichlet himself followed her the next May. (His brain joined Gauss's in the department of physiology.) Gauss's chair was now empty.

VIII. From the death of Gauss to the death of Dirichlet was four years, two months, and twelve days. In that span, Riemann lost not only the two colleagues he had esteemed above all other mathematicians, but also his father, his brother, two of his sisters, and the vicarage at Quickborn—the one place that had been a home and refuge to him since his infancy.

While his emotional life had been visited by these traumas, Riemann's star in the world of mathematics had been rising. By the end of the 1850s, the brilliance and originality of his work were known, at some level, to mathematicians all over Europe. The painfully shy young student who had shown up to begin his doctoral studies 10 years earlier was now a mathematician of note, and Göttingen, which had entered the 1850s as the home of Gauss, was beginning to be spoken of as the home of Gauss, Dirichlet, and Riemann. (Though not of Dedekind, whose best work was still in front of him. Dedekind had, in fact, left Göttingen to take up a post in Zürich in the fall of 1858.)

It was, therefore, not very surprising that the authorities selected Riemann as the second successor of Gauss. On July 30, 1859, he was given a full professorship, an assured livelihood, and—probably as an acknowledgment of his need to support his two surviving sisters—Gauss's apartments at the observatory. Other honors soon followed. The first came on August 11, when he was appointed a corresponding member of the Berlin Academy. Riemann returned to Berlin a little more than 10 years after he had left, but now he came with a modest set of laurels on his brow, to be received with honor by the great names of German mathematics: Kummer, Kronecker, Weierstrass, Borchardt.

To crown his triumph, Riemann gave the Academy his paper on "the number of primes less than a given quantity." In the paper's first sentence he acknowledged the two men, both now dead, with whose aid—though given much more willingly in Dirichlet's case than in Gauss's—he had scaled the heights. In the second sentence he showed the Golden Key. In the third he named the zeta function. Here, in fact, are the first three sentences of Riemann's 1859 paper.

> For the consideration which the Academy has shown to me by admitting me as one of its corresponding members, I believe I can best express my thanks by availing myself at once of the privilege thereby given me to communicate an inquiry into the frequency of prime numbers; a subject which, through the interest shown in it by Gauss and Dirichlet over a long period, appears not altogether unworthy of such a communication.
>
> I take as my starting-point for this inquiry Euler's observation that the product
>
> $$\prod \frac{1}{1-\frac{1}{p^s}} = \sum \frac{1}{n^s}$$
>
> for all prime numbers p and all whole numbers n. The function of a complex variable s which both these expressions stand for, so long as they converge, I signify by $\zeta(s)$.

The Riemann Hypothesis, which appears on the fourth page of that paper, asserts a certain fact about the zeta function. To advance in our understanding of the Hypothesis, we must now go deeper into the zeta function.

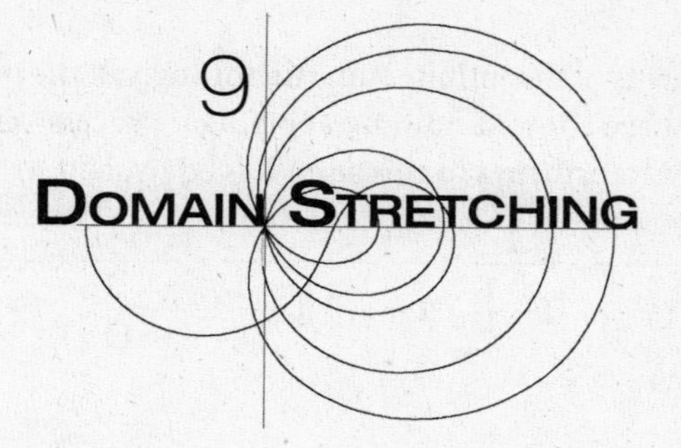

9 Domain Stretching

I. We are starting to close in on the Riemann Hypothesis. Let me state it again, just as a refresher.

The Riemann Hypothesis
All non-trivial zeros of the zeta function
have real part one-half.

Well, we've got a handle on the zeta function. If s is some number bigger than 1, the zeta function is as shown in Expression 9-1.

$$\zeta(s)=1+\frac{1}{2^s}+\frac{1}{3^s}+\frac{1}{4^s}+\frac{1}{5^s}+\frac{1}{6^s}+\frac{1}{7^s}+\frac{1}{8^s}+\frac{1}{9^s}+\frac{1}{10^s}+\frac{1}{11^s}+\cdots$$

Expression 9-1

or, to be somewhat more sophisticated about it

$$\zeta(s)=\sum_n n^{-s}$$

where the terms of the infinite sum run through all the positive whole numbers. I have showed how, by applying a process very much like the sieve of Eratosthenes to this sum, it is equivalent to

$$\zeta(s)=\frac{1}{1-\frac{1}{2^s}}\times\frac{1}{1-\frac{1}{3^s}}\times\frac{1}{1-\frac{1}{5^s}}\times\frac{1}{1-\frac{1}{7^s}}\times\frac{1}{1-\frac{1}{11^s}}\times\frac{1}{1-\frac{1}{13^s}}\times\cdots$$

that is,

$$\zeta(s)=\prod_p\left(1-p^{-s}\right)^{-1}$$

where the terms of the infinite product run through all the primes.

And so

$$\sum_n n^{-s}=\prod_p\left(1-p^{-s}\right)^{-1}$$

which I have called the Golden Key.

So far, so good, but what is this about non-trivial zeros? What is a zero of a function? What are the zeros of the zeta function? And when are they non-trivial? Onward and upward.

II. Forget about the zeta function for a moment. Here is a completely different infinite sum.

$$S(x)=1+x+x^2+x^3+x^4+x^5+x^6+\cdots$$

Does this ever converge? Sure. If x is $\frac{1}{2}$, the sum is just Expression 1-1 in Chapter 1.iv, because $\left(\frac{1}{2}\right)^2=\frac{1}{4}$, $\left(\frac{1}{2}\right)^3=\frac{1}{8}$, etc. Therefore, $S\left(\frac{1}{2}\right)=2$, because that's what that sum converged to. What's more, if you think about the rule of signs, $\left(-\frac{1}{2}\right)^2=\frac{1}{4}$, $\left(-\frac{1}{2}\right)^3=-\frac{1}{8}$, etc. Therefore, $S\left(-\frac{1}{2}\right)=\frac{2}{3}$, from Expression 1-2 in Chapter 1.v. Similarly, Expression 1-3 means that $S\left(\frac{1}{3}\right)=1\frac{1}{2}$, while Expression 1-4 gives

$S\left(-\frac{1}{3}\right)=\frac{3}{4}$. Another easy value for this function is $S(0) = 1$, since zero squared, zero cubed, and so on are all zero, and only the initial 1 is left standing.

If x is 1, however, $S(1)$ is $1 + 1 + 1 + 1 + \ldots$, which diverges. If x is 2, the divergence is even more obvious, $1 + 2 + 4 + 8 + 16 + \ldots$. When x is -1 a weird thing happens. By the rule of signs, the sum becomes $1 - 1 + 1 - 1 + 1 - 1 + \ldots$. This adds up to zero if you take an even number of terms, to one if you take an odd number. This is definitely not going off to infinity, but it isn't converging, either. Mathematicians consider it a form of divergence. For -2 things are even worse. The sum is $1 - 2 + 4 - 8 + 16 - \ldots$, which seems to go off to infinity in two different directions at once. Again, you definitely can't call this convergence, and if you call it divergence, nobody will argue with you.

In short, $S(x)$ has values only when x is between -1 and 1, exclusive. Elsewhere it has no values. Table 9-1 shows values of $S(x)$ for arguments x between -1 and 1.

TABLE 9-1 Values of $S(x) = 1 + x + x^2 + x^3 + \cdots$

x	$S(x)$
−1 or below	(No values)
−0.5	0.6666…
−0.3333…	0.75
0	1
0.3333…	1.5
0.5	2
1 or above	(No values)

That's all you can get from the infinite sum. If you make a graph, it looks like Figure 9-1, with no values at all for the function west of -1 or east of 1. If you remember the term of art, the *domain* of this function is from -1 to 1, exclusive.

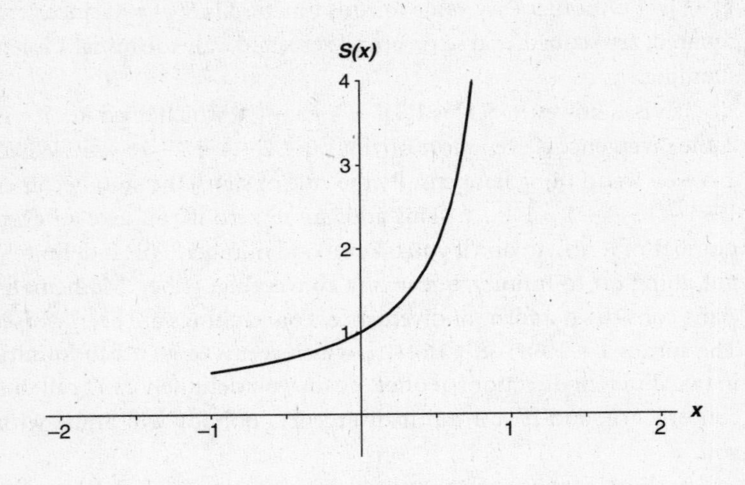

FIGURE 9-1 The function $S(x) = 1 + x + x^2 + x^3 + \cdots$

III. But look, I can rewrite that sum

$$S(x) = 1 + x + x^2 + x^3 + x^4 + x^5 + \cdots$$

like this

$$S(x) = 1 + x(1 + x + x^2 + x^3 + x^4 + \cdots)$$

Now, that series in the parenthesis is just $S(x)$. Every term that is in the one is also in the other. That means they are the same.

In other words, $S(x) = 1 + xS(x)$. Bringing the rightmost term over to the left of the equals sign, $S(x) - xS(x) = 1$, which is to say $(1 - x)\ S(x) = 1$. Therefore, $S(x) = 1/(1 - x)$. Can it be that behind that infinite sum is the perfectly simple function $1/(1 - x)$? Can it be that Expression 9-2 is true?

$$\frac{1}{1-x} = 1 + x + x^2 + x^3 + x^4 + x^5 + x^6 + \cdots$$

Expression 9-2

It certainly can. If $x = \frac{1}{2}$, for example, then $1/(1-x)$ is $1/(1-\frac{1}{2})$, which is 2. If $x=0$, $1/(1-x)$ is $1/(1-0)$, which is 1. If $x=-\frac{1}{2}$, $1/(1-x)$ is $1/(1-(-\frac{1}{2}))$, which is $1/1\frac{1}{2}$, which is $\frac{2}{3}$. If $x=\frac{1}{3}$, $1/(1-x)$ is $1/(1-\frac{1}{3})$, which is $1/\frac{2}{3}$, which is $1\frac{1}{2}$. If $x=-\frac{1}{3}$, $1/(1-x)$ is $1/(1-(-\frac{1}{3}))$, which is $1/1\frac{1}{3}$, which is $\frac{3}{4}$. It all checks out. For all the arguments $-\frac{1}{2}, -\frac{1}{3}, 0, \frac{1}{3}, \frac{1}{2}$, for which we know a function value, the value is the same for the infinite series $S(x)$ as it is for the function $1/(1-x)$. Looks like they are actually the same thing.

But they are not the same thing, because *they have different domains*, as Figures 9-1 and 9-2 illustrate. $S(x)$ only has values between −1 and 1, exclusive. By contrast, $1/(1-x)$ has values everywhere, except at $x=1$. If $x=2$, it has the value $1/(1-2)$, which is −1. If $x=10$, it has the value $1/(1-10)$, which is $-\frac{1}{9}$. If $x=-2$, it has the value $1/(1-(-2))$, which is $\frac{1}{3}$. I can draw a graph of $1/(1-x)$. You see that it is the same as the previous graph between −1 and 1, but now it has values east of 1 and west of (and including) −1, too.

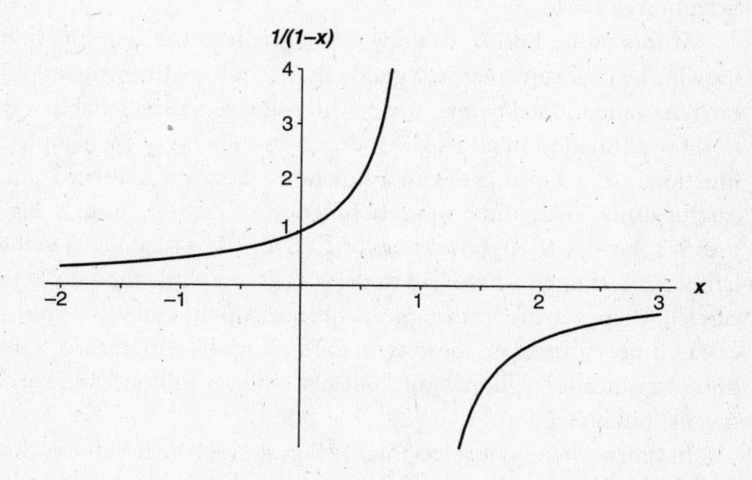

FIGURE 9-2 The function $1/(1-x)$.

The moral of the story is that an infinite series might define only part of a function; or, to put it in proper mathematical terms, an infinite series may define a function over only part of its domain. The rest of the function might be lurking away somewhere, waiting to be discovered by some trick like the one I did with $S(x)$.

IV. That raises the obvious question: is this the case with the zeta function? Does the infinite sum I've been using for the zeta function—Expression 9-1—describe only part of it? With more yet to be discovered? Is it possible that the domain of the zeta function

$$\zeta(s)=1+\frac{1}{2^s}+\frac{1}{3^s}+\frac{1}{4^s}+\frac{1}{5^s}+\frac{1}{6^s}+\frac{1}{7^s}+\frac{1}{8^s}+\frac{1}{9^s}+\frac{1}{10^s}+\frac{1}{11^s}+\cdots$$

is bigger than just "all numbers greater than 1"?

Of course it is. Why would I be going to all this trouble otherwise? Yes, the zeta function has values for arguments less than 1. In fact, like $1/(1-x)$, it has a value at *every* number with the single exception of $s = 1$.

At this point, I'd like to draw you a graph of the zeta function showing all its features across a good range of values. Unfortunately, I can't. As I mentioned before, there is no really good and reliable way to show a function in all its glory, except in the case of the simplest functions. To get intimate with a function takes time, patience, and careful study. I can graph the zeta function piecemeal, though. Figures 9-3 through 9-10 show values of $\zeta(s)$ for some arguments to the left of $s = 1$, though I have had to draw each to a different scale. You can tell where you are by the argument (horizontal) and value (vertical) numbers printed on the axes. In the scale marks, "m" means "million," "tr" means "trillion," "mtr" means "million trillion," and "btr" means "billion trillion."

In short, when s is just less than 1 (Figure 9-3), the function value is very large but negative—as if, when you cross the line $s = 1$ heading west, the value suddenly flips from infinity to minus infinity. If you continue traveling west along Figure 9-3—that is, bringing s closer and closer to zero—the rate of climb slows down dramatically. When s is zero, $\zeta(s)$ is $-\frac{1}{2}$. At $s = -2$ the curve crosses the s-axis—that is, $\zeta(s)$ is zero.

It then (we are still headed west, and now in Figure 9-4) climbs up to a modest height (actually 0.009159890…) before turning down and crossing the axis again at $s = -4$. The graph drops down to a shallow trough (−0.003986441…) before rising again to cross the axis at $s = -6$. Another low peak (0.004194), a drop to cross the axis at $s = -8$, a slightly deeper trough (−0.007850880…), across the axis at −10, now a really noticeable peak (0.022730748…), across the axis at $s = -12$, a deep trough (−0.093717308…), across the axis at $s = -14$, and so on.

The zeta function is zero at every negative even number, and the successive peaks and troughs now (Figures 9-5 to 9-10) get rapidly more and more dramatic as you head west. The last trough I show, which occurs at $s = -49.587622654…$, has a depth of about 305,507,128,402,512,980,000,000. You see the difficulty of graphing the zeta function all in one piece.

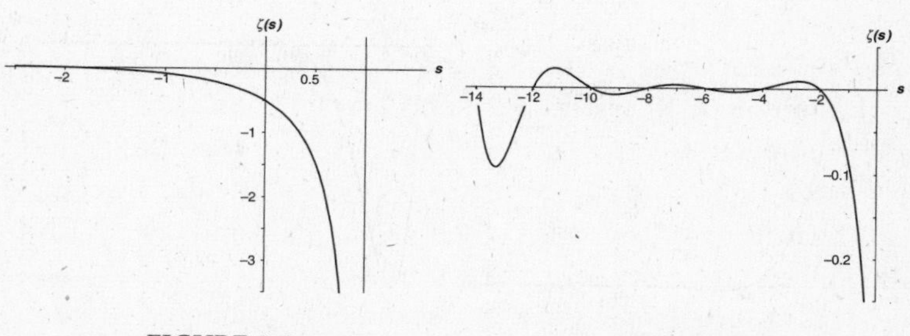

FIGURE 9-3 FIGURE 9-4

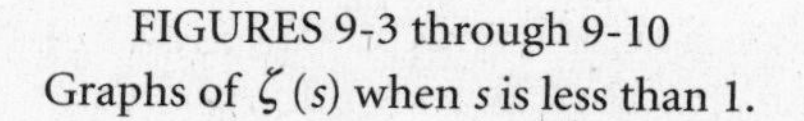

FIGURES 9-3 through 9-10
Graphs of $\zeta(s)$ when s is less than 1.

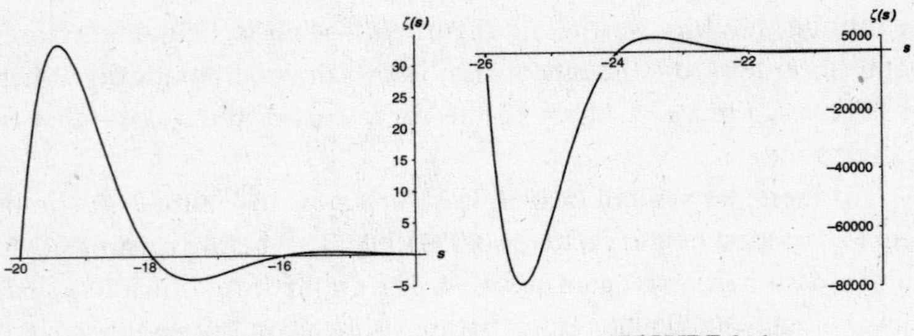

FIGURE 9-5

FIGURE 9-6

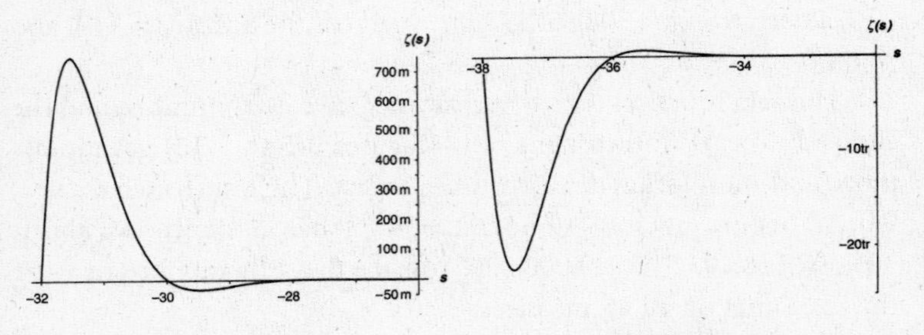

FIGURE 9-7

FIGURE 9-8

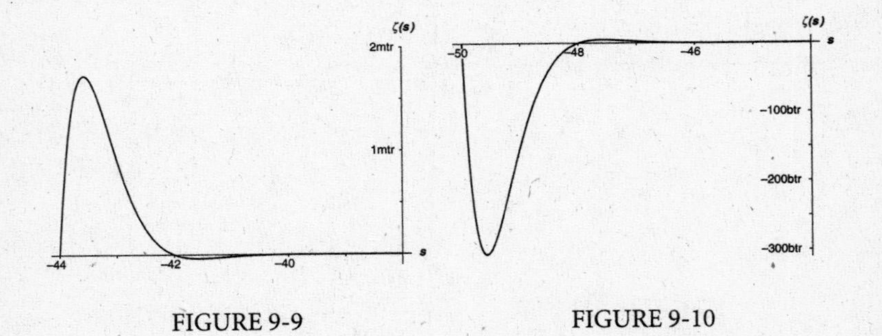

FIGURE 9-9

FIGURE 9-10

V. But how do I get these values for $\zeta(s)$ when s is less than 1? I've already shown that the infinite series in Expression 9-1 doesn't work. What *does* work? If, to save my life, I had to calculate the value of $\zeta(-7.5)$, how would I set about it?

This I can't fully explain, because it needs way too much calculus. I can give the general idea, though. First, let me define a new function, using an infinite series slightly different from the one in Expression 9-1. This is the η function; "η" is "eta," the seventh letter of the Greek alphabet, and I define the eta function as

$$\eta(s)=1-\frac{1}{2^s}+\frac{1}{3^s}-\frac{1}{4^s}+\frac{1}{5^s}-\frac{1}{6^s}+\frac{1}{7^s}-\frac{1}{8^s}+\frac{1}{9^s}-\frac{1}{10^s}+\frac{1}{11^s}-\cdots$$

In a rough sort of way, you can see that this has a better prospect of converging than Expression 9-1. Instead of relentlessly adding numbers, we are alternately adding, then subtracting, so each number will to some extent cancel out the effect of the previous number. So it happens. Mathematicians can prove, in fact—though I'm not going to prove it here—that this new infinite series converges whenever s is greater than zero. This is a big improvement on Expression 9-1, which converges only for s greater than 1.

What use is that for telling us anything about the zeta function? Well, first note the elementary fact of algebra that $A - B + C - D + E - F + G - H + \ldots$ is equal to $(A + B + C + D + E + F + G + H + \ldots)$ minus $2 \times (B + D + F + H + \ldots)$.

So I can rewrite $\eta(s)$ as

$$\left(1+\frac{1}{2^s}+\frac{1}{3^s}+\frac{1}{4^s}+\frac{1}{5^s}+\frac{1}{6^s}+\frac{1}{7^s}+\frac{1}{8^s}+\frac{1}{9^s}+\frac{1}{10^s}+\cdots\right)$$

minus

$$2\times\left(\frac{1}{2^s}+\frac{1}{4^s}+\frac{1}{6^s}+\frac{1}{8^s}+\frac{1}{10^s}+\cdots\right)$$

The first parenthesis is of course just $\zeta(s)$. The second parenthesis can be simplified by Power Rule 7, $(ab)^n = a^n b^n$. So every one of those even numbers can be broken up like this: $\frac{1}{10^s} = \frac{1}{2^s} \times \frac{1}{5^s}$, and I can take out $\frac{1}{2^s}$ as a factor of the whole parenthesis. Leaving what inside the parenthesis? Leaving $\zeta(s)$! In a nutshell

$$\eta(s) = \left(1 - 2 \times \frac{1}{2^s}\right) \text{ times } \zeta(s)$$

or, writing it the other way round and doing a last bit of tidying

$$\zeta(s) = \eta(s) \div \left(1 - \frac{1}{2^{s-1}}\right)$$

Now, this means that if I can figure out a value for $\eta(s)$, then I can easily figure out a value for $\zeta(s)$. And since I can figure out values for $\eta(s)$ between 0 and 1, I can get a value for $\zeta(s)$ in that range, too, in spite of the fact that the "official" series for $\zeta(s)$ (Expression 9-1) doesn't converge there.

Suppose s is $\frac{1}{2}$, for example. If I add up 100 terms of $\eta\left(\frac{1}{2}\right)$ I get 0.555023639...; if I add up 10,000 I get 0.599898768.... In fact, $\eta\left(\frac{1}{2}\right)$ has the value 0.604898643421630370.... (There are shortcuts for doing this without adding up zillions of terms.) Armed with this, I can calculate a value for $\zeta\left(\frac{1}{2}\right)$; it comes out to −1.460354508..., which looks pretty much right, based on the first one in that last batch of graphs.

But hold on there a minute. How can I juggle these two infinite series at the argument $s = \frac{1}{2}$, where one of the series converges and one doesn't? Well, strictly speaking, I can't, and I have been playing a bit fast and loose with the underlying math here. I got the right answer, though, and could repeat the trick for any number between zero and 1 (exclusive), and get a correct value for $\zeta(s)$.

VI. Except for the single argument $s = 1$, where $\zeta(s)$ has no value, I can now provide a value for the zeta function at every number s greater than zero. How about arguments equal to or less than zero? Here things get really tough. One of the results in Riemann's 1859 paper proves a formula first suggested by Euler in 1749, giving

$\zeta(1-s)$ in terms of $\zeta(s)$. So if you want to know the value of, say, $\zeta(-15)$, you can just calculate $\zeta(16)$ and feed it into the formula. It's a heck of a formula, though, and I give it here just for the sake of completeness.[48]

$$\zeta(1-s) = 2^{1-s}\pi^{-s}\sin\left(\frac{1-s}{2}\pi\right)(s-1)!\,\zeta(s)$$

Here π, in both occurrences, is the magic number 3.14159265…, "sin" is the good old trigonometric sine function (with the argument in radians), and ! is the factorial function I mentioned in Chapter 8.iii. In high school math, you meet the factorial function only in relation to positive whole numbers: $2! = 1 \times 2$, $3! = 1 \times 2 \times 3$, $4! = 1 \times 2 \times 3 \times 4$, and so on. In advanced math, though, there is a way to define the factorial function for all numbers except the negative integers, by a domain-stretching exercise not unlike the one I just did. For example, $\left(\frac{1}{2}\right)!$ turns out to be 0.8862269254…(half the square root of π, in fact), $\left(-\frac{1}{4}\right)! = 1.2254167024…$, etc. The negative integers create problems in the formula, but they are not major problems, and I shall say nothing about them here. Figure 9-11 shows the full factorial function, for arguments from –4 to 4.

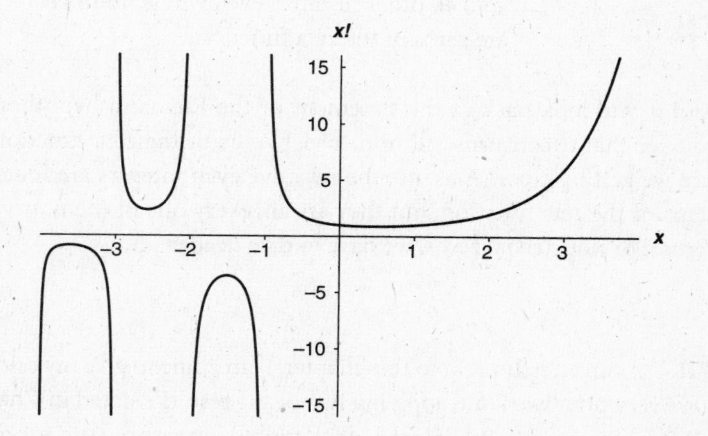

FIGURE 9-11 The full factorial function *x*!

If you find that a little over the top, just take it on faith that there is a way to get a value of $\zeta(s)$ for any number s, with the single exception of $s = 1$. Even if that last formula bounces right off your eye, at least notice this: it gives $\zeta(1 - s)$ in terms of $\zeta(s)$. That means that if you know $\zeta(16)$ you can calculate $\zeta(-15)$; if you know $\zeta(4)$ you can calculate $\zeta(-3)$; if you know $\zeta(1.2)$ you can calculate $\zeta(-0.2)$; if you know $\zeta(0.6)$ you can calculate $\zeta(0.4)$; if you know $\zeta(0.50001)$ you can calculate $\zeta(0.49999)$; and so on. The point I'm getting at is that the argument "one-half" has a special status in this relationship between $\zeta(1 - s)$ and $\zeta(s)$, because if $s = \frac{1}{2}$, then $1 - s = s$. Obviously—obviously, I mean, from glancing at Figure 5-4 and Figures 9-3 through 9-10—the zeta function is not symmetrical about the argument $\frac{1}{2}$; but the values for arguments to the left of $\frac{1}{2}$ are bound up with their mirror images on the right in an intimate, though complicated, way.

Glancing back at that last bunch of graphs, you notice something else: $\zeta(s)$ is zero whenever s is a negative even number. Now, if a certain argument gives the function a value of zero, that argument is called "a zero of" the function. So the following statement is true.

–2, –4, –6, ... and all other negative even whole numbers are zeros of the zeta function.

And if you look back at the statement of the Riemann hypothesis, you see that it concerns "all non-trivial zeros of the zeta function." Are we getting close? Alas, no, the negative even integers are indeed zeros of the zeta function; but they are all, every one of them, trivial zeros. For non-trivial zeros, we have to dive deeper yet.

VII. As an afterthought to this chapter, I am going to give my calculus a very brief workout, applying two of the results I stated in Chapter 7 to Expression 9-2. Here is that expression again, true for any number x between –1 and 1, exclusive.

$$\frac{1}{1-x} = 1+x+x^2+x^3+x^4+x^5+x^6+\cdots$$

Expression 9-2, again

All I intend to do is integrate both sides of the equals sign. Since the integral of $1/x$ is $\log x$, I hope it won't be too much of a stretch to believe—I shall not pause to prove it—that the integral of $1/(1-x)$ is $-\log(1-x)$. The right-hand side is even easier. I can just integrate term by term, using the rules for integrating powers that I gave in Table 7-2. Here is the result (which was first obtained by Sir Isaac Newton).

$$-\log(1-x) = x+\frac{x^2}{2}+\frac{x^3}{3}+\frac{x^4}{4}+\frac{x^5}{5}+\frac{x^6}{6}+\frac{x^7}{7}+\cdots$$

It will be a little handier, as you can see in Expression 9-3, if I multiply both sides by -1.

$$\log(1-x) = -x-\frac{x^2}{2}-\frac{x^3}{3}-\frac{x^4}{4}-\frac{x^5}{5}-\frac{x^6}{6}-\frac{x^7}{7}-\cdots$$

Expression 9-3

Oddly, though it makes little difference to the way I shall apply it, Expression 9-3 is true when $x=-1$, even though the expression I started with, Expression 9-2, isn't. When $x=-1$, in fact, Expression 9-3 gives the result shown in Expression 9-4.

$$\log 2 = 1-\frac{1}{2}+\frac{1}{3}-\frac{1}{4}+\frac{1}{5}-\frac{1}{6}+\frac{1}{7}-\cdots$$

Expression 9-4

Note the similarity to the harmonic series. Harmonic series … prime numbers … zeta …. This whole field is dominated by the log function.

The right-hand side of Expression 9-4 is slightly peculiar, though this is not obvious to the naked eye. It is, in fact, a textbook example of the trickiness of infinite series. It converges to log 2, which is

0.6931471805599453…, but *only if you add up the terms in this order.* If you add them up in a different order, the series might converge to something different; or it might not converge at all![49]

Consider this rearrangement, for example, $1 - \frac{1}{2} - \frac{1}{4} + \frac{1}{3} - \frac{1}{6} - \frac{1}{8} + \frac{1}{5} - \frac{1}{10} - \ldots$. Just putting in some parentheses, it is equal to $(1 - \frac{1}{2}) - \frac{1}{4} + (\frac{1}{3} - \frac{1}{6}) - \frac{1}{8} + (\frac{1}{5} - \frac{1}{10}) - \ldots$. If you now resolve the parentheses, this is $\frac{1}{2} - \frac{1}{4} + \frac{1}{6} - \frac{1}{8} + \frac{1}{10} - \ldots$, which is to say $\frac{1}{2}(1 - \frac{1}{2} + \frac{1}{3} - \frac{1}{4} + \frac{1}{5} - \ldots)$. The series thus rearranged adds up to one-half of the un-rearranged series!

The series in Expression 9-4 is not the only one with this rather alarming property. Convergent series fall into two categories: those that have this property, and those that don't. Series like this one, whose limit depends on the order in which they are summed, are called "conditionally convergent." Better-behaved series, those that converge to the same limit no matter how they are rearranged, are called "absolutely convergent." Most of the important series in analysis are absolutely convergent. There is another series that is of vital interest to us, though, that is only conditionally convergent, like the one in Expression 9-4. We shall meet that series in Chapter 21.

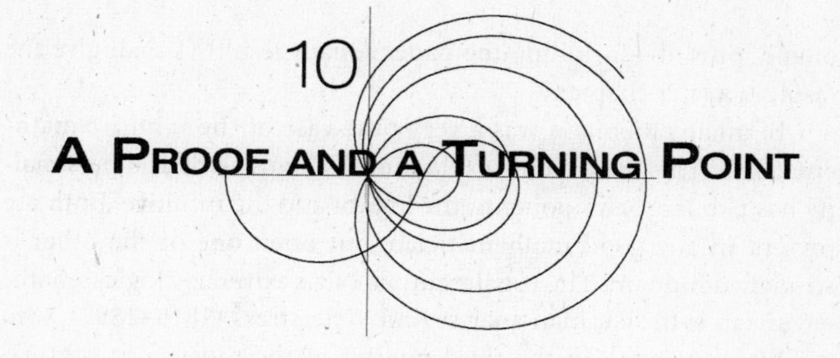

10 A Proof and a Turning Point

I. The 1859 paper "On the Number of Prime Numbers Less Than a Given Quantity" was Bernhard Riemann's only publication on number theory, and the only one of his productions that contained no geometrical ideas at all.

The paper, though dazzling and seminal, was in some respects unsatisfactory. There was, first of all, the great Hypothesis, which Riemann left hanging in the air (where it still hangs). His actual words, after making a statement that is equivalent to the Hypothesis, were

> One would, of course, like to have a rigorous proof of this, but I have put aside the search for such a proof after some fleeting vain attempts (*einigen flüchtigen vergeblichen Versuchen*) because it is not necessary for the immediate objective of my investigation.

Fair enough. Since the Hypothesis was not crucial to the ideas he was pursuing, Riemann left it unproved. That, however, is the least of the paper's deficiencies. Several other things are asserted but not thoroughly proved—including the paper's main result! (I shall give the result in a later chapter.)

Bernhard Riemann was a very pure case of the *intuitive* mathematician. This needs some explaining. The mathematical personality has two large components, the logical and the intuitive. Both are present in any good mathematician, but often one or the other is strongly dominant. The usual example of an extremely logical mathematician is the German analyst Karl Weierstrass (1815–1897), who did his great work in the third quarter of the nineteenth century. Reading Weierstrass's papers is like watching a rock climber. Every step is firmly anchored in proof before the next step is taken. Poincaré said that none of Weierstrass's books contained any diagrams. There is, in fact, just one exception to that, but certainly the precise logical progression of Weierstrass's work, with every least fact carefully justified before proceeding to the next, and no appeals to geometrical intuition at all, is representative of the logical mathematician.

Riemann is at the other pole. If Weierstrass is a rock climber, inching his way methodically up the cliff face, Riemann is a trapeze artist, launching himself boldly into space in the confidence—which to the observer often seems dangerously misplaced—that when he arrives at his destination in the middle of the sky, there will be something there for him to grab. It is plain that Riemann had a strongly visual imagination, and also that his mind leaped to results so powerful, elegant, and fruitful that he could not always force himself to pause to prove them. He was keenly interested in philosophy and physics, and notions gathered from long, deep contemplation of those two disciplines—the flow of sensations through our senses, the organizing of those sensations into forms and concepts, the flow of electricity through a conductor, the movements of liquids and gases—can be glimpsed beneath the surface of his mathematics.

The 1859 paper is therefore revered not for its logical purity, and certainly not for its clarity, but for the sheer originality of the methods Riemann used, and for the great scope and power of his results, which have provided, and will yet provide, Riemann's fellow mathematicians with decades of research.

In his book on the zeta function,[50] Harold Edwards has this to say about what followed that 1859 paper.

> For the first 30 years after Riemann's paper was published, there was virtually no progress in the field. It was as if it took the mathematical world that much time to digest Riemann's ideas. Then, in a space of less than 10 years, Hadamard, von Mangoldt and de la Vallée Poussin succeeded in proving both Riemann's main formula for $\pi(x)$ and the prime number theorem, as well as a number of other related theorems. In all these proofs Riemann's ideas were crucial.

II. Riemann's "On the Number of Prime Numbers Less Than a Given Quantity" had a direct bearing on efforts to prove the Prime Number Theorem (PNT). If the Riemann Hypothesis were true, the PNT would follow as a consequence. However, the Hypothesis is a much stronger result than the PNT, and the latter could be proved from weaker premises. The main significance of Riemann's paper for the proof of the PNT is that it provided the tools—the deep insights into analytic number theory that showed the way to a proof.

That proof came in 1896. The landmarks between Riemann's paper and the proof of the PNT were as follows.

- There was an increase in the practical knowledge of prime numbers. Longer tables of primes were published, notably Kulik's, deposited at the Vienna Academy in 1867, which provided factors for all numbers up to 100,330,200. Ernst Meissel developed a clever way to work out $\pi(x)$, the prime counting function. In 1871 he produced a correct value for $\pi(100{,}000{,}000)$. In 1885 he computed a value for

π (1,000,000,000), which was short by 56 (though this was not discovered until 70 years later).

- In 1874, Franz Mertens proved a modest result about the series of reciprocals of primes, using methods that owed something to both Riemann and Chebyshev. That series, by the way, the series $\frac{1}{2}+\frac{1}{3}+\frac{1}{5}+\frac{1}{7}+\frac{1}{11}+\frac{1}{13}+\frac{1}{17}+\cdots+\frac{1}{p}+\ldots$, diverges, though even more slowly than the harmonic series. It is ~log(log p).
- In 1881, J.J. Sylvester at Johns Hopkins University in the United States improved Chebyshev's limits (see Chapter 8.iii) from 10 percent to 4 percent.
- In 1884 the Danish mathematician Jørgen Gram published a paper titled "Investigations of the Number of Primes Less Than a Given Number" and won a prize for it from a Danish mathematical society. (The paper made no important advances but laid the groundwork for Gram's later efforts, which we shall examine in due course.)
- In 1885 the Dutch mathematician Thomas Stieltjes claimed to have a proof of the Riemann Hypothesis. More on this shortly.
- In 1890 the French Académie des Sciences announced that a grand prize would be awarded for a paper on the topic "Determination of the number of prime numbers less than a given quantity." The deadline for presentation was June 1892. It was made plain in the announcement that the Académie was soliciting work that would supply some of the proofs missing from Riemann's 1859 paper. The young French mathematician Jacques Hadamard submitted a paper concerning the representation of certain kinds of functions in terms of their zeros. Riemann had relied on this result to get a formula for $\pi(x)$; it is on this point—I shall explain the math in more detail later—that the connection between prime numbers and the zeros of the zeta function hinges. Riemann, however, had left it unproved. The key ideas in Hadamard's paper were drawn from

his own doctoral thesis, which he defended that same year. He won the prize.

- In 1895 the German mathematician Hans von Mangoldt proved the main result of Riemann's paper, which states the connection between $\pi(x)$ and the zeta function, and recast it in a simpler form. It was then plain that if a certain theorem much weaker than the Riemann Hypothesis could be proved, the application of the result to von Mangoldt's formula would prove the PNT.
- In 1896 two mathematicians working independently, the aforementioned Jacques Hadamard and the Belgian Charles de la Vallée Poussin, proved that weaker result and, therefore, the PNT.

It had been said that whoever proved the PNT would attain immortality. This prediction very nearly came true. Charles de la Vallée Poussin died five months short of his 96th birthday; Jacques Hadamard two months short of his 98th.[51] They did not know—not until late in the proceedings, anyway—that they were in competition with each other; and since both published in the same year, mathematicians consider it invidious to credit either with having got the result first. As with the ascent of Everest, the honor is shared.

In fact, de la Vallée Poussin seems to have been slightly earlier to press. Hadamard's paper—its title was *Sur la distribution des zéros de la fonction* $\zeta(s)$ *et ses conséquences arithmétiques*—appeared in the bulletin of the Mathematical Society of France. Hadamard appended a note saying that while going over the galley proofs of the paper he had learned of de la Vallée Poussin's result. He adds, "However, I believe no one will deny that my method has the advantage of simplicity."

Nobody ever has denied it. Hadamard's proof is simpler; and his knowing this before his paper went to press implies that he had not only heard of de la Vallée Poussin's result but had had the chance to examine it. However, since the two men's work was plainly independent, and since there has never been any slightest suspicion of hanky-panky, and since both Hadamard and de la Vallée Poussin were perfect gentlemen, these simultaneous proofs have never generated any rancor or controversy. I am content to say, along with the whole world of mathematics, that in 1896, Jacques Hadamard of France and Charles de la Vallée Poussin of Belgium, working independently, proved the PNT.

III. The proving of the PNT is a great turning point in our story, so much so that I have divided my book into two parts on this point. In the first place, both of the 1896 proofs depended on getting a Hypothesis-style result. If either Hadamard or de la Vallée Poussin could have proved the truth of the Hypothesis, the PNT would have followed at once. They couldn't of course, but they didn't need to. If the PNT is a nut, the Riemann Hypothesis is a sledge hammer. The PNT follows from a much weaker result (which has no name):

All non-trivial zeros of the zeta function
have real part less than one.

If you can prove this, then you can use von Mangoldt's 1895 version of Riemann's main result to prove the PNT. That is what our two scholars did in 1896.

In the second place, with the PNT out of the way, the Hypothesis came into plain view. It was the next great open issue in analytic number theory; and as mathematicians turned their attention to it, it soon became plain that if the Hypothesis could be shown to be true, a great many things would follow. If the PNT was the great white whale of number theory in the nineteenth century, the Riemann Hypothesis was to take its place in the twentieth. More than take its place, in fact, for it cast its fascination not only on number theorists, but on mathematicians of all kinds, and even, as we shall see, on physicists and philosophers.

And in the third place—apparently trivial, but these things have a way of fixing themselves in people's minds—there is the neat coincidence of the PNT being first thought of at the end of one century (Gauss, 1792), then being proved at the end of the next (Hadamard and de la Vallée Poussin, 1896). Once that theorem had been disposed of, the attention of mathematicians turned to the Riemann Hypothesis, which occupied them for the following century—which came to its end without any proof being arrived at. And that led inquisitive generalists to write books about the PNT and the Hypothesis at the beginning of the next century!

I am going to fill out the social, historical, and mathematical background to the bullet points given above by offering a sketch of the career of Jacques Hadamard; partly because he was the most important of the various players, and partly because I find him an appealing and sympathetic personality.

IV. Politically, France did not have a good nineteenth century. If Napoleon's "100 days" are included (and if you will excuse a small rounding error), the constitutional arrangements of that ancient nation from 1800 to 1899 went as follows:

First Republic (4½ years)
First Empire (10 years)
Kingdom restored (1 year)
Empire restored (3 months)
Kingdom re-restored (33 years)
Second Republic (5 years)
Second Empire (18 years)
Third Republic (29 years)

... and even that 33 years of monarchy was interrupted halfway through by a revolution and change of dynasty.

For French people of the later part of the century, the great national trauma was the defeat of their armies by Prussia in 1870, followed by the Prussian siege of Paris in the winter of 1870–1871, then by a peace treaty that involved the cession of two provinces and a huge cash indemnity. The treaty itself triggered a brief but vicious civil war. The consequences of all this for France were, of course, very great. The nation went into the Franco-Prussian War an empire and came out a republic.

The French army was particularly affected. For the rest of the century and beyond, that proud institution not only had to bear the humiliation of the 1870 defeats; it also had to embody the hopes of the nation for revenge and for recovery of the lost territories. The army also became a focus for old-fashioned French patriotism, with young men from aristocratic, clerical, and high-bourgeois families joining the officer corps in large numbers. This tipped the officer class toward the old "Throne and Altar" style of French conservatism, and to some degree cut it off from the mainstream of French life in these decades. The mainstream was all in the direction of a bustling, open-minded, commercial, and industrial republic, a leader in the arts and sciences, a center of brilliance, wit, and gaiety—the wonderful, glittering France of the Belle Epoque, one of the great high points of western civilization.

Jacques Hadamard lived through the siege of Paris as an infant, and the house his family occupied was burned down in the Civil War. He had been born in December 1865 to French-Jewish parents. His father was a high-school teacher, his mother gave piano lessons. (Among her pupils was Paul Dukas, who wrote that *Sorcerer's Apprentice* symphonic poem so well known to Disney fans.) After a degree and a brief spell of school teaching, Hadamard got his doctorate in 1892. He married that same year. In 1893 he moved with his wife to Bordeaux, where he took a position as lecturer at the university. The Hadamards' first child, Pierre, was born in October 1894, and they began raising one of those close, loving, busy bourgeois families in which everyone was expected to play a musical instrument and to enter business, academia, or the professions.

France was, then as today, a highly centralized nation. To get a lecturing position in Paris was extraordinarily difficult, and it was understood that young academics should serve an apprenticeship in the provinces for a few years. Hadamard's Paris opportunity came in 1897. He moved back to the capital in that year, quitting his professorship in Bordeaux—he had advanced from lecturer to full professor in just two years—to become an assistant lecturer at the Collège de France—from the point of view of academic prestige, a move upward.

Those six years 1892–1897 laid the foundations of Hadamard's career and fame. He was a mathematician of considerable scope, producing original work in several different areas. Undergraduate students of math generally first encounter his name attached to the Three Circles Theorem in complex function theory, a result Hadamard obtained in 1896, and which you can look up in any good encyclopedia of mathematics.[52]

You will see it written that Hadamard was the last of the universal mathematicians—the last, that is, to encompass the whole of the subject, before it became so large that this was impossible. However, you will also see this said of Hilbert, Poincaré, Klein, and perhaps of one or two other mathematicians of the period. I don't know to whom the title most properly belongs, though I suspect the answer is actually Gauss.

V. It is to the Bordeaux period that Hadamard's proof of the PNT belongs. Permit me to step back a little and look at the immediate mathematical environment of the proof.

The senior figure in French mathematics at this time was Charles Hermite (1822–1901), professor of analysis at the Sorbonne until he retired in 1897. One of his creations will play an important part later in our story (Chapter 17.v).

From 1882 onward Hermite had been conducting a mathematical correspondence with a younger mathematician, a Dutchman named Thomas Stieltjes.[53] In 1885, Stieltjes published a note in the *Comptes Rendus*[54] of the Paris Academy of Sciences, claiming to have proved my Theorem 15-1—a result stronger than the Riemann Hypothesis, from which, if Stieltjes had indeed proved it, the truth of the Hypothesis would follow (but whose falsehood would not disprove the Hypothesis—see Chapter 15.v). Stieltjes did not, however, include his proof in that note. He wrote to Hermite at about the same time, making the same claim, but adding, "My proof is very arduous; I shall try to simplify it further when I resume my research on these questions." Now, Stieltjes was an honest man and a serious and respected mathematician—there is a type of integral named after him. No one had any reason to doubt that he did, in fact, have a proof, and in all probability Stieltjes himself thought he did.

Meanwhile, Riemann's 1859 paper was being scrutinized, and its arguments tidied up. Hadamard's prize result of 1892 was a great step forward. Then, in 1895 in Berlin, the German (Germany was by this time an Empire under Kaiser Wilhelm I) mathematician Hans von Mangoldt cleared away most of the remaining underbrush and proved Riemann's main result linking the prime counting function $\pi(x)$ to the zeros of the zeta function.

Only two large points remained, the Hypothesis and the PNT. By this time everyone concerned understood that the Hypothesis was the stronger proposition. If the Hypothesis (sledgehammer) could be proved true, the PNT (nut) would follow as a consequence, with no need for further effort; but the PNT could be established from weaker results without invoking the Hypothesis, and a proof of the PNT would not imply the truth of the Hypothesis.

So, what was a mathematician to do, given that it was widely believed that Stieltjes had disposed of both matters? Start work on proving the lesser result—to which, thanks to the brush-clearing work of

Hadamard and von Mangoldt, the way was now pretty clear? Was it worth the trouble, considering that Stieltjes's superior result on the Hypothesis might appear while your own work was still in progress? On the other hand, by the mid-1890s it had been 10 years since Stieltjes's announcement, and a lot of people must have been entertaining doubts. Not doubts about Stieltjes's character; it is a very common thing for a mathematician to believe he has proved a result, only to find, going over his arguments (or more commonly, having them peer-reviewed), that there is a logical flaw in them. This happened with Andrew Wiles's first proof of Fermat's Last Theorem in 1993. It happens somewhat more dramatically to the narrator of Philibert Schogt's 2000 novel *The Wild Numbers.* Nobody would have thought the worse of Stieltjes if this had been the case, this being much too common an event in mathematical careers. But where was that proof?

Both Charles de la Vallée Poussin at the University of Louvain in Belgium and Jacques Hadamard in Bordeaux took up the lesser challenge and soon got the result. They proved the PNT. Both must have wondered, though, whether there was any point to their efforts, since, even if their papers were to be published before Stieltjes's, their lesser results would be overshadowed by his much greater one. Hadamard actually states in his paper: "Stieltjes has proved that all the imaginary zeros of $\zeta(s)$ are (conforming to Riemann's prediction) of the form $\frac{1}{2}+ti$, t being real; but his proof has never been published. I simply intend to show that $\zeta(s)$ cannot have zeros with real part equal to 1."

Stieltjes's proof never did appear; and, in fact, Stieltjes had died in Toulouse on the last day of 1894. This fact must surely have been known to Hadamard, working on his paper in 1895–1896, so presumably he was expecting the proof to turn up in unpublished papers among Stieltjes's effects. It never has. Until quite recently it was thought possible that Stieltjes might, nonetheless, have proved the Hypothesis. Then, in 1985, Andrew Odlyzko and Herman te Riele proved a result that casts serious doubt on Theorem 15-1. Belief in Stieltjes's lost proof of Riemann's Hypothesis has now, I think, pretty much evaporated.

VI. One consequence of the great national trauma of 1870–1871 was, as I have pointed out above, the reinforcing of the element of social conservatism in the officer class of the French army, and a certain distancing of that class from the main current of French society. This had one enormous consequence in the last years of the nineteenth century, the Dreyfus affair.

To attempt to do justice to "The Affair" in a couple of paragraphs is hopeless. It was a central issue in French public life for over a decade and can ignite a shouting match even today. There is a vast literature on it, along with movies, novels, and at least one TV miniseries (in French). As briefly as it can be stated: Alfred Dreyfus, an officer on the general staff of the French Army, from a wealthy Jewish-bourgeois family, was arrested and charged with treason at the end of 1894. Court-martialed *in camera*, he was condemned, degraded, and transported for life to Devil's Island. Dreyfus loudly protested his innocence and had no apparent motive for treason, having always been impeccably patriotic and without any need for money.

In March 1896 Colonel Georges Picquart of French military intelligence happened to notice that the document that had been the principal item of evidence against Dreyfus was in fact in the handwriting not of Dreyfus but of another officer, Major Esterhazy, a man of erratic character and extravagant habits, chronically beset by gambling debts. Picquart informed his superiors. He was told to say nothing further about the matter and transferred to a frontier post in French North Africa. The following year, 1897, Dreyfus's brother Mathieu learned of Picquart's discovery and demanded that Esterhazy be tried. Esterhazy was acquitted by a military tribunal in January 1898. The novelist Émile Zola promptly published an open letter, the famous *J'accuse*, to the President of the Republic, Félix Faure, denouncing the various people involved in Dreyfus's conviction as participants in a monstrous injustice and cover-up. Zola was indicted on a charge of criminal libel against the Minister of War.

The Affair then metastasized, consuming the attention of French society until Dreyfus's innocence was finally and officially proclaimed in July 1906. There were impassioned trials, dramatic reversals, the suicide of one of the conspirators, and numerous other colorful events. (Perhaps the most colorful, not arising directly from the Affair but influencing its course, was the death of President Faure while *in flagrante delicto* with his mistress in a back bedroom at the Elysée Palace. He suffered a massive stroke and in his death agony seized the poor woman by the hair with such force she was unable to separate herself from him. Her screams brought the Palace servants, who disengaged the lady, dressed her, and hustled her out a side door.)

It happened that Jacques Hadamard was a second cousin of Alfred Dreyfus's wife Lucie, *née* Lucie Hadamard. The Affair was, therefore, of direct personal concern to him. In addition to this personal connection, it confronted all French Jews with deep questions about identity and loyalty. Before the Affair, most of the Jewish-French bourgeoisie—people like the Hadamards and the Dreyfuses—had thought themselves perfectly assimilated—patriotic French people who happened to be Jewish. Anti-Semitism had been lurking below the surface, however, and not only in the army. An anti-Semitic polemical book, *La France Juive*, had been a huge publishing success in 1886, and an anti-Semitic newspaper, *La Libre Parole*, was widely read. The Affair brought all this to the surface and made French Jews wonder if they had been living in a fool's paradise. But even without the anti-Semitism factor a gross injustice had been done, and the ranks of the Dreyfusards—those agitating on behalf of the disgraced captain—included countless gentile citizens outraged by the army's deceit and the failure of the political authorities to act.

Before the Affair, Hadamard seems to have been an apolitical and unworldly man, rather the absent-minded professor type that is very common among great mathematical talents. Much is made of this stereotype, and there is in fact something to it. Because of the purely abstract nature of the material they work with and the need to concentrate on that material for long hours at a time, mathematicians tend to be somewhat detached from more earthly matters. It is not impossible for a mathematician to be worldly, and there are many counterexamples. René Descartes was a soldier and a courtier. (He survived the first but not the second.) Karl Weierstrass spent his years at university drinking and fighting and left without a degree. John von Neumann, one of the greatest of twentieth-century mathematicians, was quite a *boulevardier*, fond of pretty women and fast cars.

Jacques Hadamard, on the evidence, was not one of those counterexamples. Even discounting the apocrypha that develop around any great man, it seems plain that Hadamard could not knot his tie without assistance. His daughter claimed he could not count beyond four, "After that came *n*." His involvement in the Dreyfus Affair, therefore, speaks to the depths of the feeling aroused by that incident, stirring even such a detached soul as this. Once he had become involved, Hadamard was a passionate Dreyfusard. He became active in the League for Human Rights, founded in 1898 during the trial of Zola. Hadamard's third son, born in February 1899, was named Mathieu-Georges, "Mathieu" after Dreyfus's brother and most tireless champion, "Georges" after the remarkable Colonel Picquart, whose iron integrity and quiet insistence on telling the truth were key factors in the eventual vindication of Dreyfus (whom Picquart personally detested).

Hadamard remained a public man for the rest of his life, which, as well as being exceptionally long, was more than usually productive and busy. It was also deeply marked with tragedy. The great wars of the twentieth century took all three of his sons. The older two died at Verdun within three months of each other; Mathieu-Georges was killed in 1944 while serving with the Free French forces in North Africa. In grief and despair after the First World War, Hadamard turned to pacifism and the League of Nations. He worked to help elect the Popular Front government of 1936–1938. Like many more worldly than himself, he was to some degree taken in by communism and the Soviet Union.[55] Driven from Paris by the German advance in 1940, he taught at Columbia for four years. He traveled and lectured everywhere and met everyone. He was a keen naturalist, with museum-grade collections of ferns and fungi. He was an early supporter of the

Hebrew University of Jerusalem (founded in 1925). His many books included *The Psychology of Invention in the Mathematical Field* (1945), still well worth reading for its insights into the thought processes of mathematicians; I have used some of its ideas for this book. He organized an amateur orchestra at his home; Albert Einstein—a lifelong friend—was a visiting violinist. He was married for 68 years to the same woman. When she died, Jacques was 94 years old. He struggled on for two years; but then the death of his beloved grandson in a climbing accident robbed him of his spirit and he died a few months later, a little short of his 98th birthday.

VII. In concentrating on Jacques Hadamard, I have indulged my personal fondness for an attractive personality and fine mathematical talent, intending no disrespect to the other mathematicians who participated in the clarification of Riemann's great paper and the proof of the PNT.[56] By the later nineteenth century the world of mathematics had passed out of the era when really great strides could be made by a single mind working alone. Mathematics had become a collegial enterprise in which the work of even the most brilliant scholars was built upon, and nourished by, that of living colleagues.

One recognition of this fact was the establishment of periodic International Congresses of Mathematicians. The first such gathering was held in Zürich in August 1897. (Hadamard's wife was expecting their second child, so he did not attend. He sent a paper to be read by his friend Émile Picard. It is interesting to note that the First Zionist Congress was taking place at the same time, 40 miles away in Basel, and inspired in part by issues arising out of the Dreyfus Affair.)

There was a second Congress in Paris in the summer of 1900, and the idea was to have a Congress every four years. History had other plans, however. There was no Congress in 1916, nor in 1940, 1944, or 1948. The system started up again in 1950 in Cambridge, Massachusetts. Hadamard was, of course, invited; but because of his pro-Soviet leanings, he was at first denied a U.S. visa. It took a petition by his fellow mathematicians, and the personal intervention of President Truman, to get him to Harvard. At the time of writing, early in 2002, preparations are under way for the 24th Congress, to be held in Beijing this summer, only the second outside the West (defined as Europe, Russia, and North America).

VIII. The first of the twentieth-century Congresses was that one held in Paris from August 6 to 12, 1900, and this is the one everyone remembers. The Paris Congress will forever be linked with the name of David Hilbert, a German mathematician working at Göttingen, the university of Gauss, Dirichlet, and Riemann. Though only 38 years old, Hilbert was well established as one of the foremost mathematicians of his time.

On the morning of August 8, in a lecture hall at the Sorbonne, Hilbert stood up before the 200-odd delegates to the Congress, Jacques Hadamard among them, and delivered an address on "Mathematical Problems." His aim was to concentrate the minds of his fellow mathematicians on the challenges facing them in the new century. To effect this goal, he directed their attention to a handful of the most important topics needing investigation, and problems needing solution. He organized these topics and problems under 23 headings; number 8 was the Riemann Hypothesis.

With that address, twentieth-century mathematics began in earnest.

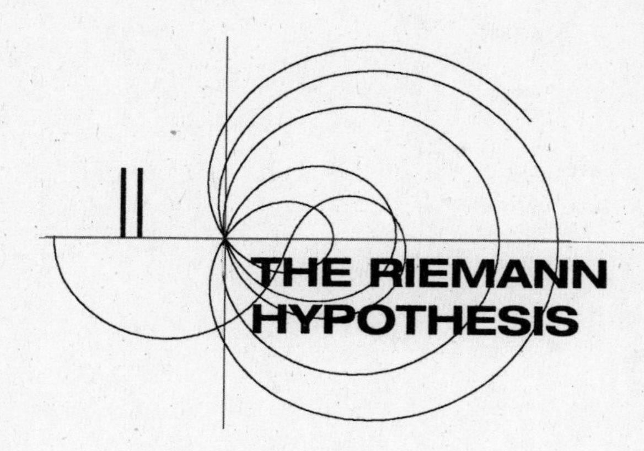
II
THE RIEMANN
HYPOTHESIS

11
Nine Zulu Queens Ruled China

I. In Chapter 9.vi I showed some zeros of the zeta function. I said that every negative even number is a zero of the zeta function: $\zeta(-2) = 0$, $\zeta(-4) = 0$, $\zeta(-6) = 0$, and so on. That gets us a certain way toward understanding the Riemann Hypothesis, which, just to remind you, says that

The Riemann Hypothesis
All non-trivial zeros of the zeta function
have real part one-half.

Unfortunately, all those negative even numbers are trivial zeros. So … where are these non-trivial ones? To answer that, I must take you into the realm of complex and imaginary numbers.

A lot of people are alarmed by this topic. They believe imaginary numbers are scary, or fantastic, or impossible—have leaked into mathematics from science fiction somehow. This is all nonsense. Complex numbers (of which imaginary numbers are a special case) came into math from very practical considerations. They were useful in helping mathematicians solve problems they couldn't solve otherwise. They are no more imaginary than any other kind of number. When was the last time you stubbed your toe on a seven?

The irrational numbers (like $\sqrt{2}$ and π) are actually more mysterious, more intellectually intimidating, and, yes, even scarier, than the square root of minus one. Indeed the irrational numbers have given—and, in the form of the so-called Continuum Hypothesis (see David Hilbert's address in Chapter 12.ii) continue to give—philosophers of mathematics far more trouble than inoffensive, handy little $\sqrt{-1}$ ever did. There have been determined attempts to reject irrational numbers, even in modern times, and even by important professional mathematicians: Kronecker in the late nineteenth century, Brouwer and Weyl in the early twentieth. For some further remarks on this topic, see Section V in this chapter.

II. To get a balanced view of complex numbers, you really need to understand how a modern mathematician thinks of numbers in general. I'm going to try to give an account of this, including complex numbers in my account. Don't worry too much about what they are right now; I'll go into more detail a bit later. I include the complex numbers in these next few paragraphs just for the sake of completeness.

So how does a modern mathematician see numbers? As hollow letters, that's how. As $\mathbb{N}$, $\mathbb{Z}$, $\mathbb{Q}$, $\mathbb{R}$, and $\mathbb{C}$. I have been trying to think of a good, memorably daft, mnemonic for keeping these letters in mind but have so far been unable to come up with anything better than "Nine Zulu Queens Ruled China."

Perhaps I'm getting ahead of things a bit. Here's an alternative answer to that question. Mathematicians think of numbers as a set of nested Russian dolls.

- Innermost doll: The *natural numbers* 1, 2, 3, 4, ….

- Next doll: The *integers.* That is, the natural numbers together with zero and the negative whole numbers (for example, –12).
- Next doll: The *rational numbers.* That is, the integers together with all positive and negative fractions (for example, numbers like $\frac{3}{2}, -\frac{1}{917635}, -\frac{1000000000001}{6}$).
- Next doll: The *real numbers.* That is, the rational numbers together with the irrational numbers, like $\sqrt{2}$, π, *e.* (Recall from note 11 to Chapter 3.vi the discovery by the ancient Greeks that there are numbers that are neither integers nor fractions—*irrational* numbers.)
- Outermost doll: The *complex numbers.*

There are several things to notice about this arrangement. The first is that there is a characteristic way to write numbers in each doll.

- Natural numbers tend to be written like this: "257."
- Integers frequently have a sign in front like this: "–34."
- Rational numbers are most often written as fractions. For purposes of writing them in fraction form, rational numbers come in two varieties. Those whose size (that is, ignoring the sign) is less than 1 are called "proper fractions," while the others are "improper." A proper fraction is written like this: $\frac{14}{37}$. An improper fraction can be written in two styles, "vulgar" ($\frac{13}{9}$) or "mixed" ($1\frac{4}{9}$).
- The most important real numbers have special symbols, like π or *e.* Many others can be expressed with "closed forms" like $\sqrt[5]{7+\sqrt{2}}$ or $\pi^2/6$. Failing all else, or to give an idea of the actual numerical value of a real number, we write it as a decimal, generally with three trailing dots to mean "this isn't the whole thing, I could supply more digits if I really had to": –549.5393169816448223.... Alternatively, we can round it to "five decimal places" (–549.53932) or "five significant digits" (–549.54) or any other level of precision.

- Complex numbers look like this: $-13.052 + 2.477i$. More on that later.

The next thing to note is that the inhabitants of each Russian doll are honorary inhabitants of the next one out and can, if there is some good reason for it, be written in the style appropriate to the outer one.

- Natural numbers (e.g., 257) are honorary integers, and can be written with a plus sign, like this: +257. When you see an integer with a plus sign in front, you think "natural number."
- Integers (e.g., –27) are honorary rational numbers, and can be written as fractions whose denominator is 1, like this: $-\frac{27}{1}$. When you see a rational number with denominator 1, you think "integer."
- Rational numbers (e.g., $\frac{1}{3}$) are honorary real numbers, and can be written out as decimals, like this: 0.33333333.... It is an interesting thing about rational numbers that if you write a rational number in decimal form, the decimal digits always repeat themselves sooner or later (unless they just come to a dead stop, like $\frac{7}{8} = 0.875$). The rational number $\frac{65463}{27100}$, for example, if written as a decimal, looks like this:

  2.41560885608856088856088....

  All rational numbers repeat like that, no irrational numbers ever do. Which is not to say that an irrational number can't have some pattern to its digits. The number

  0.12345678910111213141516171819202...

  has a clear pattern, and I could tell you in advance what the hundredth digit is, or the millionth, or the trillionth. (Wanna bet? They are 5, 1, and 1, respectively.) This number is, however, irrational. When you see a real number whose decimal repeats, you think "rational number."

- Any real number can be written as a complex number. Here is $\sqrt{2}$ written as a complex number: $\sqrt{2}+0i$. More later.

(You can jump over steps in the above list of bullet points and write, for example, a natural number as a real number: 257.0000000000....)

Each family of numbers, each Russian doll, is denoted by a hollow letter. $\mathbb{N}$ is the family of all natural numbers; $\mathbb{Z}$ is the integers; $\mathbb{Q}$ the rationals; and $\mathbb{R}$ the reals. Each family is, in a sense, contained in the next one. Each expands the power of math. It lets us do something we couldn't do with the previous doll. For example, $\mathbb{Z}$ allows us to subtract any two numbers and get an answer, which we couldn't do with $\mathbb{N}$ ($7 - 12 = ?$). Likewise, $\mathbb{Q}$ lets us divide by any number (except zero) and get an answer, which we couldn't do in $\mathbb{Z}$ ($-7 \div -12 = ?$). And $\mathbb{R}$ opens the door to analysis, the mathematics of limits, because any convergent sequence of numbers in $\mathbb{R}$ has a limit in $\mathbb{R}$, a thing not true in $\mathbb{Q}$.

(Recall those sequences and series at the end of Chapter 1. All consisted of rational numbers. Some of them converged to 2, or $\frac{2}{3}$, or $1\frac{1}{2}$—that is, their limits were also rational. Others, however, converged to $\sqrt{2}$, or π, or *e*—irrational numbers. Thus, an infinite sequence of numbers in $\mathbb{Q}$ may converge to a limit *not* in $\mathbb{Q}$. The mathematical term of art is: $\mathbb{Q}$ is not *complete*. $\mathbb{R}$, however, is complete, and so is $\mathbb{C}$. This idea of *completing* $\mathbb{Q}$ will assume new importance when I talk about *p*-adic numbers in Chapter 20.v.)

There are other categorizations of number within, or cutting across, the $\mathbb{N}$, $\mathbb{Z}$, $\mathbb{Q}$, $\mathbb{R}$, and $\mathbb{C}$ schema. Prime numbers, to take an obvious case, are a subset of $\mathbb{N}$. They are very occasionally referred to collectively as $\mathbb{P}$. There is a very important subset of $\mathbb{C}$ called the *algebraic* numbers, sometimes also given a hollow letter of its own, $\mathbb{A}$. An algebraic number is a number that is a zero of some polynomial with coefficients all in $\mathbb{Z}$, for example, $2x^7 - 11x^6 - 4x^5 + 19x^3 - 35x^2 + 8x - 3$. Among the real numbers, every rational number—and, therefore, every integer and natural number—is algebraic; $\frac{39541}{24565}$ is a zero of $24565x - 39541$ (or a solution of $24565x - 39541 = 0$, if you prefer the language of equations and solutions to the language of functions and zeros). An irrational number might or might not be algebraic. Those that are not are called *transcendental.* Both e and π are transcendental, as proved by, respectively, Hermite in 1873 and Ferdinand von Lindemann in 1882.

III. You can get another perspective on the matter from the following history of numbers I have made up. "Made up" as in "invented out of whole cloth"—it is entirely false.

John Derbyshire's Bogus History of Numbers

Human beings have always known how to count. We have had $\mathbb{N}$—the system of natural numbers—since prehistoric times. But $\mathbb{N}$ comes with a prohibition, an impossibility. *You can't subtract a greater number from a lesser one.* As technology developed, this became a stumbling block. The temperature was 5 degrees; it fell 12 degrees; what's the temperature now? There's no answer available in $\mathbb{N}$. At this point, negative numbers were invented. Oh, and someone thought up zero, too.

Negative numbers, positive numbers, and zero were gathered together in a new system, $\mathbb{Z}$, the integers. But $\mathbb{Z}$ comes with a new impossibility. *You can't divide a number by a number that isn't a factor of it.* You can divide 12 by 3 (answer: 4), or even by –3 (answer: –4), but you can't divide 12 by 7. $\mathbb{Z}$ has no answer for such an operation. As the science of measuring developed, this became a stumbling block. For finer and finer work, you need finer and finer measurements. You can finesse this for a while by just inventing new units. Need something finer than a yard? OK, here's a foot, they go three to the yard. Need something finer still? OK, here's an inch.... There is a limit to how much of this you can do, though, and the need for a general way to express fragments of a unit became pressing. So fractions were invented.

Fractions, together with all the integers, were gathered together in a new system, $\mathbb{Q}$—the rational numbers. Alas, $\mathbb{Q}$ comes with its own impossibility. *You can't always find the limit of a convergent sequence.* I gave three examples of such sequences in Chapter 1.vii. As science advanced to the point where it needed calculus, this became a stumbling block, because all of calculus rests on the idea of a limit. For calculus to develop, irrational numbers had to be invented.

Irrational numbers were gathered together with the rationals (including, of course, all the integers) to form a new system, $\mathbb{R}$—the real numbers. Yet the real numbers still contain an impossibility. *You can't take the square root of a negative number.* By the end of the sixteenth century, math had advanced to the point where this was a stumbling block. So imaginary numbers were invented. An imaginary number is the square root of a negative number.

Imaginary numbers, together with all the real numbers, were gathered together in a grand new synthesis: $\mathbb{C}$, the complex numbers. With the complex numbers, nothing is impossible, and history comes to an end.

That account is, I emphasize, totally bogus. Our understanding of numbers did not develop like that at all. Even the order is all wrong. It should be $\mathbb{N}$, $\mathbb{Q}$, $\mathbb{R}$, $\mathbb{Z}$, $\mathbb{C}$. Natural numbers were certainly known in prehistoric times. The Egyptians invented fractions early in the third millennium B.C.E. Pythagoras (or one of his disciples) discovered irrational numbers around 600 B.C.E. Negative numbers came in during the Renaissance, by way of accounting (though zero had shown up somewhat earlier). Complex numbers appeared in the seventeenth century. It all grew up haphazardly, chaotically, in the way of most human things. Nor is it true that history has ended. History never ends; as soon as one chess game has been won, another begins immediately.

My little bogus history does show how the Russian dolls fit together, though, and I hope it offers some insight into why mathematicians do not regard imaginary and complex numbers as anything very peculiar. They are just one more Russian doll, created for practical reasons—to solve problems that could not be solved otherwise.

IV. It is tedious to have to keep writing $\sqrt{-1}$, so mathematicians substitute the letter *i* for this quantity. Since *i* is the square root of minus one, $i^2 = -1$. If you multiply both sides of that by *i*, it follows that $i^3 = -i$. Repeat that process, and you get $i^4 = 1$.

What about $\sqrt{-2}$, $\sqrt{-3}$, $\sqrt{-4}$, and so on? Don't we need symbols for them, too? No. By ordinary rules for multiplying integers, $-3 = -1 \times 3$. Since $\sqrt{x}$ is just $x^{\frac{1}{2}}$, Power Rule 7 tells me that $\sqrt{a \times b} = \sqrt{a} \times \sqrt{b}$. (For example, $\sqrt{9 \times 4} = \sqrt{9} \times \sqrt{4}$, a fancy way of writing $6 = 2 \times 3$.) So $\sqrt{-3} = \sqrt{-1} \times \sqrt{3}$. Now, $\sqrt{3}$ is, of course, a perfectly ordinary real number, with a value of 1.732050807568877.... To three places of decimals, therefore, $\sqrt{-3} = 1.732i$. (In its closed form, this is usually written as $i\sqrt{3}$.) The same is true of the square root of any other negative number. You don't need a whole mass of them; you need just *i*.

Now, *i* is a very proud number. It is aloof and doesn't care to mix much with other numbers. If I add 3 to 4 I get 7; the 3-ness of the 3 and the 4-ness of the 4 disappear, absorbed into the 7-ness of the 7. If, by contrast, you add 3 to *i*, you get ... $3 + i$. It's the same with multiplication. When you multiply 5 by 2 the 5-ness and 2-ness are swallowed up by the 10-ness of the result, and vanish without trace. Multiply 5 by *i*, and you get ... $5i$. It's as if the *i* can't bear to let go of its identity; or perhaps as if the real numbers know that *i* just isn't the same kind of thing as they are.

The result is that once you introduce *i* into the scheme of things, it spawns a whole new class of numbers like $2 + 5i$, $-1-i$, $47.242-101.958i$, $\sqrt{2} + \pi i$, and every other possible $a + bi$, where *a* and *b* are any real numbers at all. These are the *complex numbers*. Each complex number has two parts, the *real* part and the *imaginary* part. The real part of $a + bi$ is *a*; the imaginary part is *b*.

As is the case with the other Russian dolls $\mathbb{N}$, $\mathbb{Z}$, $\mathbb{Q}$, and $\mathbb{R}$, numbers belonging to any inner doll are honorary complex numbers. The natural number 257, for example, is the complex number $257 + 0i$; the real number $\sqrt{7}$ is the complex number $\sqrt{7} + 0i$. A real number is just a complex number with zero imaginary part.

What about complex numbers with no real part? They are called *imaginary numbers.* Examples of imaginary numbers are: $2i$, $-1479i$, πi, $0.0000000577i$. An imaginary number can, of course, be written as a full complex number, if you want to make a point: $2i$ can be written as $0 + 2i$. If you square an imaginary number, you get a negative real number. Note that this is true even for negative imaginaries. The square of $2i$ is -4, and the square of $-2i$ is also -4, by the rule of signs.

Adding two complex numbers is a breeze. You just add the real parts, then add the imaginary parts; $-2 + 7i$ plus $5 + 12i$ would be $3 + 19i$. Subtraction likewise; if you subtract instead of adding, the answer is $-7 - 5i$. For multiplication, you must remember how to multiply out brackets, and keep in mind that $i^2 = -1$. So $(-2 + 7i) \times (5 + 12i)$ is $-10 - 24i + 35i + 84i^2$, which reduces to $-94 + 11i$. In general, $(a + bi) \times (c + di) = (ac - bd) + (bc + ad)$i.

Division depends on a simple trick. What is $2 \div i$? Answer: write it as a fraction, $2/i$. The wonderful thing about fractions is that if you multiply both the top and bottom of a fraction by the same number (not zero), its value does not change: $\frac{3}{4}$, $\frac{6}{8}$, $\frac{15}{20}$ and $\frac{12000}{16000}$ are all ways of writing the same fraction. So multiply top and bottom of $2/i$ by $-i$. Two times $-i$ is, of course, $-2i$. And i times $-i$ is $-i^2$, which is $-(-1)$, which is 1. Therefore, $2/i$ is just $-2i/1$, which is $-2i$.

There is always a way to do this, turning the bottom of a fraction into a real number. Since dividing by real numbers is no mystery, we are home and dry. How do I divide two full-blown complex numbers, say $(-7 - 4i)/(-2 + 5i)$? I multiply top and bottom of the fraction by $-2 - 5i$, that's how. Multiply out the top, $(-7 - 4i) \times (-2 - 5i) = -6 + 43i$. Multiply out the bottom, $(-2 + 5i) \times (-2 - 5i) = 29$. Answer:

$-\frac{6}{29}+\frac{43}{29}i$. You can always turn the bottom of $(a + bi)/(c + di)$ into a real number; just multiply by $c - di$. The general rule, in fact, is

$$(a+bi)\div(c+di)=\frac{ac+bd}{c^2+d^2}+\frac{bc-ad}{c^2+d^2}i.$$

What is the square root of *i*? Don't we have to define a whole other class of numbers to take in $\sqrt{i}$? And so on for ever? Answer: Multiply out the brackets $(1 + i) \times (1 + i)$. You will see that the result is $2i$. So the square root of $2i$ is $1 + i$. Scaling down, the square root of i must be $1/\sqrt{2}+i/\sqrt{2}$, which indeed it is.

Complex numbers are wonderful. You can do anything with them. You can even raise them to complex powers, if you know what you are doing. For example, $(-7 - 4i)^{-2+5i}$ is approximately $-7611.976356 + 206.350419i$. That, however, is something I shall explain more fully elsewhere.

V. The thing you *can't* do with complex numbers is lay them out on a line, as you can with real numbers.

You can visualize the family of real numbers, $\mathbb{R}$ (which of course includes within itself $\mathbb{Q}$, $\mathbb{Z}$, and $\mathbb{N}$) very easily. Just lay it out on a straight line. This way of illustrating the real numbers is called "the real line," as shown in Figure 11-1.

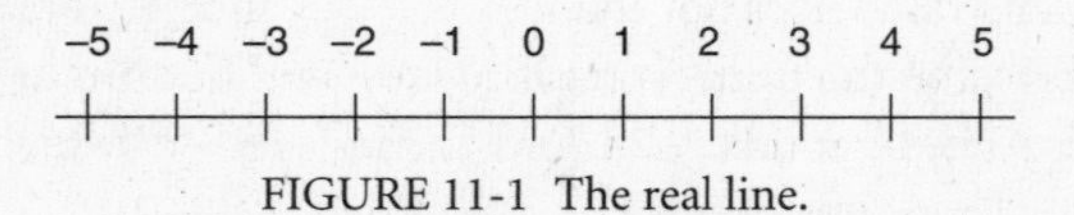

FIGURE 11-1 The real line.

Every real number is on there somewhere. For example, $\sqrt{2}$ is a little way east of 1, not quite half way to 2, $-\pi$ is just slightly west of –3, and 1,000,000 is off in the next county somewhere. I can, of course, show only part of the line on a finite sheet of paper. You must use your imagination.

The real line looks obvious, but in fact it is a very deep and mysterious affair. The rational numbers, for example, are "everywhere dense" on it. That means that between any two rational numbers, you can always find another one. And that means that between any two rational numbers you can find an *infinity* of other ones. (Look: if, between *a* and *b*, I am guaranteed to find *c*, then between *a* and *c*, and between *c* and *b*, I am guaranteed to find a *d* and an *e* ... and so on forever.) That, you can just about visualize. But where do the irrational numbers go? It seems they have to somehow squish in between the rational numbers, which, as I've just said, are themselves everywhere dense! While yet managing to be not complete!!

Take that sequence from Chapter 1.vii that closes in on $\sqrt{2}$, for example, $\frac{1}{1}$, $\frac{3}{2}$, $\frac{7}{5}$, $\frac{17}{12}$, $\frac{41}{29}$, $\frac{99}{70}$, $\frac{239}{169}$, $\frac{577}{408}$, $\frac{1393}{985}$, $\frac{3363}{2378}$, ... The terms are alternately less than and greater than $\sqrt{2}$, so that $\frac{1393}{985}$ is short of $\sqrt{2}$ by about 0.00000036440355 and $\frac{3363}{2378}$ exceeds it by about 0.00000006252177. Squeezed in between those two fractions, though, is an infinity of other fractions ... and still there is room in there somewhere for $\sqrt{2}$. And not only for $\sqrt{2}$, either, but for an infinity of other irrationals!

For the amazing thing is that not only is there an infinity of irrationals, and not only are they, too, everywhere dense; but there is a precise mathematical sense in which there are *far more* irrationals than rationals. This was shown by Georg Cantor in 1874. The number of rational numbers is infinite, and the number of irrational numbers is infinite; but the second infinity is bigger than the first. How on earth do they all fit on the real line? How does such an inconceivably vast number of irrationals squeeze in among the rationals, if the rationals themselves are everywhere dense?

I have no space to go into such things here. My advice is not to think about these matters too much. That way lies madness. (In fact, Cantor ended his life in an asylum, although this was more a consequence of a congenital disposition toward depression aggravated by difficulty in getting his theories accepted than a result of thinking too much about the real line. Those theories are not now seriously doubted.) Just accept that all real numbers are there on the line somewhere.

But now, where on earth are we going to put the complex numbers? The real line is jammed up—and then some!—with rationals and irrationals. Yet for any real number a, there's a whole infinity of complex numbers $a + bi$, with b roaming freely up and down the real line. Where shall we put them all?

That last remark suggests the answer. For each real number, we need a line, and since there is an infinity of reals, we need an infinity of lines, side by side. That means a flat plane. While the real numbers can be spread out for inspection on a line, the complex numbers need a plane—which of course we call "the complex plane." Every complex number is illustrated by a point somewhere in the plane.

In the complex plane as usually drawn (see Figure 11-2), the real line stretches west-east as usual. Set at right angles to it is a new line going south-north, containing all the imaginary numbers: i, $2i$, $3i$, and so on. To get to the number $a + bi$, you go a distance a to the east (west if negative), then b to the north (south if negative). The real line and the imaginary line—they are more commonly called "the real axis" and "the imaginary axis"—cross at zero. Points on the real line have imaginary part zero; points on the imaginary line have real part zero. The point where they cross, the point that is on both, has both real and imaginary parts zero. It is $0 + 0i$, that is, zero.

Let me introduce three terms of art. The *modulus* of a complex number is its straight-line distance from zero. The symbol is $|z|$, pronounced "mod z." By Pythagoras's Theorem, the modulus of $a + bi$ is $\sqrt{a^2 + b^2}$. It is always a positive number or zero. The *amplitude* of a complex number is the angle it makes with the positive real line, measured in radians. (One radian is 57.29577951308232... degrees; 180 degrees is π radians.) The amplitude is conventionally taken to be an angle between $-\pi$ (exclusive) and π (inclusive) radians, and its symbol is $Am(z)$.[57] Positive real numbers have amplitude zero; negative real numbers have amplitude π; positive imaginary numbers have amplitude $\pi/2$; negative imaginaries have amplitude $-\pi/2$.

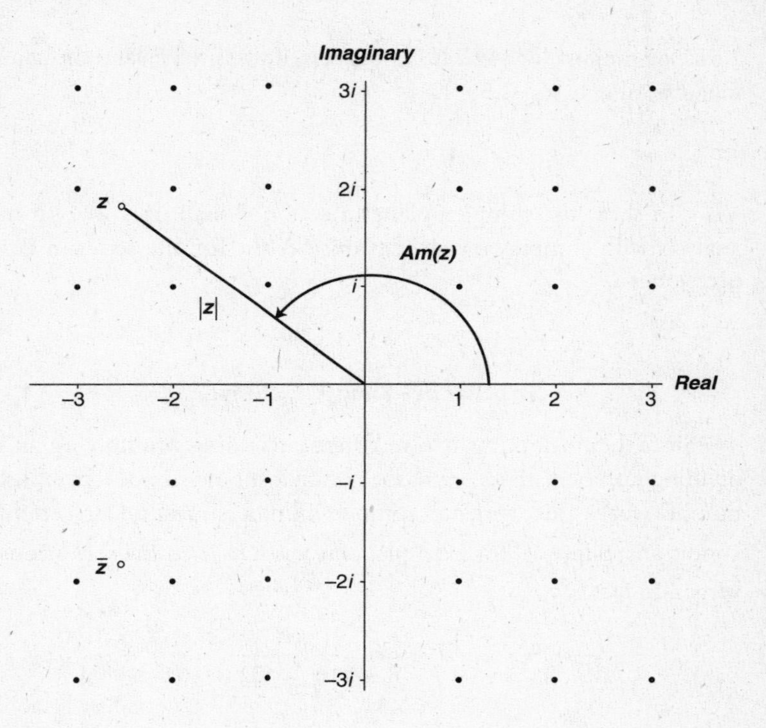

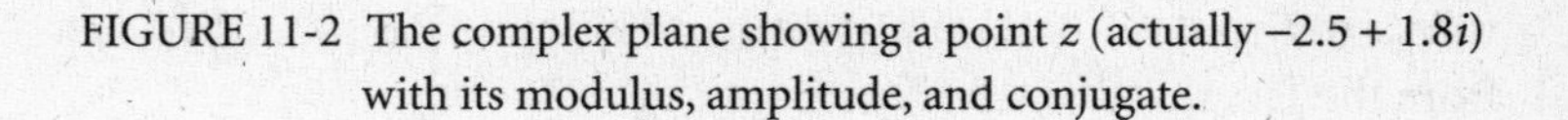
FIGURE 11-2 The complex plane showing a point z (actually $-2.5 + 1.8i$) with its modulus, amplitude, and conjugate.

Finally, the *complex conjugate* of a complex number is its mirror image in the real line. The complex conjugate of $a + bi$ is $a - bi$. Its symbol is $\bar{z}$, pronounced "z bar." If you multiply a complex number by its conjugate, you get a real number: $(a + bi) \times (a - bi) = a^2 + b^2$, which is, in fact, the modulus of $a + bi$, squared. That's what makes the trick for division work. In proper symbols it is $z \times \bar{z} = |z|^2$, and the division trick is just $z / w = (z \times \overline{w}) / |w|^2$.

For the complex number $-2.5 + 1.8i$, shown in Figure 11-2, the modulus is, $\sqrt{9.49}$, that is, about 3.080584, the amplitude is about

2.517569 radians (or 144.246113 degrees, if you prefer), and the conjugate is, of course, $-2.5 - 1.8i$.

VI. To show the complex plane in action, I shall do a wee bit of analysis with complex numbers. Consider the infinite series in Expression 9-2.

$$\frac{1}{1-x} = 1 + x + x^2 + x^3 + x^4 + x^5 + x^6 + \cdots$$
$$(x \text{ between } -1 \text{ and } 1, \text{ exclusive})$$

Since there's nothing involved there but adding, multiplying, and dividing numbers, there seems no reason x should not be a complex number. Does this work for complex numbers? Yes, under certain conditions. Suppose, for example, that x is $\frac{1}{2}i$. Then the series converges. In fact,

$$\frac{1}{1-\frac{1}{2}i} = 1 + \frac{1}{2}i + \frac{1}{4}i^2 + \frac{1}{8}i^3 + \frac{1}{16}i^4 + \frac{1}{32}i^5 + \frac{1}{64}i^6 + \cdots$$

The left-hand side, if you do the trick I described above for division, works out to $0.8 + 0.4i$. The right-hand side can be simplified just from the fact that $i^2 = -1$.

$$0.8 + 0.4i = 1 + \frac{1}{2}i - \frac{1}{4} - \frac{1}{8}i + \frac{1}{16} + \frac{1}{32}i - \frac{1}{64} - \cdots$$

You can actually walk out the right-hand side on the complex plane; Figure 11-3 gives the general idea. Start at the point 1 (which is on the real line, of course); then go north to add the $\frac{1}{2}i$; then go west $\frac{1}{4}$; then go south $\frac{1}{8}i$ … and so on. You get a neat spiral, closing in on the complex number $0.8 + 0.4i$. Analysis in action, an infinite series closing in on its limit.

Notice that while we lost the simplicity of one dimension when we moved to complex numbers, we gained some imaginative power.

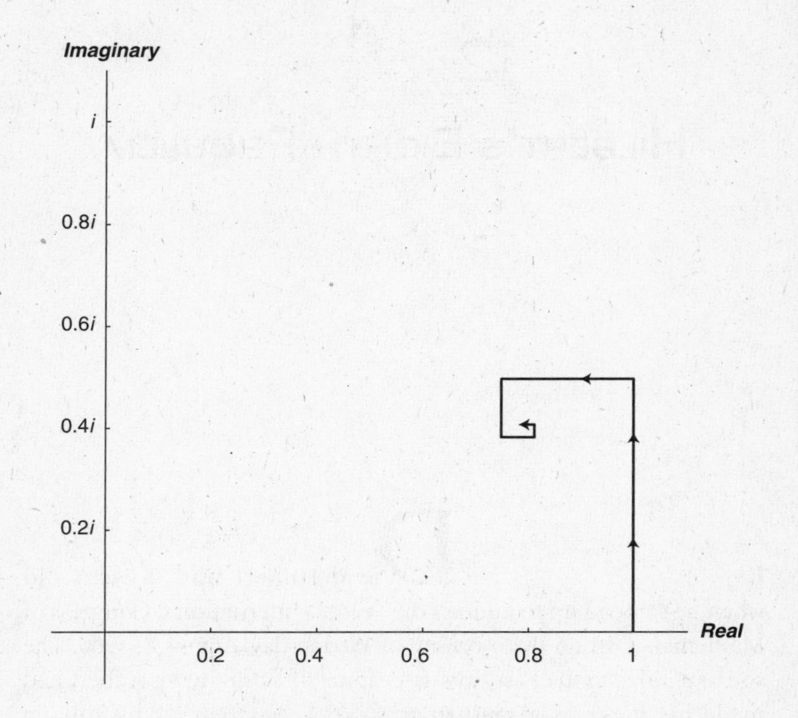

FIGURE 11-3 Analysis in the complex plane.

With two dimensions to play with, you can show mathematical results the way I just did, as striking visual patterns or pictures. This is part of the appeal, for me anyway, of complex analysis. In Chapter 13, I shall actually show you Riemann's zeta function, and the great hypothesis itself!, laid out as elegant patterns on the complex plane.

12 Hilbert's Eighth Problem

I. David Hilbert was 38 years old when he stepped up to address the Second International Congress of Mathematicians on the morning of Wednesday, August 8, 1900. The son of a judge in the East Prussian capital of Königsberg, Hilbert had made his name as a mathematician 12 years earlier by solving Gordan's Problem, in the theory of algebraic invariants.

This had been not only a *succès d'estime*, but also, in a minor way, a *succès de scandale*. Gordan's Problem concerned the existence of a certain class of objects. Hilbert had proved that the objects exist but had not produced them, nor even suggested any method for constructing them. Mathematicians refer to this kind of thing as an "existence proof." Hilbert used the following everyday example in his lectures. "There is at least one student in this class—let us name him 'X'—for whom the following statement is true: no other student in the class has more hairs on his head than X. Which student is it? That we shall never know; but of his existence we can be absolutely certain." Existence proofs are rather common in modern mathematics and are nowadays not particularly controversial. Matters were different in the Germany of 1888. Just one year previously, Leopold Kronecker, a respected member of the Berlin Academy, had issued his manifesto *On the Concept of Number*, which attempted to banish from mathematics what he regarded as unnecessary levels of abstraction—anything, in his view, that could not be derived from the integers in a finite number of steps. Gordan himself famously remarked of Hilbert's existence proof: "This is not mathematics. This is theology."

The generality of mathematicians, however, acknowledged the validity of Hilbert's solution. Hilbert then went on to do important work in the theory of algebraic numbers and in the foundations of geometry. He created brilliant new proofs—three and one-half pages for *both*—of the facts that π and e are transcendental. (When, in 1882, von Lindemann had been the first to prove π transcendental, the aforementioned Kronecker[58] had complimented him on the elegance of his argument but added that it proved nothing, since transcendental numbers did not exist!) In 1895 Hilbert was given a chair at Göttingen, where he remained until his retirement in 1930.

The names "Hilbert" and "Göttingen" are yoked together in the minds of modern mathematicians as closely as, in other spheres, are "Joyce" and "Dublin," or "Johnson" and "London." Hilbert and Göttingen dominated mathematics during the first third of the twentieth century—not merely German mathematics, but all mathematics. The Swiss physicist Paul Scherrer, arriving at Göttingen as a student in 1913, reported finding there "an intellectual life of unsurpassed intensity." An astonishing proportion of important mathematicians and physicists of the first half of the century had studied either at Göttingen, or under someone who had studied there.

Of Hilbert's personality, mixed reports have come down to us. By no means antisocial, he was a keen dancer and a popular lecturer. He was also something of a skirt-chaser, to the very limited degree that was possible in the ambience of provincial Wilhelmine Germany. (It is not likely that anything very improper took place.) He had an irreverent streak and seems to have been impatient with the stuffiness of university life, the customs, regulations, and social proscriptions.

One old professor's wife was scandalized to hear that Hilbert had been seen in the back room of a restaurant in the town, playing billiards with his junior lecturers. When, during World War I, the university refused to give Emmy Noether a regular lecturing position on the grounds that she was a female, Hilbert simply announced a course of lectures to be given by himself, then let Noether deliver them. He seems to have been a soft examiner, always ready to give a candidate the benefit of the doubt.

It is hard to avoid the impression, though, that Hilbert was a man who did not suffer fools gladly, and that he classed rather a large part of humanity as fools. This was particularly unfortunate in Hilbert's case because Franz, his only child, was afflicted with serious mental problems. Unable to learn anything much, or to hold down any kind of job, Franz also suffered occasional lapses into paranoia, following which he had to be kept in a mental hospital for a while. Hilbert is recorded as saying, at the time of the first of these incarcerations: "From now on I must consider myself as not having a son."

Hilbert was, at any rate, revered by his students and mathematical colleagues. There is a vast number of anecdotes about him, mostly of an affectionate sort. Here are just three. The first, which touches on the Riemann Hypothesis, I have taken from Constance Reid's English-language biography.

> Hilbert had a student who one day presented him with a paper purporting to prove the Riemann Hypothesis. Hilbert studied the paper carefully and was really impressed by the depth of the argument; but unfortunately he found an error in it which even he could not eliminate. The following year the student died. Hilbert asked the grieving parents if he might be permitted to make a funeral oration. While the student's relatives and friends were weeping beside the grave in the rain, Hilbert came forward. He began by saying what a tragedy it was that such a gifted young man had died before he had had an opportunity to show what he could accomplish. But,

> he continued, in spite of the fact that this young man's proof of the Riemann Hypothesis contained an error, it was still possible that some day a proof of the famous problem would be obtained along the lines which the deceased had indicated. "In fact," he continued with enthusiasm, standing there in the rain by the dead student's grave, "let us consider a function of a complex variable...."

The second I have borrowed from Martin Davis's book *The Universal Computer.*

> Hilbert was seen day after day in torn trousers, a source of embarrassment to many. The task of tactfully informing Hilbert of the situation was delegated to his assistant, Richard Courant. Knowing the pleasure Hilbert took in strolls in the countryside while talking mathematics, Courant invited him for a walk. Courant managed matters so that the pair walked through some thorny bushes, at which point Courant informed Hilbert that he had evidently torn his pants on one of the bushes. "Oh no," Hilbert replied, "they've been that way for weeks, but nobody notices."

The third is apocryphal, though quite possibly true.

> One of Hilbert's students stopped showing up to classes. On enquiring the reason, Hilbert was told that the student had left the university to become a poet. Hilbert: "I can't say I'm surprised. I never thought he had enough imagination to be a mathematician."

Hilbert was not, by the way, Jewish, though his given name, unusual among German gentiles, brought him under suspicion in the Hitler years. His paternal ancestors belonged to a fundamentalist Protestant sect called Pietists, who favored Old Testament and hortatory names. Hilbert's grandfather rejoiced in the names David Fürchtegott Leberecht (i.e., *Fear God Live Right*) Hilbert.

II. Constance Reid describes Hilbert at the 1900 Congress thus:

> The man who came to the rostrum that morning was not quite forty, of middle height and build, wiry, quick, with a noticeably high forehead, bald except for wisps of still reddish hair. Glasses were set firmly on a strong nose. There was a small beard, a still somewhat straggly moustache, and under it a mouth surprisingly wide and generous for the delicate chin. Bright blue eyes looked innocently but firmly out from behind shining lenses.

Hilbert delivered his address, in German, in a stuffy lecture hall at the Sorbonne. Total attendance at the Congress was 250, but it is not likely that all of them were present to hear Hilbert speak on the morning of August 8.

The title of the address was "Mathematical Problems." Its opening words became as familiar to twentieth-century mathematicians as those of the Gettysburg Address are to American schoolchildren. "Who of us would not be glad to lift the veil behind which the future lies hidden; to cast a glance at the next advances of our science and at the secrets of its development during future centuries?"[59] Hilbert went on to speak of the importance of difficult problems in concentrating the attention of mathematicians, inspiring new developments and new symbols, and in pushing mathematics to higher and higher levels of generalization. He ended with a list of 23 particular problems "from the discussion of which an advancement of science may be expected."

I should like to take you on a tour of Hilbert's 23 problems.[60] To do so, however, would make this book unacceptably long. Besides, there is a considerable literature, pitched at many different levels of understanding, providing such tours.[61] I shall only note in passing that the very first of Hilbert's problems was that of the Continuum Hypothesis, which I mentioned in my previous chapter, and which goes to the heart of the knotty issue of the nature of the real numbers, and of Kronecker's objections to them. There is a large literature on the Continuum Hypothesis, too. A good library, or a good internet search engine, will satisfy the curiosity of anyone who wants to look into this fascinating issue.[62]

Only one of Hilbert's problems is of direct concern to the topic of this book, and that is the eighth. Here it is, as translated for the *Bulletin of the American Mathematical Society* by Mary Winston Newson.

8. Problems of Prime Numbers

> Essential progress in the theory of the distribution of prime numbers has lately been made by Hadamard, de la Vallée Poussin, von Mangoldt and others. For the complete solution, however, of the problems set us by Riemann's paper "Über die Anzahl der Primzahlen unter einer gegebenen Grösse," it still remains to prove the correctness of an exceedingly important statement of Riemann, viz., *that the zero points of the function* $\zeta(s)$ *defined by the series*
>
> $$\zeta(s)=1+\frac{1}{2^s}+\frac{1}{3^s}+\frac{1}{4^s}+\cdots$$
>
> *all have the real part* $\frac{1}{2}$, except the well-known negative integral real zeros. As soon as this proof has been successfully established, the next problem would consist in testing more exactly Riemann's infinite series for the number of primes below a given number and, especially, *to decide whether the difference between the number of primes below a number* x *and the integral logarithm of* x *does in fact become infinite of an order not greater than* $\frac{1}{2}$ *in* x. Further, we should determine whether the occasional condensation of prime numbers which has been noticed in counting primes is really due to those terms of Riemann's formula which depend upon the first complex zeros of the function $\zeta(s)$.

Some parts of this will be understood by readers who have followed me this far. I hope all of it will make sense by the time I have finished. The main point to note here is that the Riemann Hypothesis was regarded as one of 23 large, difficult issues or problems facing mathematics in the twentieth century, and it was so regarded by David Hilbert, probably the greatest mathematician doing productive work in 1900.[63]

III. I briefly mentioned, in Chapter 10.iii, the reason for the prominence of the Riemann Hypothesis at the turn of the century. The main factor was that the Prime Number Theorem had now been proved. Since 1896 it was known, with mathematical certainty, that, yes indeed, $\pi(N) \sim Li(N)$. Everyone's attention now focused on that twiddle sign. OK, so as N gets larger and larger without limit, $\pi(N)$ gets proportionally closer and closer to $Li(N)$. But what is the nature of that closeness? Is a better approximation possible? How approximate is the approximation anyway? What is the "error term"?

Free—now that the proof of the PNT was in the bag—to think about these secondary matters, mathematicians found their eyes being drawn to the Riemann Hypothesis. Bernhard Riemann's 1859 paper had not, of course, proved the PNT, but it had mightily suggested that it should be true, and even further had suggested an expression for the error term. That expression involved all the non-trivial zeros of the zeta function. Knowing where, precisely, those zeros lie thus became a matter of pressing importance.

The mathematics of all this will become clearer as we go along, but I think you will not be at all surprised to hear that those non-trivial zeros are all complex numbers. In 1900 the following things were known, with mathematical certainty, about the location—the location on the complex plane, that is—of the non-trivial zeros.

- There is an infinity of them, all having real parts between 0 and 1 (exclusive). Using the complex plane to visualize this (see Figure 12-1), mathematicians say that all non-trivial zeros are known to lie in *the critical strip*. The Riemann Hypothesis makes a much stronger assertion, that they all lie on

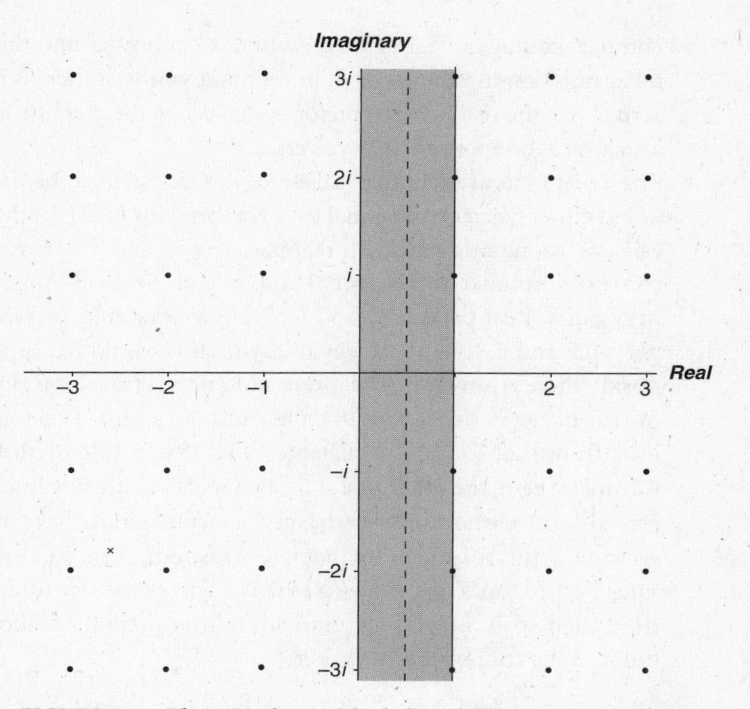

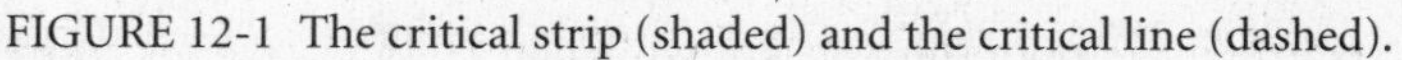
FIGURE 12-1 The critical strip (shaded) and the critical line (dashed).

the line whose real part is one-half, that is, on *the critical line.* "Critical strip" and "critical line" are common terms of art in discussions of the Riemann Hypothesis, and from now on I shall use them quite freely.

The Riemann Hypothesis (stated geometrically)
All non-trivial zeros of the zeta function
lie on the critical line.

- The zeros occur in conjugate pairs. That is, if $a + bi$ is a zero, then so is $a - bi$. In other words, if z is a zero, then so is its complex conjugate $\bar{z}$. I defined "complex conjugate" and the z-bar notation in Chapter 11.v. In yet other words, if there is a zero above the real line, its mirror image below the real line is also a zero (and, of course, vice versa).

- Their real parts are symmetrical about the critical line; that is, a zero either has real part equal to $\frac{1}{2}$ (in line with the Hypothesis), or is one of a pair with real parts $\frac{1}{2}+\alpha$ and $\frac{1}{2}-\alpha$, for some real number α between 0 and $\frac{1}{2}$, and identical imaginary parts. Real parts 0.43 and 0.57 are an example, or real parts 0.2 and 0.8. Another way of saying this would be: supposing there is any non-trivial zero not *on* the critical line, its mirror image *in* the critical line must also be a zero. This follows from that formula in Chapter 9.vi. If one side of that formula is zero, the other side must be too. Leaving aside integer values of *s*, where other terms in the formula misbehave or go to zero, this formula says that if $\zeta(s)$ is zero, then $\zeta(1-s)$ must be zero too. Thus, if $\left(\frac{1}{2}+\alpha\right)+it$ is a zero of the zeta function, then so is $\left(\frac{1}{2}-\alpha\right)-it$, and so, by the previous bullet point, is the conjugate $\left(\frac{1}{2}-\alpha\right)+it$.

Little more than this was known when Hilbert gave his address. Riemann had suggested another twiddle formula for the approximate number of zeros with imaginary part between zero and some large number *T* (see Chapter 16.iv). However, this formula was not actually proved until 1905, by von Mangoldt. The Hypothesis had not been entirely ignored. It turns up as a discussion topic in some mathematical literature of the 1890s, for example in the French problem journal *L'Intermediaire de Mathematiciens.* To all intents and purposes, though, the mathematicians of the nineteenth century left it to those of the twentieth to take on Bernhard Riemann's tremendous and subtle conjecture.

IV. The twentieth century was a very ... *busy* century. A great deal happened, in all spheres of human endeavor. This makes it seem, in retrospect, awfully long, far longer than the mere one-and-a-half standard lifetimes that a century actually is. Mathematics, however, moves at a stately pace, and the deep problems tackled by modern mathematicians yield up their secrets only very slowly and reluctantly. The world of any given mathematical specialty is, too, a small one, with its own heroes, folklore, and oral traditions binding the community together in both time and space. From speaking with living mathematicians to gather material for this book, I came to feel that the twentieth century was not such a very long span of time after all, the great names of its early years almost within hailing distance.

I am writing these words, for example, just a week after talking to Hugh Montgomery, a key figure in developments of the 1970s and 1980s (which I shall tell you about in the proper place). Hugh did postgraduate work at Trinity College, Cambridge, in the later 1960s. Among the faculty members he knew personally was John Edensor Littlewood, 1885–1977, who obtained one of the earliest major advances toward the understanding of the Riemann Hypothesis in 1914. "He tried to persuade me to take snuff," reports Hugh, who still has in his possession handwritten notes from Littlewood. Littlewood could in theory have met and talked mathematics with Riemann's friend Richard Dedekind, who lived until 1916 and was mathematically active almost to the end of his life ... and who had studied under Gauss! (I have not been able to discover if any such meeting took place. It is not actually very likely. Dedekind retired from his professorship at the Brunswick Polytechnic in 1894 and thereafter, according to George Pólya[64] "lived in a quiet way, seeing very few people.")

Because of this strong impression of continuity across the period, I am tempted to abandon a strictly chronological approach to the twentieth century. That temptation is strengthened by the nature of developments through that century. The story of the Riemann Hypothesis in the twentieth century is not a single linear narrative, but a number of threads, sometimes crossing, sometimes tangling with each other. This needs a little preliminary explanation; and the explanation itself needs a preamble, a note about how mathematics developed from 1900 to 2000.

V. Aside from having been distinguished by Hilbert's Paris address, the year 1900 is, of course, an arbitrary mark. Mathematics has developed steadily and continuously across the modern period. Mathematicians did not go home from their New Years' parties in the small hours of January 1, 1900 (or 1901, if you like—see Chapter 6.ii) thinking, "It's the twentieth century! We must move to a higher level of abstraction!" any more than Europeans woke up on the morning of May 30, 1453, thinking, "The Middle Ages are over! We'd better start disseminating printed books, challenging the authority of the Pope, and discovering the New World!" I should hate to have to stand before a jury of my peers and justify the term "twentieth-century mathematics."

It is nonetheless true that the mathematics of the last few decades has had a distinctive flavor, quite different from the flavor of mathematics as practiced by Gauss, Dirichlet, Riemann, Hermite, and Hadamard. As well as it can be captured in a word, that flavor is *algebraic.* Here is the beginning of the first proposition in Alain Connes' *Noncommutative Geometry* (1990), a pretty typical higher-math text of the later twentieth century.

> The classes of bounded random operators $(q_l)_{l \in X}$ modulo equality almost everywhere, endowed with the following algebraic rules, form a von Neumann algebra $W(V,F)$....

Algebraic ... *algebra* ... And this in a book about geometry! (The 11th word in the statement of the book's final theorem, by the way, is "Riemannian.")

What has been happening these past few decades, very roughly, is this. For most of its development, mathematics has been firmly rooted in number. Most of nineteenth-century math was concerned with numbers: whole numbers, rational numbers, real numbers, complex numbers. In the course of this development, new mathematical objects were created, or the scope of existing ones extended—functions, spaces, matrices—and powerful new tools devised for the manipulation of these objects. Still, it was all about numbers. A function maps one set of numbers into another set. The squaring function maps 3, 4, 5 into 9, 16, 25; Riemann's zeta function maps 0, $1 + i$, $2 + 2i$ into $-\frac{1}{2}$, $0.58216 - 0.92685i$, $0.86735 - 0.27513i$. Similarly, a space is a set of points, known by their coordinates, which are numbers. A matrix is an array of numbers, and so on. (I shall introduce matrices in Chapter 17.iv.)

In twentieth-century math the objects that had been invented to encapsulate important facts about number *themselves became the objects of inquiry*, and the techniques that had been developed for investigating numbers and sets of numbers were turned on those objects themselves. Mathematics broke free, as it were, from its mooring in number and soared up to a new level of abstraction.

Classical analysis, for example, concerns itself with the limit of an infinite sequence of numbers or points (with "point" defined by coordinates, which are numbers). A typical product of the twentieth century, by contrast, was "functional analysis," where the fundamental object of study is sequences of *functions*, which might or might not converge, and where a function is itself liable to be treated as a "point" in a space of infinitely many dimensions.

Mathematics even turned on itself to such a degree that the very techniques of investigation and proof became objects of inquiry. Some of the most important theorems of twentieth-century mathematics were concerned with the completeness of mathematical systems (Kurt Gödel, 1931) and the decidability of mathematical propositions (Alonzo Church, 1936).

These momentous developments have not yet, even at the opening of the twenty-first century, been reflected in mathematics education, at least up to college-entrance level. Perhaps they cannot be. Mathematics is a cumulative subject. Every new discovery adds to the body of knowledge, and nothing is ever subtracted. When a mathematical truth has been discovered it is there forever, and every succeeding generation of students must learn it. It never (well, hardly ever) becomes untrue or irrelevant—though it might become unfashionable, or be subsumed as a particular case of some more general theory. (And note that in mathematics, "more general" does not necessarily mean "more difficult." There is a theorem in projective geometry, Desargues' Theorem, which is easier to prove in three dimensions than in two. Chapter 7 of H.S.M. Coxeter's *Regular Polytopes* contains a theorem[65] that is easier to prove in four dimensions than in three!)

A bright young American turning up for a first class as a college math major learns math pretty much as it was known to the young Gauss, with perhaps a few forward excursions. Since I am pitching my book to readers at about that level, the mathematics you are reading here has a strong nineteenth-century flavor to it. I shall cover all developments down to the present day in these narrative chapters, explaining them as best I can, but my mathematical chapters do not often go beyond 1900.

VI. The story of the Riemann Hypothesis in the twentieth century is the story of an obsession that gripped most of the great mathematicians of the age sooner or later. Instances of this obsession are abundant, as will become clear over the next few chapters. Here I shall just give a single example.

David Hilbert, as I have already described, listed the Riemann Hypothesis eighth in his list of 23 problems for mathematicians of the twentieth century to concentrate their efforts on. That was in

1900, before the obsession took hold. His state of mind a few years later is revealed in the following story, told by his younger colleague George Pólya.

> The thirteenth-century German emperor Frederick Barbarossa, who died while on a crusade, was popularly supposed by Germans to be still alive, asleep in a cave deep in the Kyffhäuser Mountains, ready to awake and emerge when Germany needed him. Someone asked Hilbert what he would do if, like Barbarossa, he could be revived after a sleep of several centuries. Hilbert: "I would ask whether anyone had proved the Riemann Hypothesis."

And this was not an era short of challenging problems. Fermat's Last Theorem (that there are no whole-number solutions to the equation $x^n + y^n = z^n$ when n is greater than 2, proved in 1994) was still open; so was the Four Color Theorem (that four colors are sufficient to color any map in the plane, no two adjacent regions having the same color, proved in 1976); so was Goldbach's Conjecture (that every even number greater than 2 is the sum of two primes, still unproved); so were many lesser but long-standing problems, conjectures, and conundrums. The Riemann Hypothesis soon came to tower over them all.

The obsession took different mathematicians in different ways, according to their mathematical inclinations. Thus a number of threads developed during the course of the century—different approaches to investigating the Hypothesis, each originated by some one person, then carried forward by others, the threads sometimes crossing and tangling with each other. There was, for example, the *computational* thread, in which mathematicians set about actually calculating the value of more and more zeros, and developing better methods for doing so. There was an *algebraic* thread, started by Emil Artin in 1921, attempting to take the Riemann Hypothesis by a flanking movement through an algebraic topic called Field Theory. Later in the century, as a result of a remarkable encounter I shall write about in due course, a *physical* thread emerged, linking the Hypothesis to the mathematics of particle physics. While all this was going on, *analytic* number theorists were still working steadily away, continuing the tradition begun by Riemann himself, tackling the Hypothesis with the tools of complex function theory.

And research into the primes themselves went on, too, without any particular application to the Hypothesis but still, very often, with the hope that new insights into the distribution of the primes might throw light on why the Hypothesis is true—or, as the case may be, false. Key advances here were the development of a probabilistic model for the distribution of primes in the 1930s, and Selberg's "elementary" proof of the Prime Number Theorem in 1949, which I described in Chapter 8.iii.

In covering these developments I shall try to make it clear at every point which thread I am talking about, though sometimes skipping carelessly from one to another to maintain the overall chronological narrative. Let me begin with a brief introductory remark about the computational thread, since that is the easiest for a non-mathematician to understand. What are the actual values, as numbers, of the non-trivial zeros? How can they be calculated? What are their overall statistical properties, taken as a collectivity?

VII. The first concrete information about the zeros was provided by the Danish mathematician Jørgen Gram, to whom I gave a passing mention in Chapter 10. An amateur mathematician with no university position—his day job was, like the poet Wallace Stevens's, as an insurance company executive—Gram seems to have been doodling for some years with methods of actually calculating the location of the non-trivial zeros (this was long before the age of computers, of course). In 1903, after settling on a fairly efficient method, he published a list of the "first" 15 zeros—the ones closest to, and above, the real line. Gram's zeros are shown dotted along the critical line in

Figure 12-2. His list, which contained some slight inaccuracies in the right-most digits, begins

$$\tfrac{1}{2}+14.134725\,i,\ \tfrac{1}{2}+21.022040\,i,\ \tfrac{1}{2}+25.010856\,i\ldots.$$

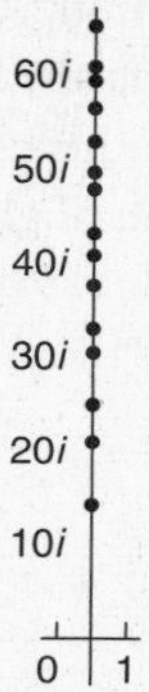

FIGURE 12-2 Gram's zeros.

Every one of these numbers, as you can see, has real part one-half.[66] (And the existence of each one, of course, implies a conjugate one below the real axis: $\tfrac{1}{2}-14.134725\,i$, and so on. I shall take this as understood and not mention it again until it becomes important, in Chapter 21.) Therefore, as far as they go, they confirm the truth of the Riemann Hypothesis. But of course they don't go very far. The number of zeros was known to be infinite—that was implicit in Riemann's 1859 paper. Do they all have real part one-half? Riemann thought so. That was his mighty Hypothesis. At this point, however, no one had a clue.

When Gram's list appeared, mathematicians must have looked on it in fascinated awe. The secret of the distribution of prime numbers, which had engaged the attention of mathematicians since the days of the legendary Gauss, was locked up somehow in this string of numbers: $\tfrac{1}{2}+14.134725\,i$, $\tfrac{1}{2}+21.022040\,i$, $\tfrac{1}{2}+25.010856\,i$, ... But how? Their real parts were certainly one-half, as Riemann had hypothesized; but the imaginary parts showed no apparent order or pattern.

I said "mathematicians must have...." I really should have said "a few continental mathematicians must have...." The obsession with Riemann's Hypothesis that seized mathematicians during the twentieth century was only just beginning to gather strength in 1905. In some parts of the world, it was hardly known. In the next part of my historical narrative I shall take the reader to England, in the high Edwardian summer of her imperial glory. But first let me show you what the zeta function actually looks like.

The giants, and those who nourished them.

Leonhard Euler

Peter the Great of Russia

Carl Friedrich Gauss

Carl Wilhelm Ferdinand, Duke of Brunswick

Plate 2

Bernhard Riemann, his mentor, and his friend.

Riemann, early 1850s

Riemann, 1863

Lejeune Dirichlet

Richard Dedekind

The Prime Number Theorem.

Charles de la Vallée Poussin

Jacques Hadamard

Pafnuty Lvovich Chebyshev

Atle Selberg

Twentieth-century pioneers.

David Hilbert

Edmund Landau

G.H. Hardy

J.E. Littlewood

The computational thread.

Jørgen Pedersen Gram

Carl Siegel

Alan Turing

Andrew Odlyzko

The algebraists.

Emil Artin

André Weil

Pierre Deligne

Alain Connes

The physical thread.

George Pólya

Freeman Dyson

Hugh Montgomery

Sir Michael Berry

The Lindelöf Hypothesis and Cramér's model.

Ernst Lindelöf

Harald Cramér

Counting vs measuring.

The author and family with Taiye, who is arithmetically 97 years old, but analytically only 95.522....

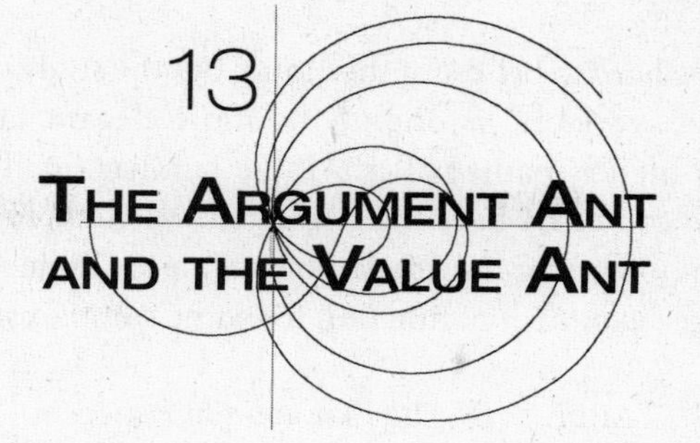

13 The Argument Ant and the Value Ant

I. Supposing, as I have tried to persuade you, that complex numbers are a perfectly straightforward extension of ordinary real numbers, obeying all the normal rules of arithmetic with the single extra one that $i^2 = -1$; and recalling that a function just turns one range of numbers—its *domain*—into another; is there any reason there should not be functions of complex numbers? No reason at all.

The squaring function, for example, works just fine for complex numbers, following the rule for multiplication. The square of $-4 + 7i$, for example, is $(-4 + 7i) \times (-4 + 7i)$, which is $16 - 28i - 28i + 49i^2$, i.e., $-33 - 56i$. Table 13-1 shows a sample of the squaring function for some random complex numbers.[67]

TABLE 13-1 The Squaring Function.

z	z^2
$-4 + 7i$	$-33 - 56i$
$1 + i$	$2i$
i	-1
$0.174 - 1.083i$	$-1.143 - 0.377i$

It may be hard to believe at this point, but the study of "functions of a complex variable" is one of the most elegant and beautiful branches of higher mathematics. All the familiar functions of high school math can easily have their domains extended to cover all, or most, of the complex numbers. For example, Table 13-2 gives a glimpse of the exponential function for some complex numbers.

TABLE 13-2 The Exponential Function.

z	e^z
$-1 + 2.141593i$	$-0.198766 + 0.30956i$
$3.141593i$	-1
$1 + 4.141593i$	$-1.46869 - 2.28736i$
$2 + 5.141593i$	$3.07493 - 6.71885i$
$3 + 6.141593i$	$19.885 - 2.83447i$

Note that, just as before, when I choose the arguments to go up by addition—as of course I do, in this case adding $1 + i$ each time—the function values go up by multiplication, in this case by $1.46869 + 2.28736i$. If I had picked the arguments to go up by adding 1 each time, then of course the values would have multiplied by e. Note also that I slipped into this table one of the most beautiful identities in all of math.

$$e^{\pi i} = -1$$

Gauss is supposed to have said—and I wouldn't put it past him—that if this was not immediately apparent to you on being told it, you would never be a first-class mathematician.

How on earth is it possible to define a complex power for e, or any other number? By a series, that's how. Expression 13-1 shows the actual definition of e^z for any number z whatsoever, real or complex.

$$e^z = 1 + z + \frac{z^2}{1\times 2} + \frac{z^3}{1\times 2\times 3} + \frac{z^4}{1\times 2\times 3\times 4} + \cdots$$

Expression 13-1

Miraculously (it seems to me) this infinite sum converges for every number. The denominators grow so fast they eventually swamp any power of any number. Equally miraculously, if z is a natural number, the infinite sum works out to exactly what the basic meaning of "power" would lead you to expect, though from just looking at Expression 13-1, there is no obvious reason why it should. If z is 4, it works out to exactly the same as $e \times e \times e \times e$, which is what e^4 is *supposed* to mean.

Let me just feed πi into Expression 13-1, to show how it converges. If z is πi, then z^2 is $-\pi^2$, z^3 is $-\pi^3 i$, z^4 is π^4, z^5 is $\pi^5 i$, and so on. Feeding these into the infinite sum, and calculating the actual powers of π (to just six decimal places for simplicity's sake), the sum is

$$e^{\pi i} = 1 + 3.141592i - \frac{9.869604}{2} - \frac{31.006277i}{6} + \frac{97.409091}{24} + \frac{306.019685i}{120} - \cdots$$

If you add up the first 10 terms of this, you have $-1.001829104 + 0.006925270i$. If you add up the first 20, you have $-0.9999999999243491 - 0.000000000528919i$. Sure enough, it is converging on -1. The real part is closing in on -1, and the imaginary part is disappearing.

Can the log function be extended to complex numbers, too? Yes, it can. It is, of course, just the inverse of the exponential function. If $e^z = w$, then $z = \log w$. Unfortunately, as with square roots, you run into the many-valued function quicksand unless you take precautions. This is because, in the complex world, the exponential function sometimes gives the same value for different arguments. The cube of -1, for example, is, by the rule of signs, -1; so if you cube both sides of $e^{\pi i} = -1$, you get $e^{3\pi i} = -1$; so the arguments πi and $3\pi i$ both yield the same function value of -1, just as -2 and $+2$ both yield value 4 under the squaring function. So what is $\log(-1)$? Is it πi? Or $3\pi i$?

It's πi. To stay out of trouble, we restrict the imaginary part of the function value to between $-\pi$, exclusive, and π, inclusive. Then every non-zero complex number has a log, and $\log(-1) = \pi i$. In fact, in the symbols I introduced in Chapter 11.v, $\log z = \log |z| + i\,Am(z)$, with $Am(z)$ measured in radians, of course. Table 13-3 is a sample of the log function, using six decimal places.

TABLE 13-3 The Log Function.

z	$\log z$
$-0.5i$	$-0.693147 - 1.570796i$
$0.5 - 0.5i$	$-0.346574 - 0.785398i$
1	0
$1 + i$	$0.346574 + 0.785398i$
$2i$	$0.693147 + 1.570796i$
$-2 + 2i$	$1.039721 + 2.356194i$
-4	$1.386295 + 3.141592i$
$-4 - 4i$	$1.732868 - 2.356194i$

Here the arguments go up by multiplication (each row is $1 + i$ times the previous row) while the function values go up by addition (of $0.346574 + 0.785398i$ each time). So, it's a log function. The only wrinkle is, when the imaginary part of the function value gets bigger than π, as it does in going from argument -4 to argument $-4 - 4i$, you have to subtract $2\pi i$ to keep it in range, 2π radians being 360 degrees. (Recall from Chapter 11.v that radians are just mathematicians' favorite way to measure angles.) This doesn't cause any problems in practice.

II. Since there is an exponential function for complex numbers, and a log function, there doesn't seem to be any reason we can't raise any complex number to any complex power. By Power Rule 8 in Chapter 5.ii, any real number a is just $e^{\log a}$, so by Power Rule 3, a^x is just $e^{x \log a}$. Can't we just extend this idea into the realm of complex numbers, and say that for any two complex numbers z and w, z^w just means $e^{w \log z}$?

We certainly can, and do. If you wanted to raise $-4+7i$ to the power of $2-3i$, you'd first calculate the log of $-4+7i$, which turns out to be around $2.08719+2.08994i$. Then you'd multiply that by $2-3i$, getting answer $10.4442-2.08169i$. Then you'd raise e to that power, giving final result $-16793.46-29959.40i$. So

$$(-4+7i)^{2-3i}=-16793.46-29959.40i$$

Piece of cake. As another example, since $e^{\pi i}=-1$, taking the square root of both sides gives $i=e^{\frac{\pi i}{2}}$. If you now raise both sides to the power of i, remembering Power Rule 3 again, you get $i^i=e^{-\frac{\pi}{2}}$. Note that this is a real number, equal to 0.2078795763….

Since I can raise any complex number to the power of any complex number, it should be easy to raise a *real* number to a complex power. Given a complex number z, I can, therefore, calculate 2^z, 3^z, 4^z, and so on. You can see where this is leading. Can we extend the domain of the zeta function

$$\zeta(s)=1+\frac{1}{2^s}+\frac{1}{3^s}+\frac{1}{4^s}+\frac{1}{5^s}+\frac{1}{6^s}+\frac{1}{7^s}+\frac{1}{8^s}+\cdots$$

into the world of complex numbers? Of course we can. I tell you, with complex numbers you can do anything.

III. Since the formula for zeta is still an infinite sum, the question of convergence arises. It turns out that the sum converges for any complex number whose real part is greater than one. Mathematicians say "in the half-plane $\mathrm{Re}(s)>1$," where $\mathrm{Re}(s)$ is understood to mean "the real part of s."

As with the zeta function for real arguments, though, mathematical tricks can be used to extend the domain of the zeta function back into regions where the infinite sum doesn't converge. After applying those tricks, you have the complete zeta function, whose domain is all the complex numbers, with a single exception at $s = 1$. There, just as I started out showing with that deck of cards in Chapter 1, the zeta function has no value. Everywhere else, it has a single, definite value. There are some places, of course, where that value is zero. We already know that. Those graphs in Chapter 9.iv show the zeta function taking the value zero at all the negative even numbers $-2, -4, -6, -8, \ldots$. I have already dismissed these arguments as not being very important. They are the trivial zeros of the zeta function. Could it be that there are some *complex* arguments for which the value of the zeta function is zero? And that these are the non-trivial zeros mentioned in the Hypothesis? You bet; but I am getting a little ahead of the story.

IV. Forty years ago the brilliant but eccentric Theodor Estermann[68] wrote a textbook titled *Complex Numbers and Functions,* which contains just two diagrams. "I ... have avoided any appeal to geometric intuition," announced the author in his preface. There has been a small number of kindred spirits, but the generality of mathematicians do not follow Estermann's approach. They tackle the theory of complex functions in a strongly visual way. Most of us feel that complex functions are easier to get to grips with if you have some visual aids.

How then can complex functions be visualized? Let's take the simplest non-trivial complex function, the squaring function. Is there any way to get a handle on what it looks like?

In the first place, ordinary graphs are no help. In the world of real numbers you can graph a function like this. Draw a line to represent the arguments (remember the real numbers live on a line). Draw another line at right angles to represent the function values. To represent the fact that this function turns the number x into the number y, go east from argument zero a distance x (west if x is negative); then go north from value zero a distance y (south if y is negative). Mark the spot. Repeat for as many function values as you care to compute. This gives you a graph of the function. Figure 13-1 shows an example.

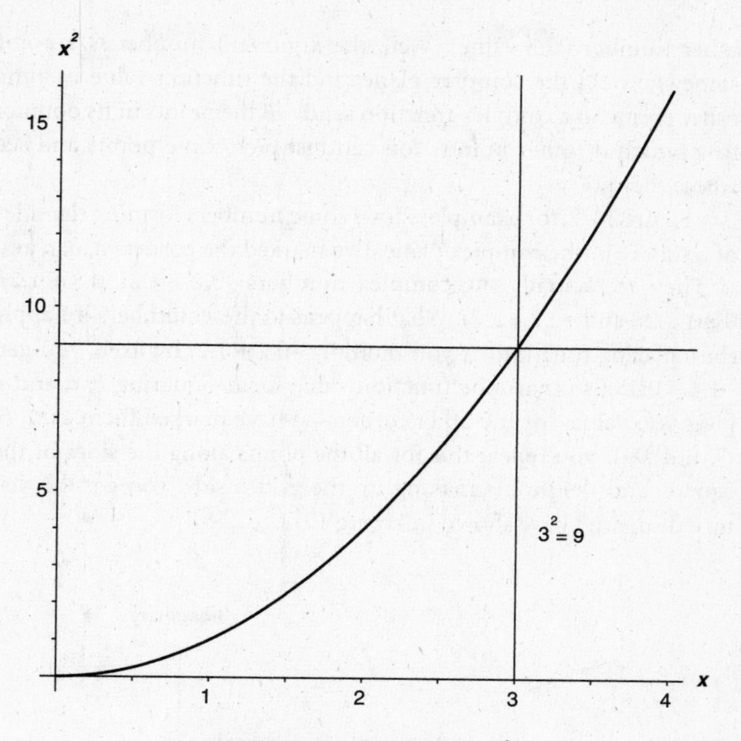

FIGURE 13-1 The function x^2.

This can't be made to work for complex functions. The arguments need a two-dimensional plane to be laid out on. The function values need another two-dimensional plane. So to get a graph, you need four dimensions of space to draw it in: two for the argument, two for the function value. (In four dimensions of space, believe it or not, two flat two-dimensional planes can intersect in a single point. Compare the fact, utterly inconceivable to the inhabitants of a two-dimensional universe, that in three dimensions, two non-parallel straight lines need not intersect at all.)

To compensate us for this disappointment, there are things you can do to make pictures of complex functions. Remember the basic thing about a function: it turns one number (the argument) into another number (the value). Well, the argument number is a point somewhere on the complex plane; and the function value is some other point. So a complex function sends all the points in its domain to a bunch of other points. You can just pick some points and see where they go.

Figure 13-2, for example, shows some numbers forming the sides of a square in the complex plane. I've marked the corners *a*, *b*, *c*, and *d*. They are actually the complex numbers $-0.2 + 1.2i$, $0.8 + 1.2i$, $0.8 + 2.2i$, and $-0.2 + 2.2i$. What happens to these numbers if I apply the squaring function? If you multiply $-0.2 + 1.2i$ by itself, you get $-1.4 - 0.48i$; so that's the function value for *a*. Squaring *b*, *c*, and *d* gives you values for the other corners—I have marked them as *A*, *B*, *C*, and *D*. If you repeat this for all the points along the sides of the square, and the points making up the grid inside, you get the distorted square I have shown in Figure 13-2.

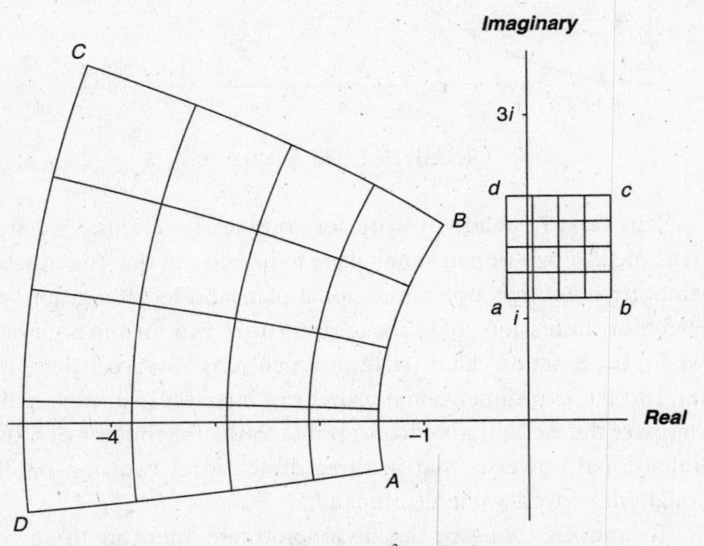

FIGURE 13-2 The function z^2 applied to a square.

V. It helps with complex functions to think of the complex plane as an infinitely stretchable sheet of rubber and ask what a function does to this sheet. You can see from Figure 13-2 that the squaring function stretches the sheet counter-clockwise round the zero point, while simultaneously stretching it outward from that point for the numbers I've shown. The number $2i$, for example, whose natural home is on the positive-imaginary (north) axis, when you square it, goes to –4, which is on the negative-real (west) axis, and twice as far from the zero point. In turn, –4, when you square it, gets stretched round to 16, on the positive-real (east) axis, and even further from zero. By the rule of signs, $-2i$, down on the negative-imaginary (south) axis, gets winched all the way round to –4. Because of the rule of signs, in fact, every function value turns up twice, from two arguments. Remember that –4 is not only the square of $2i$, it is also the square of $-2i$.

Bernhard Riemann, who seems to have had a very powerful visual imagination, conceived of the matter like this. Take the entire complex plane. Make a cut along the negative real (west) axis, stopping at the zero point. Now grab the top half of that cut and pull it round counter-clockwise, using the zero point as a hinge. Stretch it right round through 360 degrees. Now it's over the stretched sheet, with the other side of the cut under the sheet. Pass it through the sheet (you have to imagine that the complex plane is not only infinitely stretchable, but also is made of a sort of misty substance that can pass through itself) and rejoin the original cut. Your mental picture now looks something like Figure 13-3. That is what the squaring function does to the complex plane.

This is not a fanciful or trivial exercise. From it, Riemann developed a whole theory, called the theory of Riemann surfaces. It contains some powerful results and gives deep insights into the behavior of complex functions. It also yokes function theory to algebra and topology, two key growth areas of twentieth-century math. It is, in fact, a typical product of Riemann's bold, fearless, and ever-original imagination—a fruit of one of the greatest minds that ever existed.

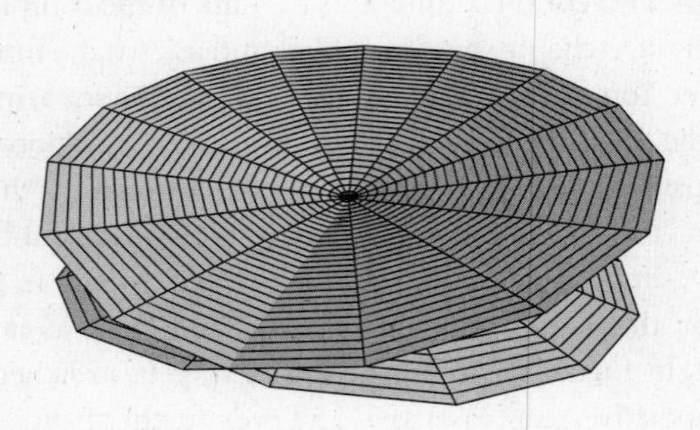

FIGURE 13-3 The Riemann surface corresponding to the function z^2.

VI. I am going to take a much simpler approach to illustrating complex functions. I'd like you to meet my pal the argument ant, shown in Figure 13-4.

FIGURE 13-4 The argument ant.

The argument ant is awfully hard to see, because she is infinitesimal in size. If you *could* see her, however, she would look just like a regular ant—a *Camponotus japonicus* worker, to be precise—with the regulation number of appendages, antennae, etc. In one of the frontmost appendages, which for convenience we may call a "hand," the argument ant holds a small gadget rather like a beeper, or a mobile telephone, or one of those global positioning devices that can tell you exactly where you are. This gadget (Figure 13-5) has three displays. The first display, labeled "Function," shows the name of some function: z^2, log z, or whatever the gadget might be set to. The second display, labeled "Argument," shows the point—the complex number—the argument ant is currently standing on. The third display, labeled "Function value," shows the value of the function at that argument. So the argument ant always knows exactly where she is; and, for any given function, she knows where the point she's standing on gets sent to by the function.

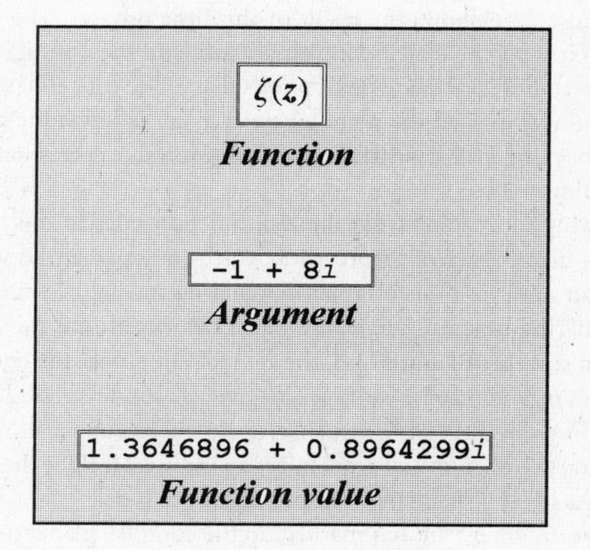

FIGURE 13-5 The ant's gadget.

I have set the gadget to show the zeta function, and I am going to let the argument ant wander freely over the complex plane. When "Function value" shows zero, she will be standing on a point ("Argument") that is a zero of the zeta function. I can have her mark those points for us with a magic marker she carries in a small pouch under her thorax. Then we shall know where the zeros of the zeta function are.

In fact, I am going to have the argument ant do a little bit more work than that. I am going to have her mark *all arguments that give a pure-real or pure-imaginary function value.* An argument whose value is 2 or –2, or $2i$ or $-2i$, will be marked; a point whose value is $3 - 7i$ will not. To put it another way, all those points that zeta will send to the real line or the imaginary line will be marked. And, of course, since the real line and the imaginary line cross at zero, the arguments where these lines cross, will be zeros of the zeta function. In this way, I can get some kind of picture of the zeta function.

Figure 13-6 shows the result of this little odyssey. The straight lines in it show the real and imaginary axes and the critical strip. All the curved lines are made up of points that are sent to either the real or imaginary axis. At the point where each curve leaves the diagram at left or right, I have written in the function value corresponding to that point.

Trying to imagine what the zeta function does to the complex plane—in the sense of Figure 13-3, which shows what the squaring function does to it—is a rather demanding mental exercise. While the squaring function wraps the plane over itself into the double-sheeted surface of Figure 13-3, the zeta function does the equivalent thing an *infinite* number of times, to give an infinite-sheeted surface. If you find this difficult to visualize, don't feel bad about it. You need long practice over several years to get an intuitive feel for these functions. As I said, I shall take a simpler approach here.

The argument ant has marked up the complex plane to give the patterns of Figure 13-6. Now I shall set her to wandering along some of those curves. Let's suppose she starts out standing on the point –2. Since this is a zero of the zeta function—one of the trivial zeros—the "Function value" display reads 0. Now she starts heading west along the real axis. The function value begins to creep up from zero.

Shortly after she passes the point –2.717262829, heading west, "Function value" reaches the number 0.009159890.... Then it starts

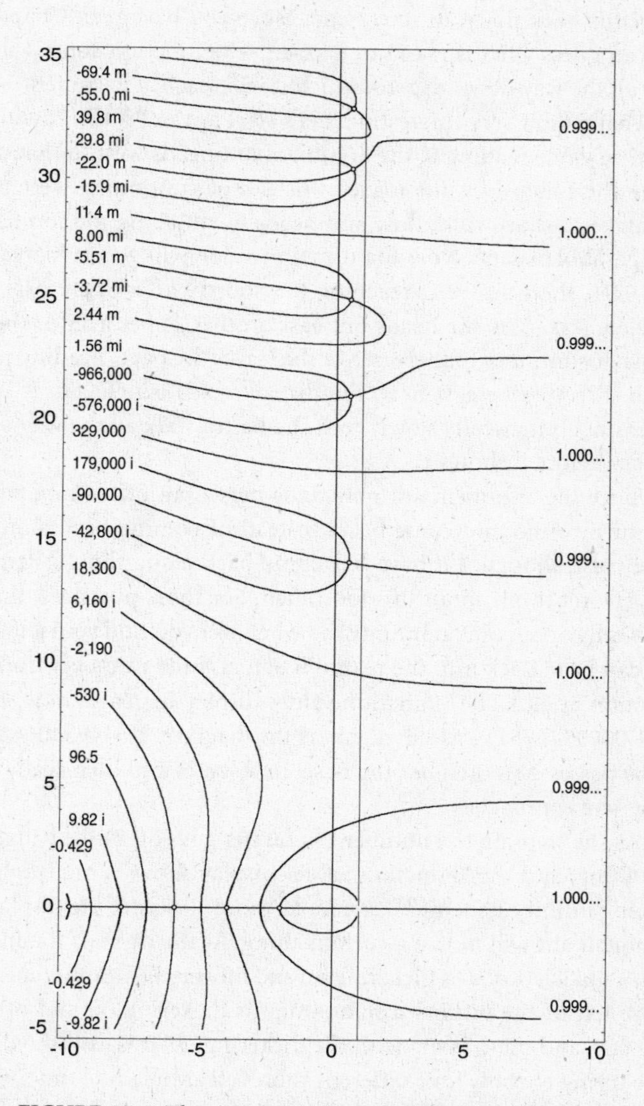

FIGURE 13-6 The argument plane, showing points that zeta "sends to" the real and imaginary axes.

to decline back down to zero again. Since you have read Chapter 9, you can guess what is going to happen. The function value will decline all the way down to zero, which it will reach at argument –4.

That wasn't very interesting. Let's start again. From –2, with the function value reading 0, the argument ant heads west to that point where the function value maxed. Instead of continuing west to –4, she makes a sharp right turn and heads north along the top half of that parabola shape. Now the function value will go on increasing, past 0.01, then past 0.1, reaching 0.5 shortly after she crosses the imaginary axis. As she heads out east on that upper arm of the parabola, it continues to increase. As she leaves the page, heading pretty much directly due east now, the display reads 0.9990286. It is still increasing, but awfully slowly, and she has to walk all the way out to infinity before it shows 1.

Since the argument ant now finds herself at infinity, she may as well turn round and come back. Instead of coming back along the same path, though, I'll have her come back along the positive real axis. (Don't think about this too much. For these purposes, there is really only one "point at infinity," so whenever you find yourself there, you can head back into the realm of actual finite numbers from any direction at all.) The "Function value" display increases now, showing 1.0009945751... as she re-enters the diagram, 1.644934066848... as she passes 2 (remember the Basel problem?) and then really soaring as she approaches 1.

As she steps on the number 1, a buzzer goes off in the gadget she is holding, and the "Function value" display shows a big bright red flashing infinity sign, "∞." If she looks more closely at the display, the argument ant will notice a curious thing. At the right of the infinity sign, a small letter "*i*" is flickering on and off very fast. Simultaneously, to the left of the infinity a minus sign is flickering on and off, also very fast, and out of sync with the flickering "*i*." It is as if the display were trying to show four different values all at the same time: ∞, –∞, ∞*i*, and –∞*i*. Curious!

The reason is that the argument ant now has three choices (other than to go back the way she came). If she just goes forward, heading west along the real axis until she comes home to the zero at argument −2, she will see the function values turn into large negative numbers, like minus 1 trillion, then rise very fast to moderate-sized negative numbers (minus 1,000, minus 100) eventually coming up to −1, then to −0.5 as she steps on the zero point (because $\zeta(0) = -0.5$), and eventually back to zero at argument −2.

If, on the other hand, she takes a sharp right turn northwards at 1 and traverses the top half of that oval shape around the zero point, she will find from the display that the function values are ascending the negative imaginary axis, from numbers like $-1{,}000{,}000i$, up through $-1{,}000i$ to $-10i$, $-5i$, $-2i$, then to $-i$. Shortly before she crosses the imaginary axis the display reads $-0.5i$. Then, as she heads to the zero at −2, the function value rises to, of course, zero.

Just to help you keep your bearings, and to anchor this firmly in the world of functions (which I first introduced in Chapter 3 by way of tables),Table 13-4 shows that last walk, counterclockwise round the top of the oval shape. I have picked the arguments for this table to have the following amplitudes (in degrees, not radians): 0°, 30°, 60°, 90°, 120°, 150°, and 180°. All numbers are rounded to four decimal places in Table 13-4.

TABLE 13-4 The Argument Ant Traverses the Top of the Oval in Figure 13-6.

z	$\zeta(z)$
1	$-\infty i$
$0.8505 + 0.4910i$	$-1.8273i$
$0.4799 + 0.8312i$	$-0.7998i$
$0.9935i$	$-0.4187i$
$-0.5737 + 0.9937i$	$-0.2025i$
$-1.3206 + 0.7625i$	$-0.0629i$
-2	0

If the ant had taken a left turn at 1, the function values would have come back to zero down the *positive* imaginary axis instead, through 1.8273*i*, 0.7998*i*, and so on.

VII. The argument ant can start her walk from any other zero of the function. I have shown them all in Figure 13-6 with teeny circles. To help the ant know where she is going, I have shown the actual values that are on the "Function value" display at the moment she leaves the diagram along any particular line. (To save space, I have written "m" for "million" in these values. "*i*," of course, just means *i*.) Notice the pattern as she goes up the left-hand edge of the diagram, that is, through arguments whose real part is −10. The first line to leave the diagram at this edge is one that maps into the negative real axis. The next maps into the positive imaginary axis; the next, into the positive real axis; the next, into the negative imaginary axis, … and so on, this pattern repeating itself.

The lines that leave the diagram on the right-hand edge, by contrast, are all mapping into the positive real axis. To the right of the critical strip, in fact, this is a pretty dull function. This whole vast eastern region maps into a tiny area around the 1 point. It is not as "busy" as the left-hand western region; and that western region is not as interesting as the critical strip. With the zeta function, all the interesting stuff happens in the critical strip. (For another illustration of this general truth, see my account of the Lindelöf Hypothesis in the Appendix.)

Figure 13-6 is really the heart of this book. There you actually *see* the Riemann zeta function, as well as a complex function *can* be seen. I urge you to spend some time in silent contemplation of this diagram, and to venture out on a few of those ant-walk exercises. The functions of higher mathematics are very wonderful things. They don't yield up their secrets easily. Some, like this one, can offer a lifetime of study. I can by no means claim to be an expert on the zeta function. I don't have a comprehensive collection of zeta-function literature, having relied mainly on university libraries and personal acquaintances for the facts in this book. Still, even without trying hard, I seem to have acquired my own copies of E.C. Titchmarsh's *The Theory of the Riemann Zeta-function* (412 pages), S.J. Patterson's *An Introduction to the Theory of the Riemann Zeta-Function* (156 pages), and Harold Edwards's indispensable *Riemann's Zeta Function* (316 pages, and I have three copies of this one—that's a long story), as well as a thick folder of photocopied articles from various journals and periodicals. There must be a score of other full-length books plumbing the mysteries of this function, and thousands of articles. This is serious math.

And, best of all, you can see in that diagram the Riemann Hypothesis shining clear. Look!—the non-trivial zeros actually do all lie on the critical line. I have not shown the critical line in Figure 13-6, but obviously it lies halfway down the critical strip, like a highway median.

VIII. Before leaving the topic of visualizing zeta, just a couple more pictures. First, note that the general pattern you see in Figure 13-6 continues all the way up, for as far as we know.

To illustrate this, Figure 13-7 shows a block of zeros up around $\frac{1}{2} + 100i$. You will notice that they are packed closer together than the ones in Figure 13-6. In fact the average spacing between the eight zeros shown here is 2.096673119.... For the five zeros shown in Figure 13-6, the average spacing was 4.7000841.... So up here around $100i$ on the imaginary axis, the zeros are packed more than twice as densely as down around $20i$.

There is in fact a rule for the average spacing of zeros at height T in the critical strip. It is $\sim 2\pi / \log(T/2\pi)$. If T is 20 this works out to 5.4265725.... If T is 100 it is 2.270516724.... You can see that the rule is not very precise, though, as the twiddle sign tells you, it gets better

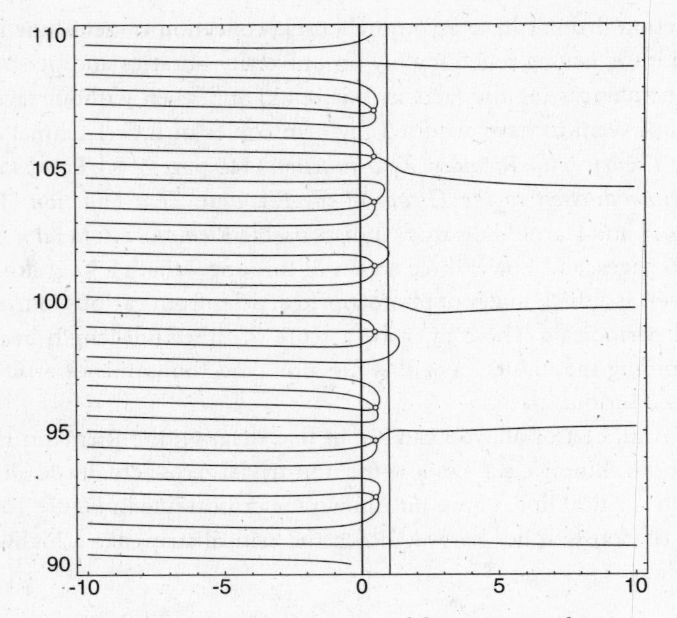

FIGURE 13-7 A higher region of the argument plane.

for bigger numbers. Andrew Odlyzko has published a list of 10,000 zeros up in the neighborhood of $\frac{1}{2} + 1{,}370{,}919{,}909{,}931{,}995{,}308{,}897i$. In that neck of the woods, $2\pi/\log(T/2\pi)$ is worth about 0.13416467. The actual average of the 9,999 spaces is 0.13417894.... Not bad.

Next, note a point that will be of some importance later in the book. There is a certain symmetry about the real (i.e., east-west) axis. If I extended Figure 13-6 down south of the real axis, the lines would be mirror images of what they are north of it. The only difference is that while the real numbers I have written in on Figure 13-6 are just the same south as they are north, the imaginary numbers have their signs flipped. To put it mathematically, if $\zeta(a+bi) = u+vi$, then $\zeta(a-bi) = u-vi$. In proper complex-number symbols, $\zeta(\overline{z}) = \overline{\zeta(z)}$.

The important thing that follows is: If $a + bi$ is a zero of the zeta function, then so is $a - bi$.

IX. Finally, a pictorial representation of the Riemann Hypothesis—or at any rate, of the fact that there are lots of zeros on the critical line.

To understand Figure 13-8, you should remember that Figures 13-6 and 13-7 were pictures of the *argument* plane. A function of a complex variable sends one set of complex numbers, the arguments, to another set, the values. Since the complex numbers can be laid out as points on a plane, you can think of a function as sending points of one plane, the argument plane, to points of another plane, the value plane. The zeta function sends the point $\frac{1}{2} + 14.134725i$ in the argument plane to the point 0 in the value plane. Look back at Figure 13-2. There I showed both the argument plane and the value plane together, as if they were transparencies, one laid on top of the other.

Figures 13-6 and 13-7 are pictures of the argument plane, showing which arguments are sent to interesting values. The argument ant lives on the argument plane—whence her name. I had her wandering over the argument plane, noting where the argument points are sent to by the zeta function. I actually had her wandering along strange curves and loops, made up of points that are sent to (i.e., whose function values are equal to) pure real or pure imaginary numbers. I shall call these "'sent to' pictures of the argument plane."

An alternative way to show a function is with a "comes from" picture of the *value* plane.[69] Instead of showing, as I did in Figures 13-6 and 13-7, which arguments are sent to interesting values (in those cases, pure-real and pure-imaginary numbers), I can present a picture of the value plane, showing which value points *come from* interesting arguments.

Let us imagine that the argument ant has a twin sister who lives on the value plane. This sister is, of course, the value ant. Let's further suppose that the two sisters are in instantaneous radio communication; and that by this means they synchronize their movements, so that whichever argument the argument ant is standing on at any moment, the value ant is standing on the corresponding value in the value plane. If the argument ant is standing on $\frac{1}{2} + 14.134725i$, for example, with her gadget set to the zeta function, then the value ant is standing on 0 in her plane, the value plane.

Now suppose that the argument ant, instead of following those fancy loops and whorls in Figure 13-6 (which send the value ant on dull hikes up and down the real and imaginary axes), takes a walk straight up the critical line, heading due north from argument $\frac{1}{2}$. What path will the value ant follow? Figure 13-8 shows you. Her path starts out at $\zeta(\frac{1}{2})$, which, as I showed in Chapter 9.v, is −1.4603545088095…. Then she does a sort of half-circle counterclockwise below the zero point, then turns and loops clockwise

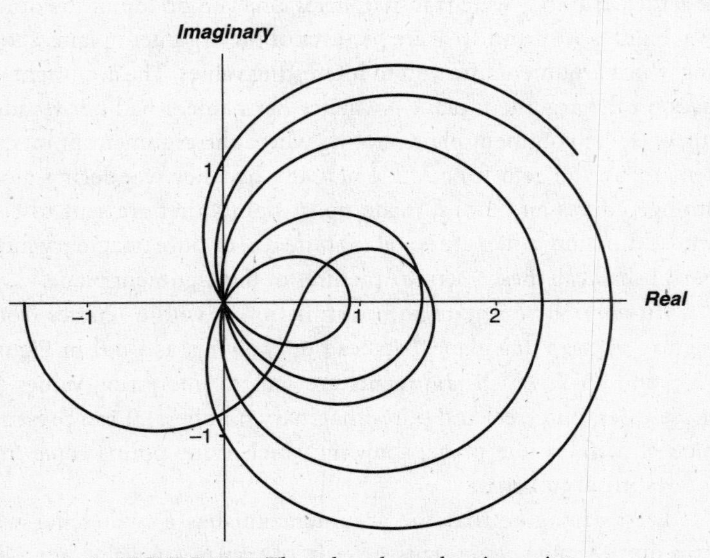

FIGURE 13-8 The value plane, showing points that come from the critical line.

around 1. She heads to zero and passes through it (that's the first zero—the argument ant has just passed $\frac{1}{2} + 14.134725i$). Then she keeps going round in clockwise loops, passing through the zero point every so often—whenever her twin on the argument plane steps on a zero of the zeta function. I stopped her walk when the argument ant reached $\frac{1}{2} + 35i$, because that's as far as Figure 13-6 goes. By that point, the curve has passed through zero five times, corresponding to the five non-trivial zeros in Figure 13-6. Notice that points on the critical line have a strong tendency to map to points with positive real part.

Once again, Figure 13-8 shows the *value* plane. It is not a "sent to" diagram like Figures 13-6 and 13-7; it is a "comes from" diagram, showing what the zeta function does to the critical line, just as Figure 13-2 showed what the squaring function does to that little checkered box. If you want to be properly mathematical about it, that looping curve in Figure 13-8 is ζ (*critical line*), the set of all points that come from points on the critical line. The curves in Figures 13-6 and 13-7 are ζ^{-1}(*real and imaginary axes*), the set of all points that are sent to the real and imaginary axes. The notation "ζ (*critical line*)" means "all zeta function values for arguments on the critical line." Conversely, "ζ^{-1}(*real and imaginary axes*)" means "all arguments whose zeta function values are on the real or imaginary axis." Note that the expression "ζ^{-1}" is used here in the special function-theory sense of "inverse function." Don't confuse it with a^{-1} as in Power Rule 8, which has the meaning $1/a$, the arithmetic reciprocal of a. This is a different usage—another case of overloading math symbols, like the use of π for both 3.14159... and the prime counting function.

Speaking very generally, "sent to" pictures of the argument plane are better tools for understanding a function in its broad properties (e.g., where its zeros are). "Comes from" pictures of the value plane are more useful for exploring particular aspects or curious features of the function.[70]

The Riemann Hypothesis states that all the non-trivial zeros of the zeta function lie on the critical line—the line of complex numbers with real part one-half. All the non-trivial zeros I have shown in this chapter do indeed lie on that line, as you can see from Figures 13-6, 13-7, and 13-8. Of course, that doesn't prove anything. The zeta function has an infinite number of non-trivial zeros, and no diagram could show them all. How do we know that the trillionth one, or the trillion trillionth, or the trillion trillion trillion trillion trillion trillionth lies on the critical line? We don't, not from drawing diagrams anyway. What's it all got to do with prime numbers? To answer that, I shall have to turn the Golden Key.

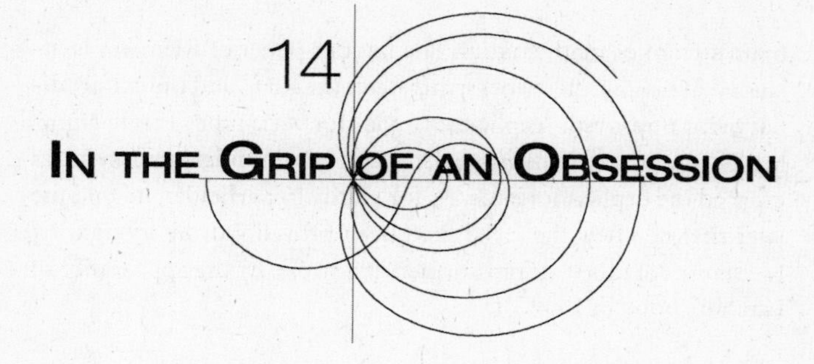

14

In the Grip of an Obsession

I. Göttingen was not, of course, the only place where first-class mathematics was being done in the early years of the twentieth century. Here is the English mathematician John Edensor Littlewood, 60-odd years before he offered snuff to Hugh Montgomery. As a young mathematician at Trinity College, Cambridge, in 1907, Littlewood was casting around for a good meaty problem on which to do postgraduate research.

> Barnes[71] was now encouraged to suggest a new problem: "Prove the Riemann Hypothesis." As a matter of fact this heroic suggestion was not without result; but I must begin by sketching the background of $\zeta(s)$ and prime numbers in 1907, especially so far as I was myself concerned. I had met $\zeta(s)$ in Lindelöf,[72] but there is nothing there about primes, nor had I the faintest idea there was any connexion; for me the R.H. was famous, but only as a problem in integral functions; and all this took place in the long vacation when I had no access to literature, had I suspected there was any. (As for people better instructed, only some had heard of Hadamard's paper, and fewer still knew of de la Vallée Poussin's in a Belgian journal. In any case, the work was considered very sophisticated and outside the

main stream of mathematics. The famous paper of Riemann is included in his collected works; this states the R.H., and the extraordinary, but unproved, "explicit formula" for $\pi(x)$; the "Prime Number Theorem" is not mentioned, though it is doubtless an easy guess granted the explicit formula. As for Hardy in particular, he told me later that he knew the P.N.T. had been proved, but he thought by Riemann. All this was transformed at a stroke by the appearance of Landau's book in 1909.)

I have taken that passage from *Littlewood's Miscellany*, a quirky collection of autobiographical fragments, jokes, math puzzles, and character sketches, first published (under a slightly different title) in 1953. The other *dramatis personae* in the extract are the older English mathematician Godfrey Harold Hardy, 1877–1947, and the German Edmund Landau, 1877–1938. These three men, half a generation after Hilbert, were all pioneers in the early assaults on the Riemann Hypothesis.

II. British mathematics in the nineteenth century had been oddly asymmetrical in its development and achievements. Great advances were made by British mathematicians in the *least* abstract areas of math, those most closely connected with physics. This was something I noticed during my own higher-mathematical education in London. We would sit through a class in real analysis, or complex function theory, or number theory, or algebra, and the names attached to the theorems would come rolling in across the English Channel from the Continent: Cauchy, Hadamard, Jacobi, Chebyshev, Riemann, Hermite, Banach, Hilbert.... Then we would have a Methods lecture (i.e., on mathematical methods used in physics), and suddenly we were back in Victoria's islands: Green's Theorem (1828), Stokes's Formula (1842), the Reynolds Number (1883), Maxwell's Equations (1855), the Hamiltonian (1834)....

Such other activity as took place in Britain was concentrated in the *most* abstract areas of math. Arthur Cayley, with J.J. Sylvester, invented matrices (more about them later), and the theory of algebraic invariants. George Boole opened up the whole territory of "foundations"—that is, mathematical logic, which he called "the laws of thought." (You can get an argument going about whether this is really at the high end of the abstraction scale. Boole himself declared that his intention was to make logic a branch of applied mathematics. However, I think mathematical logic is sufficiently abstract for most of us mortals.) It is curious to note that the week before Hilbert addressed the Paris Congress, the same lecture rooms at the Sorbonne had been booked for an International Congress of Philosophy. One of the papers read was "The Idea of Order and Absolute Position in Space and Time." Its author was a young British logician, also a Trinity man, named Bertrand Russell, who 10 years later, with Alfred North Whitehead, produced the classic of mathematical logic (to be more precise, of logicized mathematics), *Principia Mathematica.*

The least abstract math, and the most, but the great middle ground of abstraction—function theory, number theory, most of algebra—was yielded to the Continentals. In analysis, the most fertile field of nineteenth-century mathematics, the British were nearly invisible. At the end of the century they were in fact barely visible even in their strong areas. Only seven British mathematicians showed up at the Paris Congress, ranking Britain below France (90), Germany (25), the U.S.A. (17), Italy (15), Belgium (13), Russia (9), Austria, and Switzerland (8 each). Mathematically, Britain in 1900 was a backwater.

Even a backwater, of course, has some pockets of vitality. Trinity College, Cambridge, where Littlewood was in residence, maintained a strong mathematical tradition. It had been Sir Isaac Newton's college, 1661–1693, and counted several geniuses of mathematics and physics among its nineteenth-century alumni: Charles Babbage, generally credited with inventing the computer; the astronomer George Airy, after whom a family of mathematical functions is named;

Augustus de Morgan, the logician; Arthur Cayley, the algebraist; James Clerk Maxwell, and some lesser lights. Bertrand Russell got his degree at Trinity in 1893, was elected a fellow[73] in 1895, and was teaching there at the time Hardy joined the faculty. The college's history in the twentieth century was somewhat more mixed. It supplied most of the personnel for the Cambridge spy ring,[74] as well as several Bloomsberries.[75] So far as mathematics was concerned in the early years of the century, though, it was first and foremost the home of G.H. Hardy—the Hardy of Littlewood's memoir. It was Hardy, more than anyone else, who awoke English pure mathematics from its long slumber.

Studying for his degree at Trinity in 1897, Hardy came across a famous textbook of the time, *Cours d'Analyse*, by the French mathematician Camille Jordan. Jordan is familiar to students of complex variable theory for Jordan's Theorem, which says, basically, that a simple closed curve in the plane, for example a circle, has an inside and an outside. This theorem is ferociously difficult to prove—Estermann describes Jordan's own proof as "an intelligent attempt." *Cours d'Analyse* seems to have had the same effect on Hardy as Chapman's Homer had on Keats. After getting his fellowship at Trinity in the summer of Hilbert's address, Hardy spent the next few years publishing papers on analysis.

One fruit of Hardy's early analytical obsession was an undergraduate textbook, *A Course of Pure Mathematics*, first published in 1908 and never subsequently out of print. I learned analysis from this book, as did most twentieth-century British undergraduates. We referred to the book simply as "Hardy." The book's title is entirely misleading, as it contains nothing but analysis—no algebra, no number theory, no geometry, no topology. Nobody has ever minded this, though. As an introduction to classical (i.e., nineteenth-century) analysis, it is as near to perfect as a textbook can be. Its influence on my own approach to math was tremendous. Looking through what I have written in this book, I see Hardy all too plainly.

III. G.H. Hardy is the kind of oddity that only nineteenth-century England could produce. In old age he wrote a very curious book titled *A Mathematician's Apology* (1940), in which he described his own life as a mathematician. It is in some ways a sad book—an elegiac book, to be precise. The reason for this is very well explained in C.P. Snow's preface to the later editions. Hardy was a Peter Pan, a boy who never grew up. Snow: "His life remained the life of a brilliant young man until he was old: so did his spirit: his games, his interests, kept the lightness of a young don's. And, like many men who keep a young man's interests into their sixties, his last years were the darker for it." Littlewood: "Until he was about 30 he looked incredibly young." Hardy's games were cricket, about which he was passionate, and real tennis (a.k.a. court tennis or *jeu de paume*), a more difficult, more intellectually challenging game than ordinary tennis.

For 12 years, 1919–1931, Hardy held a chair at Oxford, with an exchange year at Princeton, 1928–1929; the rest of his life was spent at Trinity, Cambridge. A handsome and charming man, he never married, nor had any intimate attachments of any kind, so far as anyone knows. It must be remembered that the old Oxford and Cambridge colleges were men-only institutions with a strong flavor of misogyny. Until 1882, Fellows of Trinity were not permitted to marry. In the manner of our age, there has recently been some speculation that Hardy may have been homosexual. I refer the curious reader to Robert Kanigel's biography of Hardy's protégé Srinivasa Ramanujan, *The Man Who Knew Infinity*, which contains a full discussion of this point. The answer seems to be: probably not, except perhaps in the innermost sense.

There are even more Hardy stories than there are Hilbert stories—I see that I have already told one. Here are two more, both containing the Riemann Hypothesis. The first is from his obituary in the British science journal *Nature*.

> Hardy had one ruling passion—mathematics. Apart from that his main interest was in ball-games, of which he was a skilled player

and an expert critic. An illustration of some of his interests and antipathies is given by this list of "six New-Year wishes" which he sent on a postcard to a friend (in the 1920s):

(1) prove the Riemann Hypothesis;
(2) make 211 not out in the fourth innings of the last Test Match at the Oval;
(3) find an argument for the non-existence of God which shall convince the general public;
(4) be the first man at the top of Mount Everest;
(5) be proclaimed the first president of the USSR of Great Britain and Germany;
(6) murder Mussolini.

The second illustrates another of Hardy's eccentricities. Though claiming not to believe in God, he carried on a perpetual battle of wits with Him. In the 1930s, Hardy often visited with his friend Harald Bohr, who was Professor of Mathematics at the University of Copenhagen (and younger brother of the physicist Niels Bohr). George Pólya told the following story about one of these trips.

> Hardy stayed in Denmark with Bohr until the very end of the summer vacation, and when he was obliged to return to England to start his lectures there was only a very small boat available.... The North Sea can be pretty rough, and the probability that such a small boat would sink was not exactly zero. Still, Hardy took the boat, but sent a postcard to Bohr: "I proved the Riemann Hypothesis. G.H. Hardy." If the boat sinks and Hardy drowns, everybody must believe that he has proved the Riemann Hypothesis. Yet God would not let Hardy have such a great honor and so He will not let the boat sink.

His wonderful textbook aside, Hardy is best known for two great collaborations of which he was a part. The one with Ramanujan has been better publicized, and for good reason because it is one of the most curious and affecting stories in the history of mathematics. It is told in full in the aforementioned book by Robert Kanigel. However, the Hardy-Ramanujan collaboration is of only the most incidental concern to the history of the Riemann Hypothesis, and I shall have no more to say about it.

Hardy's other great collaboration was with Littlewood, with whose memoir about his own postgraduate research I opened this chapter. Littlewood joined the Trinity faculty in 1910. His collaboration with Hardy began the following year and continued until 1946. It was conducted mostly by mail during the years that Hardy was at Oxford and Princeton, and also during World War I, when Littlewood worked on artillery matters for the British army. Collaboration by mail was not much of a departure for Hardy and Littlewood, though: they often communicated by mail when living in rooms at Trinity.

Both Hardy and Littlewood were great mathematicians, both were the sons of schoolmasters, and both were lifelong bachelors. In most other ways they were different. There is something distinctly strange about Hardy. He hated having his photograph taken, for example—there are only half a dozen extant photographs of him[76]—and when staying in a hotel or guest room, he would cover up all the mirrors. Littlewood was much more of a meat-and-potatoes man. Where Hardy was slender and finely made, Littlewood was stocky and strong, a good all-round sportsman: swimming, rowing, rock climbing, cricket. He took up skiing at age 39 and became very proficient—an unusual thing among Englishmen at that time. He loved music and dancing.

Though conforming to the old idea of a college fellow—never married, he occupied the same set of rooms at Trinity for 65 years, 1912–1977—Littlewood had at least two children. The story as his colleague Béla Bollobás tells it is that Littlewood, in his younger years, used to go for annual vacations with the family of a doctor in Cornwall, whose children grew up calling him "Uncle John." One of these children was named Ann; Littlewood referred to her as "my niece." However, after becoming close friends with Bollobás and his wife, Littlewood confessed that Ann was, in fact, his daughter. They persuaded him to stop calling her his niece and start saying "my daughter." He accordingly did so, in the faculty common room one evening, and was mortified that none of his colleagues displayed the least surprise. Then, after Littlewood's death in 1977, a middle-aged man showed up at Trinity asking about his effects, explaining that he was Littlewood's son.

IV. "Hardy and Littlewood" became such a common byline on mathematical papers in the 1910s and 1920s that jokes were circulating about Littlewood being a fiction, invented by Hardy to take the blame for his mistakes. One German mathematician was said to have crossed the English Channel solely to confirm his belief that Littlewood did not exist.

That mathematician was Edmund Landau, who was seven days younger than Hardy. Landau was an instance of that uncommon phenomenon, the scion of a wealthy family who yet had a powerful work ethic and a record of great achievement in a non-commercial field. Landau's mother Johanna, *née* Jacoby, came from a rich banking family. His father was a Professor of Gynecology in Berlin, with a successful practice. Landau Senior was also a keen supporter of Jewish causes. The family home was at Pariser Platz 6a, in the most elegant quarter of Berlin, close to the Brandenburg Gate. Edmund was appointed to a professorship at Göttingen in 1909. When people asked for directions to his house, he would reply "You can't miss it. It's the finest house in town." He followed his father's (and Jacques Hadamard's) interest in Zionism, helping to establish the Hebrew University of Jerusalem and giving the first math lecture there, in Hebrew, shortly after the university opened in April 1925.

Landau was something of a character—this was a great age for mathematical characters—and there are apocrypha about him rivaling those of Hilbert and Hardy. Perhaps the best-known story is his remark about Emmy Noether, a colleague at Göttingen. Noether was mannish and very plain. Asked if she was not an instance of a great female mathematician, Landau replied: "I can testify that Emmy is a great mathematician, but that she is female, I cannot swear." His work ethic was legendary. It is said that when one of his junior lecturers was in hospital, recuperating from a serious illness, Landau climbed a ladder and pushed a huge folder of work through the poor man's window. Littlewood: "He simply did not know what it was like to be tired." Hardy says that Landau worked from 7 A.M. until midnight every day.

Landau was a gifted and enthusiastic teacher, and an extraordinarily productive mathematician. He wrote more than 250 papers and 7 books. His main importance for our story is the first of those books, a classic of number theory, published in 1909. This is the book Littlewood was speaking of in the extract I opened this chapter with: "All this was transformed at a stroke by the appearance of Landau's book...." The book's full title was *Handbuch der Lehre von der Verteilung der Primzahlen*—"Handbook of the Theory of the Distribution of the Prime Numbers." It is generally referred to by number theorists as simply "the *Handbuch*."[77] In two volumes of more than 500 pages each, this book gathered together all that was known about the distribution of primes up to that time, with a strong emphasis on analytic number theory. The Riemann Hypothesis is stated on page 33. The *Handbuch* was not the first book on analytic number theory—Paul Bachmann had published one in 1894—but its extremely detailed and systematic presentation laid out the subject in a style both clear and attractive, and Landau's book at once became the standard in its field.

I don't think Landau's *Handbuch* has ever been translated into English. Number theorist Hugh Montgomery, the star of my Chapter 18, taught himself German by reading his way through the *Handbuch*, one finger on the dictionary. He tells the following story. The first 50-odd pages of the book are given over to a historical survey, in sections each of which is headed with the name of a great mathematician who made contributions in the field: Euclid, Legendre, Dirichlet, and so on. The last four of these sections are headed "Hadamard," "von Mangoldt," "de la Vallée Poussin," "Verfasser." Hugh was extremely impressed with the contributions of Verfasser, but was puzzled to know why he had not heard the name of this fine mathematician before. It was some time before he learned that "Verfasser" is a German word meaning "author" (ordinary nouns are capitalized in German).

V. "All this was transformed at a stroke by the appearance of Landau's book...." Both Hardy and Littlewood must have read Landau's book soon after it became available. Here is what Hardy has to say, in the obituary of Landau he wrote (with Hans Heilbronn) for the London Mathematical Society.

> The *Handbuch* was probably the most *important* book he wrote. In it the analytic theory of numbers is presented for the first time, not as a collection of a few beautiful scattered theorems, but as a systematic science. The book transformed the subject, hitherto the hunting ground of a few adventurous heroes, into one of the most fruitful fields of research of the last thirty years. Almost everything in it has been superseded, and that is the greatest tribute to the book.

It was certainly from the *Handbuch* that both Hardy and Littlewood became infected with the Riemann Hypothesis obsession. The first fruits came in 1914, not in the form of a collaboration, though they were collaborating by that time, but as two separate papers, both of major importance in the development of the theory.

Hardy's paper was titled *Sur les zéros de la fonction* $\zeta(s)$ *de Riemann* and appeared in the *Comptes Rendus* of the Paris Academy of Sciences. In it, he proved the first major result on the distribution of the non-trivial zeros.

Hardy's 1914 Result

Infinitely many of the zeta function's non-trivial zeros satisfy the Riemann Hypothesis—that is, have real part one-half.

Though a major step forward, it is important for the reader to understand that this did not settle the Hypothesis. There is an infinity of non-trivial zeros; Hardy proved that infinitely many of them have real part one-half. This leaves three possibilities still open:

- Infinitely many zeros do not have real part one-half.
- Only finitely many zeros do not have real part one-half.
- There are no zeros that do not have real part one-half—the Hypothesis!

For an analogy, consider the following statements about the even numbers greater than two, that is: 4, 6, 8, 10, 12, ...

- Infinitely many of them are divisible by 3; infinitely many are not.
- Infinitely many are greater than 11; only four are not.
- Infinitely many are the sum of two primes; there are none that are not—the Goldbach Conjecture (which is still unproven at the time of writing).

Littlewood's paper, also published in the Paris Academy's *Comptes Rendus* of that year, was titled *Sur la distribution des nombres premiers.* It proved a result as subtle and striking as Hardy's, though in a different part of the field. It needs some preamble.

VI. I have already pointed out the following general trend in thinking about the Riemann Hypothesis at the beginning of the twentieth century. The Prime Number Theorem (PNT) had been proved. It was known with mathematical certainty that indeed $\pi(x) \sim Li(x)$—to put it in words, that the relative difference between $\pi(x)$ and $Li(x)$ dwindles away to zero as x gets bigger and bigger. So now what can we say about this difference, this error term? It was in focusing on the error term that mathematicians' attention was drawn to the Riemann Hypothesis, because Riemann's 1859 paper gave an exact expression for the error term. That expression, as I shall show in due course, involves all the non-trivial zeros of the zeta function, so the key to understanding the error term is hidden in among the zeros somehow.

Let me make this concrete by showing some actual values of the error term. In Table 14-1, "absolute error" means $Li(x) - \pi(x)$, while "relative error" means that number as a proportion of $\pi(x)$—in other words, the absolute error divided by $\pi(x)$.

TABLE 14-1

		Error Term	
x	$\pi(x)$	Absolute	Relative
1,000	168	10	0.059523809524
1,000,000	78,498	130	0.001656093149
1,000,000,000	50,847,534	1,701	0.000033452950
1,000,000,000,000	37,607,912,018	38,263	0.000001017419
1,000,000,000,000,000	29,844,570,422,669	1,052,619	0.000000035270
1,000,000,000,000,000,000	24,739,954,287,740,860	21,949,555	0.000000000887

Well, the relative error is certainly dwindling away to zero, just as the PNT says it should. This is happening because the absolute error, though increasing, is not increasing anything like as fast as $\pi(x)$.

The inquiring mathematical mind now asks how, exactly, do these numbers behave? Are there rules to describe the slow increase of the absolute error, or the dwindling to zero of the relative error? To put it another way, if you drop the second and fourth columns of Table 14-1, or the second and third, and consider the resulting two-column tables to be snapshots of some functions (argument, value)—what functions are they? Can we get twiddle formulas for them, as we did for $\pi(x)$?

That is where the non-trivial zeros of the zeta function come in. They are intimately connected, in a way I shall later show you in exact mathematical detail, with the error term.

Although it is the relative error that the PNT speaks about, investigations in this area more often concentrate on the absolute error. It really makes no difference, of course, which one you consider. The relative error is just the absolute error divided by $\pi(x)$, so you can always skip easily from one to the other. So can we get any kind of result for the absolute error term, $Li(x) - \pi(x)$?

VII. Looking at Figure 7-6, and at Table 14-1, we can say with fair confidence that the the absolute difference $Li(x) - \pi(x)$ is positive and increasing. The numerical evidence for this is so strong that Gauss, when he made his own investigations, believed it to be always the case. Probably most early researchers agreed, or at least felt sure that $\pi(x)$ is always less than $Li(x)$. (Riemann's opinion on the matter is unclear.) Littlewood's 1914 paper therefore came as a sensation, for it proved that this is not so; that, on the contrary, there are numbers x for which $\pi(x)$ is greater than $Li(x)$. It actually proved much more.

Littlewood's 1914 Result

$Li(x) - \pi(x)$ changes from positive to negative and back infinitely many times.

Given that $\pi(x)$ is less than $Li(x)$ for as far as we have been able to take x, even with the most powerful computers, where is that first crossing point, the first "Littlewood violation," where $\pi(x)$ becomes equal to, and then greater than, $Li(x)$?

In situations like this, mathematicians go looking for what they call an *upper bound*, that is, a number N for which they can prove that whatever the precise answer to the question, it is at any rate definitely less than *N*. Proven upper bounds *N* of this sort are sometimes far larger than the actual answer.

That was the case with the first upper bound for the Littlewood violation. In 1933 Littlewood's student Samuel Skewes showed that if the Riemann Hypothesis is true, the crossover point must come before $e^{e^{e^{79}}}$, a number of about $10^{\text{ten billion trillion trillion}}$ digits. That's not the number; that's *the number of digits* in the number. (By way of contrast, the number of atoms in the cosmos is thought to have about eighty digits.) This monstrosity attained fame as "Skewes' number," the largest number ever to emerge naturally from a mathematical proof up to that time.[78]

In 1955 Skewes improved his result, this time without assuming the truth of the Riemann Hypothesis, to a number of a mere $10^{\text{one thousand}}$ digits. In 1966, Sherman Lehman pulled the upper bound down to a much more manageable (or at least, writable) figure, 1.165×10^{1165} (a number, that is, of a mere 1,166 digits), and established an important general theorem about the upper bound. In 1987, using Lehman's theorem, Herman te Riele reduced the upper bound still further, to 6.658×10^{370}.

At the time of writing (mid-2002), the best figure is the one established by Carter Bays and Richard Hudson in 2000, also starting from Lehman's theorem.[79] They showed that there are Littlewood violations in the vicinity of 1.39822×10^{316} and even gave some reasons for thinking that these may be the first violations. (Bays's and Hudson's paper leaves open a small possibility that lower violations might exist, perhaps even as low as 10^{176}. They also show a huge zone of violation around 1.617×10^{9608}.)

VIII. These oscillations of the error term $Li(x) - \pi(x)$ from positive to negative and back take place within fairly well-defined constraints, though. If this were not so, the PNT would not be true. Some ideas about the nature of those constraints had already emerged out of the effort to prove the PNT. De la Vallée Poussin had actually included an estimate for the constraining function in his own proof of the PNT. Five years later, in 1901, the Swedish mathematician Helge von Koch[80] had proved the following key result, which I'll state in a modern form.

Von Koch's 1901 Result

If the Riemann Hypothesis is true, then

$$\pi(x) = Li(x) + O\left(\sqrt{x}\log x\right)$$

The equation is pronounced as, "Pi of *x* equals log integral of *x* plus big oh of root *x* log *x*." Now I have to explain the "big oh" notation.

15 BIG OH AND MÖBIUS MU

I. I have given over this chapter to two mathematical topics that are related to the Riemann Hypothesis, but not otherwise to each other. The topics are the "big oh" notation and the Möbius mu function. First, big oh.

II. When the great Hungarian number theorist Paul Turán lay dying of cancer in 1976, his wife was at his bedside. She reported that his last murmured words were: "Big oh of one...." Mathematicians tell this story with awed admiration. "Doing number theory to the very end! A real mathematician!"

Big oh came into math from Landau's 1909 book, whose influence, as I have already described, was tremendous. Landau did not actually invent big oh. He candidly acknowledges, on page 883 of the *Handbuch*, that he borrowed it from Paul Bachmann's 1894 treatise. It is, therefore, very unfair that it is always referred to as "Landau's big oh," and that most mathematicians probably believe Landau did invent it. Big oh is all over the place in analytic number theory, and has leaked into other areas of math too.

Big oh is a way of setting a limit on the size of a function, as the argument goes off (usually) to infinity.

Definition of Big Oh

Function *A* is big oh of function *B* if, for large enough arguments, the size of *A* never exceeds some fixed multiple of *B*.

Let me take a cue from Paul Turán and consider big oh of one. "One," as used here, is a function, a function of the simplest kind. Its graph is a flat horizontal line, one unit above the horizontal axis. For any argument at all, the function value is ... 1. What, then, does it mean to say that some function $f(x)$ is big oh of one? By the definition I just gave, it means that as the argument x goes off to infinity, $f(x)$ never exceeds some fixed multiple of 1. To put it another way, the graph of $f(x)$ stays forever below some horizontal line. This is useful information about $f(x)$. There are lots of functions for which this is not true. It's not true for x^2, for example, or for x to any positive power, or for e^x, or even for log x.

Big oh means a bit more than that, actually. Note that in my definition I said "the size of *A*...." That means: "the value of *A*, ignoring its sign." The size of 100 is 100; the size of −100 is also 100. Big oh doesn't care about minus signs. To say that some function $f(x)$ is big oh of one is to say that $f(x)$ is forever trapped between two horizontal lines, one above the axis, one an equal distance below it.

As I said, lots of functions are not big oh of one. The simplest is the function x—that is, the function whose value is always equal to the argument. Its graph is a diagonal straight line, disappearing off the graph paper at top right. Clearly it is not contained between any pair of horizontal lines. No matter how far apart you set those horizontal lines, the function x breaks through them eventually. This remains true even if you reduce the slope. The functions $0.1x$ (shown in Figure 15-1), $0.01x$, $0.001x$, $0.0001x$ all eventually break through any fixed horizontal lines you set as bounds. None is big oh of one.

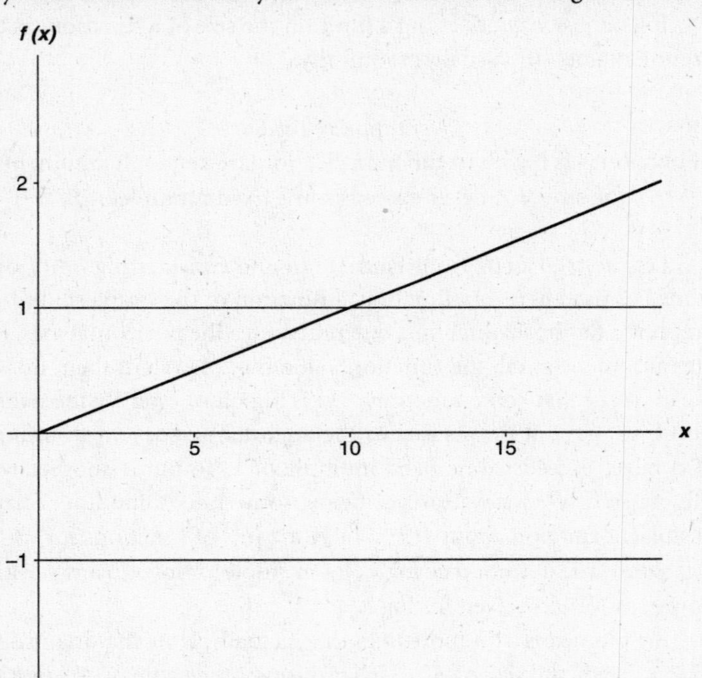

FIGURE 15-1 $0.1x$ is not $O(1)$.

Which illustrates another thing about big oh. Not only does big oh not care about signs, it doesn't care about multiples, either. If *A* is big oh of *B*, then so is ten times *A*, a hundred times *A*, a million times *A*; so is one-tenth of *A*, one-hundredth of *A*, one-millionth of *A*. Big oh doesn't tell you a precise rate of increase—we have derivatives to do that for us. It tells you the *type* of rate of increase. The function "one" has no rate of increase at all; it's dead flat. A function that is big oh of one never increases any faster than that. It might do all sorts of other stuff: dwindle to zero, oscillate indefinitely inside its bounding lines, or approach one of those bounding lines ever more closely, but it never shoots suddenly upward, or dives suddenly downward, breaking through the lines and staying outside them thereafter.

Those functions $0.1x$, $0.01x$, $0.001x$, $0.0001x$ are not big oh of one; they are all big oh of x. So is any other function that remains forever trapped in a "pie wedge" between a line ax and its mirror-image line $-ax$. Figure 15-2 is an example of a function that does not stay thus trapped. This is $0.1x^2$, the squaring function. No matter how wide you make the pie wedge—no matter how big the value of a—the graph of $0.1x^2$ eventually crashes through the upper line.

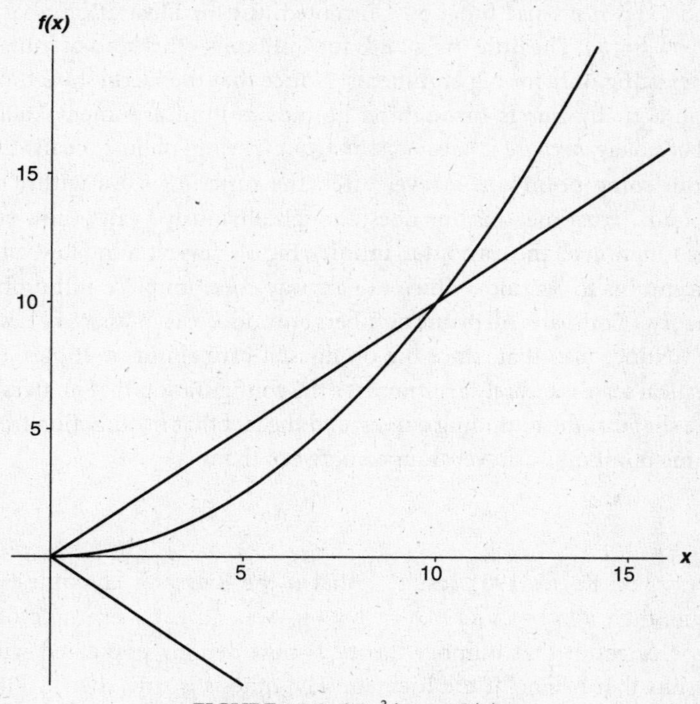

FIGURE 15-2 $0.1x^2$ is not $O(x)$.

Now you can see the meaning of von Koch's 1901 result. If the Riemann Hypothesis is true, then the absolute difference between $\pi(x)$ and $Li(x)$—either $Li(x) - \pi(x)$ or $\pi(x) - Li(x)$, it doesn't matter because big oh doesn't care about signs—stays trapped between two bounding curves as x goes off to infinity. The bounding curves are $C\sqrt{x}\log x$ and its mirror image, for some fixed number C. The error term can do what it likes between those two curves, but it can't break out from them and suddenly soar away out of their control. The difference between $\pi(x)$ and $Li(x)$ is big oh of $\sqrt{x}\log x$.

Figure 15-3 is an instance of a function that is $O(\sqrt{x}\log x)$. The graph shows (1) the curve $\sqrt{x}\log x$ (top half of the vaguely parabola shape), (2) the mirror-image curve $-\sqrt{x}\log x$ (bottom half of same), and (3) a nonsense function I invented just for illustration, that is $O(\sqrt{x}\log x)$. The little "m" stands for "million"—this kind of thing is interesting only for big arguments. Notice that the Derbyshire function actually bursts through its bounds around argument 200m. That's okay, *because it never does it again.* The big oh just means that from some point on, forever after, the function stays within its bounds. Trust me, this one does, though obviously I can't show you the function all the way out to infinity. Big oh doesn't mind low value exceptions to its rules, which are anyway commonplace in number theory. (Compare: All prime numbers are odd... except the very first.)

Notice also that, since big oh doesn't care about multiples, the vertical scale is entirely arbitrary. It's the configuration that matters—the shape of the bounding curves, and the fact that my function from some point on is forever trapped between them.

III. Von Koch's 1901 result[81]—that if the Riemann Hypothesis is true, then $\pi(x) = Li(x) + O(\sqrt{x}\log x)$—was an early example of a type of result that number theory is now densely populated with, results that begin: "If the Riemann Hypothesis is true, then...." If it turns out that the Riemann Hypothesis is not true, quite large parts of number theory will have to be rewritten.

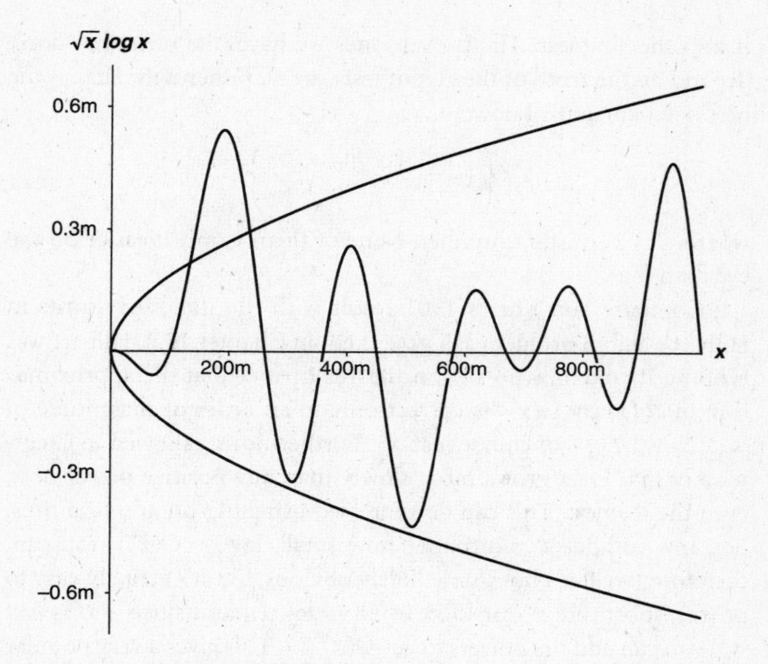

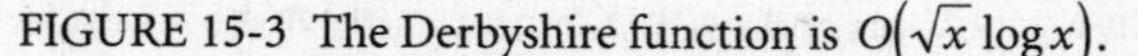
FIGURE 15-3 The Derbyshire function is $O\left(\sqrt{x}\,\log x\right)$.

Is there any big-oh type result for the error term $Li(x) - \pi(x)$ that does *not* depend on the truth of the Riemann Hypothesis? Oh, yes. It has for decades been a popular sport among analytic number theorists to find ever better big-oh formulas for the error term. None is as good as $O\left(\sqrt{x}\,\log x\right)$. That is the bee's knees, the tightest possible bound on the error term known up to the present. Since it depends on the Hypothesis being true, though, we can't be certain it applies. The ones we do know for certain are all looser than that. The corresponding parabola shape in Figure 15-3 is a tad wider, the difference getting more and more noticeable as x goes out to infinity. If the Riemann Hypothesis is true, we have the best possible—the tightest—big-oh formula for the error term, $O\left(\sqrt{x}\,\log x\right)$ so far known. It is also the simplest. The proven ones we have, the ones that don't depend on the truth of the Hypothesis, are all rather ugly. Here is the best one I currently know:

$$O\left(xe^{-C\left[(\log x)^{3/5}/(\log\log x)^{1/5}\right]}\right),$$

where C is a constant number. None of them is much easier on the eye than that.

Compare von Koch's 1901 result with the italicized words in Hilbert's eighth problem as I gave them in Chapter 12.ii. Hilbert was echoing Riemann, who says in the 1859 paper that the approximation of $\pi(x)$ by $Li(x)$ "is correct only to an order of magnitude of $x^{\frac{1}{2}}$." Now, $\sqrt{x}$ is of course just $x^{\frac{1}{2}}$. Furthermore, I showed in Chapter 5.iv that $\log x$ grows more slowly than any positive power of x, even the teeniest. This can be expressed using big-oh notation thus: For any number ε, no matter how small, $\log x = O(x^{\varepsilon})$. You can, therefore (well, it's not immediately obvious, but it's actually easy to prove), substitute x^{ε} for $\log x$ in $O(\sqrt{x}\log x)$; and since $\sqrt{x}$ is just $x^{\frac{1}{2}}$, you can add the powers to get $O(x^{\frac{1}{2}+\varepsilon})$. This gives a very popular alternative way to express von Koch's result, $\pi(x) = Li(x) + O(x^{\frac{1}{2}+\varepsilon})$. The symbol ε is so commonly used for vanishingly small numbers that the words "...for any ε, no matter how small" are understood.

Notice, however, that in making this substitution, I weakened von Koch's result slightly. "Error term = $O(\sqrt{x}\log x)$" implies "Error term = $O(x^{\frac{1}{2}+\varepsilon})$"; but the converse is not true. The two results are not precisely equivalent. This is because, as I showed in Chapter 5.iv, not only does $\log x$ increase more slowly than any power of x; so does $(\log x)^N$, for any positive N. So if von Koch's result had stated that the error term was $O(\sqrt{x}(\log x)^{100})$, we could *still* deduce the alternative form $O(x^{\frac{1}{2}+\varepsilon})$!

Writing von Koch's result in this slightly weaker form $O(x^{\frac{1}{2}+\varepsilon})$ is, though, very suggestive. Riemann was almost right, in the sense that the log function is almost x^0; the order of magnitude is not $x^{\frac{1}{2}}$, it's $x^{\frac{1}{2}+\varepsilon}$. Given the tools at his disposal, the state of knowledge in the field, and the known numerical facts at that time, Riemann's $x^{\frac{1}{2}}$ must still count as an intuition of breathtaking depth.[82]

I introduced big oh with a story, so I shall take my leave of it with another. The point of this story is that mathematicians, like other professionals, sometimes put out a cloud of squid ink to deter and confuse outsiders.

At the Courant conference in Summer 2002 (see Chapter 22), I was talking to Peter Sarnak about this book. Peter is Professor of Mathematics at Princeton University and is an expert on number theory. I mentioned that I was trying to think of a way to explain big oh to readers who weren't familiar with it. "Oh," said Peter, "You should speak to my colleague Nick" (i.e., Nicholas Katz, also a professor at Princeton, though mainly an algebraic geometer). "Nick hates big oh. Won't use it." I swallowed this and made a note of it, thinking I might find some place for it in this book. Then that evening I happened to be talking to Andrew Wiles, who knows Sarnak and Katz both very well. I mentioned Katz's not liking big oh. "That's all nonsense," said Wiles. "They're just teasing you. Nick uses it a lot." Sure enough, he used it in a lecture the next day. Funny sense of humor, mathematicians.

IV. So much for big oh. Now, the Möbius function. There are many ways to introduce the Möbius function. I am going to approach it by way of the Golden Key.

Take the Golden Key and turn it upside down, that is, take the reciprocal of each side in Expression 7-1. Obviously, if $A = B$ and neither is zero, then $1/A = 1/B$. The result is Expression 15-1.

$$\frac{1}{\zeta(s)} = \left(1-\frac{1}{2^s}\right)\left(1-\frac{1}{3^s}\right)\left(1-\frac{1}{5^s}\right)\left(1-\frac{1}{7^s}\right)\left(1-\frac{1}{11^s}\right)\left(1-\frac{1}{13^s}\right)\cdots$$

Expression 15-1

I'm now going to multiply out those parentheses on the right-hand side. At first blush this seems like a fairly ambitious thing to want to do. There are, after all, infinitely many of them. It does, in fact, demand a bit more justification and care than I'm going to give it here; but I shall get a useful and true result, so in this case, the end justifies the means.

Multiplying out parentheses is a thing you learn in basic algebra. To multiply out $(a+b)(p+q)$, you first multiply the $(p+q)$ by a to give $ap+aq$. Then you multiply $(p+q)$ by b to give $bp+bq$. Then, since the first parenthesis is a plus b, you add the two sub-results together for the final result, $ap+aq+bp+bq$.

If you have to multiply out three parentheses $(a+b)$ $(p+q)$ $(u+v)$, repeating the process gets you $apu + aqu + bpu + bqu + apv + aqv + bpv + bqv$. Multiplying out four parentheses $(a+b)(p+q)(u+v)(x+y)$ gives a result like the one in Expression 15-2.

$$apux + aqux + bpux + bqux + apvx + aqvx + bpvx + bqvx + \\ apuy + aquy + bpuy + bquy + apvy + aqvy + bpvy + bqvy$$

Expression 15-2

All of this is starting to look a bit formidable. And we have an infinity of parentheses to multiply out! The trick is to look at it with a mathematician's eyes. What is Expression 15-2 made up of? Well, it's the sum of a number of terms. What do these terms look like? Take one of them at random, let's say $aqvy$. It's got an a from the first parenthesis, a q from the second, a v from the third and a y from the fourth. *It's a product made up of one number plucked from each parenthesis.* And the whole expression is got by adding up the results of all possible combinations of plucks.

Once you have seen this, multiplying out an infinity of parentheses is a breeze. The answer is going to be a sum—an infinite sum, of course—of terms; and each term is got by plucking one number from each parenthesis and multiplying all those plucked numbers together. If you add up the result of all possible plucks, you have the result. As written, that still looks pretty daunting. It says that every term in my infinite sum is an infinite product. Yes, it is, but since every parenthesis on the right hand of Expression 15-1 contains a 1, I can finesse the situation by plucking an infinity of 1s and only a finite number of not-1s. After all, since every not-1 term in every parenthesis is a number between $-\frac{1}{2}$ and 0, if I multiplied an infinity of them, the size of the result (I mean, ignoring the sign) would certainly be no bigger than $\left(\frac{1}{2}\right)^{\infty}$ —which is zero! Watch me build the infinite sum.

First term of the infinite sum: Pluck the 1 from every parenthesis. This gives you the infinite product $1 \times 1 \times 1 \times 1 \times 1 \times 1 \times 1 \times \ldots$, whose value is of course just 1.

Second term: Pluck the 1 from every parenthesis except the first. From that one, pluck the $-\frac{1}{2^s}$. This gives the infinite product $-\frac{1}{2^s} \times 1 \times 1 \times 1 \times 1 \times 1 \times 1 \times \ldots$, which is just $-\frac{1}{2^s}$.

Third term: Pluck the 1 from every parenthesis except the second. From that one, pluck the $-\frac{1}{3^s}$. This gives the infinite product $1 \times (-\frac{1}{3^s}) 1 \times 1 \times 1 \times 1 \times 1 \times 1 \times \ldots$, which is just $-\frac{1}{3^s}$.

Fourth term.... Well, I think you can see that by plucking a 1 from every parenthesis except the *n*th, I am going to get a term equal to $-1/p^s$, where p is the *n*th prime. So the infinite sum looks like Expression 15-3.

$$1-\frac{1}{2^s}-\frac{1}{3^s}-\frac{1}{5^s}-\frac{1}{7^s}-\frac{1}{11^s}-\frac{1}{13^s}-\ldots$$

Expression 15-3

That's not the end of it, though. When you multiply out parentheses, you end up with the sum of *all possible terms* got by plucking one number out of each parenthesis. Suppose I pluck $-\frac{1}{2^s}$ from the first parenthesis, $-\frac{1}{3^s}$ from the second, and 1 from all the others. This gives me $(-\frac{1}{2^s}) \times (-\frac{1}{3^s}) \times 1 \times 1 \times 1 \times 1 \times 1 \times \ldots$, which is $\frac{1}{6^s}$. I shall get a similar term from every possible pair of not-1 plucks. Plucking $-\frac{1}{5^s}$

from the third parenthesis, $-\frac{1}{13^s}$ from the sixth, and 1 from every other, gives me a term $\frac{1}{65^s}$.

(Note that there are two simple rules of arithmetic at work here. One is the rule of signs, a minus times a minus gives a plus. The other is Power Rule 7, $(x \times y)^n = x^n \times y^n$.)

So as well as the terms I've already gathered in Expression 15-3, I have a new bunch, of which there is one for every pair of different primes—like 5 and 13—and whose signs are all positive. So now Expression 15-3 has grown to look like this.

$$1-\frac{1}{2^s}-\frac{1}{3^s}-\frac{1}{5^s}-\frac{1}{7^s}-\frac{1}{11^s}-\frac{1}{13^s}-\cdots$$

$$+\frac{1}{6^s}+\frac{1}{10^s}+\frac{1}{14^s}+\frac{1}{15^s}+\frac{1}{21^s}+\frac{1}{22^s}+\frac{1}{26^s}+\frac{1}{33^s}+\cdots$$

with every number in that second row being the product of two different primes.

And we've only just started at this business of multiplying out an infinity of parentheses. The next step is to take all possible plucks of three not-1s, with all other plucks equal to 1. An example is $1 \times (-\frac{1}{3^s}) \times 1 \times 1 \times (-\frac{1}{11^s}) \times (-\frac{1}{13^s}) \times 1 \times 1 \times 1 \times \ldots$, which comes to $-\frac{1}{429^s}$. Now the result has expanded to

$$1-\frac{1}{2^s}-\frac{1}{3^s}-\frac{1}{5^s}-\frac{1}{7^s}-\frac{1}{11^s}-\frac{1}{13^s}-\cdots$$

$$+\frac{1}{6^s}+\frac{1}{10^s}+\frac{1}{14^s}+\frac{1}{15^s}+\frac{1}{21^s}+\frac{1}{22^s}+\frac{1}{26^s}+\frac{1}{33^s}+\cdots$$

$$-\frac{1}{30^s}-\frac{1}{42^s}-\frac{1}{66^s}-\frac{1}{70^s}-\frac{1}{78^s}-\frac{1}{102^s}-\frac{1}{105^s}-\cdots$$

with every number in that third row being the product of three different primes.

Assuming that I can just keep doing this, and assuming that I can rearrange the resulting terms at will, Expression 15-1 boils down to the one shown in Expression 15-4.

$$\frac{1}{\zeta(s)}=1-\frac{1}{2^s}-\frac{1}{3^s}-\frac{1}{5^s}+\frac{1}{6^s}-\frac{1}{7^s}+\frac{1}{10^s}-\frac{1}{11^s}-\frac{1}{13^s}+\frac{1}{14^s}+\frac{1}{15^s}-\cdots$$

Expression 15-4

The natural numbers that show up on the right-hand side there are ... what? Not all the natural numbers, for sure: 4, 8, 9, and 12 are missing. Not the primes: 6, 10, 14, and 15 aren't primes. If you look back at the process I went through to multiply out that infinity of parentheses, you will see that the answer is: every natural number that is the product of an odd number (including 1) of different primes, prefixed by a minus sign, together with every natural number that is the product of an even number of different primes, prefixed by a plus sign. The numbers that are missing are those like 4, 8, 9, 12, 16, 18, 20, 24, 25, 27, 28, ... that divide by some prime squared.

Welcome to the Möbius function, named after the German mathematician and astronomer August Ferdinand Möbius (1790–1868).[83] It is universally referred to now by the Greek letter μ, pronounced "mu," the Greek equivalent of "m."[84] Here is a full definition of the Möbius function $\mu(n)$.

- Its domain is $\mathbb{N}$, that is, all the natural numbers 1, 2, 3, 4, 5, ...
- $\mu(1) = 1$.
- $\mu(n) = 0$ if n has a square factor.
- $\mu(n) = -1$ if n is a prime, or the product of an odd number of different primes.
- $\mu(n) = 1$ if n is the product of an even number of different primes.

That might seem like an awfully cumbersome function definition to you. However, the Möbius function is tremendously useful in the theory of numbers and will play a starring role later in this book. As an instance of its utility, note that all that laborious algebra I just went through boils down to the elegant result shown in Expression 15-5.

$$\frac{1}{\zeta(s)} = \sum_n \frac{\mu(n)}{n^s}$$

Expression 15-5

V. As important as $\mu(n)$ itself in the history of the Riemann Hypothesis is its cumulative value, that is, the number you get if you add up $\mu(1) + \mu(2) + \mu(3) + \ldots + \mu(k)$ for some number k. This is "Mertens's function," $M(k)$. Its first 10 values (that is, for arguments $k = 1, 2, 3, \ldots$ up to 10) are: 1, 0, −1, −1, −2, −1, −2, −2, −2, −1. $M(k)$ is a very irregular function, oscillating back and forth around zero in the manner of what mathematicians call a "random walk." For arguments 1,000, 2,000, … up to 10,000 it has the values: 2, 5, −6, −9, 2, 0, −25, −1, 1, −23. For arguments 1 million, 2 million, … up to 10 million it has values: 212, −247, 107, 192, −709, 257, −184, −189, −340, 1037. If you ignore the signs, it's pretty clear that the size of $M(k)$ increases, but nothing else is clear.

Because of Expression 15-5, the behaviors of the μ function and the M function (cumulative μ) are intimately tied up with the zeta function and, therefore, with the Riemann Hypothesis. In fact, if you could prove Theorem 15-1, it would follow that the Riemann Hypothesis is true!

$$M(k) = O\left(k^{\frac{1}{2}}\right)$$

Theorem 15-1

However, if Theorem 15-1 is not the case, it does not follow that the Hypothesis is false. Mathematicians say that Theorem 15-1 is stronger than the Hypothesis.[85] A slightly weaker version, Theorem 15-2, is precisely as strong as the Hypothesis.

$$M(k) = O\left(k^{\frac{1}{2}+\varepsilon}\right),$$

for every number ε, no matter how small.

Theorem 15-2

If Theorem 15-2 is true, the Hypothesis is true; and if it is false, the Hypothesis is false. They are exactly equivalent theorems. More on this in Chapter 20.vi.

16

CLIMBING THE CRITICAL LINE

I. In 1930 David Hilbert attained his 68th birthday. In conformity with the regulations of Göttingen University, he retired. Honors poured in. Among them was a resolution by the authorities of Königsberg to award the keys of the city to this eminent native son. The presentation was to be made at the opening session of a conference scheduled for that fall, a meeting of the Society of German Scientists and Physicians. The occasion naturally required a speech. Thus, on September 8, 1930, in Königsberg, Hilbert delivered the second great public speech of his career.

The title of the speech was "Logic and the Understanding of Nature." Hilbert's purpose was to express some opinions about the relationship between our inner lives—our mental processes, including those that help us to create and prove mathematical truths—and the physical universe. This was, of course, a topic with a long philosophical pedigree, one in which the name of another of Königsberg's native sons, the eighteenth-century philosopher Immanuel Kant, has particular prominence. It is one to which the modern understanding of the Riemann Hypothesis is, as it happens, rather especially pertinent, as I shall show in Chapter 20. This was not known at the time of Hilbert's Königsberg address, though.

It had been arranged that, following the speech, Hilbert would give a shorter version of the address over the radio—at that time, of course, a very new thing. That shorter version was recorded and was actually released as a 78 R.P.M. gramophone record. ("Celebrity mathematician" was apparently not an oxymoron in Weimar Germany.) It can now be found on the Internet. With very little effort you can now hear spoken, in Hilbert's own voice, the six words for which he is best remembered, and which appear on his memorial stone at the cemetery in Göttingen. Those words are the last in his Königsberg address.

Hilbert believed firmly in the unbounded power of the human mind to uncover the truths of Nature and mathematics. In his youth, the rather pessimistic theories of the French philosopher Emil du Bois-Reymond had been very popular. Du Bois-Reymond maintained that certain things—the nature of matter and of human consciousness, for example—are intrinsically unknowable. He coined the apothegm *ignoramus et ignorabimus*—"we are ignorant and we shall remain ignorant." Hilbert had never liked this gloomy philosophy. Now, with all the world (or at any rate the scientific-mathematical part of it) listening, he gave it a last resounding kick.

> We ought not believe those who today, with a philosophical air and a tone of superiority, prophesy the decline of culture, and are smug in their acceptance of the *Ignorabimus* principle. For us there is no *Ignorabimus*, and in my opinion there is none for the natural sciences either. In place of this foolish *Ignorabimus*, let our resolution be, to the contrary: "We must know, we shall know."

Those last six words—in German, *Wir müssen wissen, wir werden wissen*—are the most famous that Hilbert ever spoke, and among the most famous in the history of science. They express a strong optimism, all the more remarkable from a man who was heading into retirement and was furthermore unwell. (Hilbert had for some years been suffering from pernicious anemia, a disease that in the 1920s was only just beginning to yield to treatment.) Those words make a happy contrast with the rather gloomy solipsism of Hardy's *Apology*—written 10 years later when Hardy was 63, five years younger than was Hilbert at the time of the Königsberg address.

II. A happy contrast too—though now we are in the realm of hindsight—with the horrors that were soon to engulf Germany. When Hilbert retired from his professorship in 1930, Göttingen was still what it had been for 80 years, a great center of mathematical research and study, probably the greatest in the world at that point. Four years later it was an empty shell, from which the greatest minds had fled, or been driven out.

The principal events here were of course those that took place in the early months of 1933: Adolf Hitler's swearing-in as Chancellor on January 30, the Reichstag fire on February 27, the elections of March 5, in which the Nazis won 44 percent of the votes (a plurality), and the Enabling Act of March 23, which transferred key constitutional powers from legislature to executive. By April the Nazis had almost total control of Germany.

One of their first decrees, on April 7, was intended to bring about the dismissal of all Jews from the civil service. I say "was intended" because the old Field Marshal, Paul von Hindenburg, was still president of the German Republic and had to be deferred to. He insisted that there be two categories of exemption to the April 7 decree: first, any Jew who had performed military service in World War I, and second, any who had already held a civil service position before August 1914, when that war began.

University professors were civil servants and so came under the scope of the decree. Of the five professors teaching mathematics at

Göttingen, three—Edmund Landau, Richard Courant, and Felix Bernstein—were Jewish. A fourth, Hermann Weyl (who had succeeded to Hilbert's chair), had a Jewish wife. Only Gustav Herglotz was racially uncompromised. As a matter of fact, the April 7 decree did not apply to Landau or Courant, since they fell within the Hindenburg exemptions. Landau had been appointed to his professorship in 1909; Courant had performed valiant war service on the Western Front.[86]

It was not the way of the Nazis to stick to the letter of the law in such matters, though. It did not help that Göttingen at large was rather strong for Hitler. This was true of both "town" and "gown." In the 1930 elections, Göttingen had delivered twice as many votes to Hitler's party as the national average; and the Nazis had a majority in the university's student congress as far back as 1926. (That grand house of which Edmund Landau was so proud had been defaced with a painting of a gallows in 1931.) On April 26 the town newspaper, *Göttinger Tageblatt*, which was keenly pro-Nazi,[87] printed an announcement that six professors at the university were being placed on indefinite leave. The announcement came as a surprise to the six professors; they had not been notified.

Between April and November that year, Göttingen as a mathematical center was gutted. Not only Jewish faculty were involved; anyone thought to have leftist leanings came under suspicion. The mathematicians fled—most eventually finding their way to the United States. Altogether 18 faculty members left or were dismissed from the Mathematics Institute at Göttingen.

One holdout was Edmund Landau (the only Göttingen math professor, by the way, who was a member of the town's synagogue). Relying on the integrity of the law, Landau attempted to resume his calculus classes in November 1933, but the Science Students' Council learned of his intention and organized a boycott. Uniformed storm troopers prevented Landau's students from entering the lecture hall. With singular courage, Landau asked the Council leader, a 20-year-old student named Oswald Teichmüller, to write out as a letter his reasons for organizing the boycott. Teichmüller did so, and the letter somehow survived.

Teichmüller was a very intelligent man and in fact became a fine mathematician.[88] It is clear from his letter that his motivation for the boycott was ideological. He believed, wholeheartedly and sincerely, in the Nazi doctrines, including the racial ones, and felt it improper that German students should be taught by Jews. We are accustomed to think of Nazi activists as thugs, low-lifes, opportunists, and failed artists of one sort or another, which indeed most of them were. It is salutary to be reminded that they also included in their ranks some people of the highest intelligence.[89]

Landau himself then left Göttingen, brokenhearted. He went back to the family home in Berlin. There were a few overseas lecturing trips, which seemed to give him great pleasure, but he would not leave his native land to live permanently abroad and died from natural causes at his Berlin home in 1938.

Hilbert himself died in wartime Göttingen on February 14, 1943, three weeks after his 81st birthday, from complications following a fall in the street. No more than a dozen people attended the funeral service. Only two of them had much claim to mathematical honors: the physicist Arnold Sommerfeld, who had been an old friend of Hilbert's, and the above-mentioned Gustav Herglotz. Hilbert's home city of Königsberg was flattened in the war; it is now the Russian city of Kaliningrad. Göttingen is now a rather ordinary provincial German university with a strong math department.

III. Those years of the early 1930s, before the darkness fell, brought forth one of the most romantic episodes in the history of the Riemann Hypothesis, the discovery of the Riemann-Siegel formula.

Carl Ludwig Siegel, the son of a Berlin letter carrier, was a lecturer at the University of Frankfurt. An accomplished number theorist, he understood very well, as any mathematician who reads it must, that Riemann's 1859 paper was only, to employ the terminology of Erving Goffman that I introduced in Chapter 4.ii, a "front" display—a summary for formal presentation of what must have been a far greater amount of "back" work. Siegel spent such time as he could spare, going through Riemann's private mathematical papers from the period, to see if he could gain any insight into the activity of Riemann's mind when he was constructing the paper.

Siegel was by no means the first to attempt this. The papers had been deposited at the university library in 1895 by Heinrich Weber, following his second edition of Riemann's collected works. When Siegel arrived, they had been sitting there in the Göttingen archives (where they still sit—see Chapter 22.i) for 30 years. Several researchers had investigated them, but all had been defeated by the fragmentary and disorganized style of Riemann's jottings, or else they lacked the mathematical skills needed to understand them.

Siegel was made of sterner stuff. He persevered with the piles of scribbled sheets, and made an astonishing discovery, which he published in 1932 in a paper titled "Of Riemann's *Nachlass*[90] as It Relates to Analytic Number Theory." This is one of the key papers in the story of the Riemann Hypothesis. To explain the nature of Siegel's discovery, I shall have to return to the computational thread in my narrative—that is, to the effort to actually calculate the zeros of the zeta function and to verify the Riemann Hypothesis experimentally.

IV. I left the computational thread in Chapter 12 with Jørgen Gram's publication of the first 15 non-trivial zeros in 1903. Further work in this direction has continued down to the present day. At the 1996 Seattle conference on the Riemann Hypothesis, Andrew Odlyzko presented the history shown in Table 16-1.

Van de Lune went on to carry his investigation to 5 billion zeros at the end of 2000, and to 10 billion by October 2001. In the mean-

TABLE 16-1 Computational Work on the Zeta Zeros.

Researcher(s)	Publication date	Number of zeros with real part ½
J. Gram	1903	15
R.J. Backlund	1914	79
J.I. Hutchinson	1925	138
E.C. Titchmarsh et al.	1935–1936	1,041
A.M. Turing	1953	1,054
D.H. Lehmer	1956	25,000
N.A. Meller	1958	35,337
R.S. Lehman	1966	250,000
J.B. Rosser et al.	1969	3,500,000
R.P. Brent et al.	1979	81,000,001
H. te Riele, J. van de Lune et al.	1986	1,500,000,001

time, in August 2001, Sebastian Wedeniwski, using idle time on 550 office PCs at IBM Corp., Germany, began a project to advance computation yet further. The latest result posted by Wedeniwski is dated August 1, 2002, and reports that the number of non-trivial zeros with real part one-half has now been carried to 100 billion.

There are actually a number of different things going on here, and it is important to keep them distinct in one's mind.

First, there is the confusion between (a) *height up the critical line*, and (b) *number of zeros.* "Height" here just means the imaginary part of a complex number; the height of $3 + 7i$ is 7. In discussions of the zeta zeros, it is now customary to refer to this height as t or T. (Since we know that the zeros are symmetrical about the real axis, we only bother with positive t, by the way.) We have a formula for the number of zeros up to height T.

$$N(T) = \frac{T}{2\pi}\log\left(\frac{T}{2\pi}\right) - \frac{T}{2\pi} + O(\log T)$$

This is actually a very good formula—the first two terms are Riemann's—giving excellent approximations even for quite low values of *T*. Ignoring the big oh term,[91] for *T* equal to 100, 1,000, and 10,000, it gives 28.127, 647.741, and 10,142.090. The actual numbers of zeros up to these heights are 29, 649 and 10,142. To get a value of $N(T)$ equal to Wedeniwski's 100 billion, you need *T* to be 29,538,618,432.236 . . . , and that is the height Wedeniwski has carried his work to.

And then there is the confusion about what is actually being calculated. It should not be assumed that Wedeniwski can show us all 100 billion of those zeros, to a high (or even medium) degree of accuracy. The aim of most of this kind of work is to confirm the Riemann Hypothesis, and this can be done without very precise computations of the zeros. There is a piece of theory that lets you compute how many zeros are in the critical strip between heights T_1 and T_2—that is, inside a rectangle whose bottom and top edges are imaginary T_1 and T_2, and whose left and right edges are real 0 and 1, as illustrated in Figure 16-1. There is another piece of theory that lets you compute how many zeros are on the critical line between these heights.[92] If the two computations give the same result, you have confirmed the Riemann Hypothesis in that range. You can do this with only a rough knowledge of where the zeros actually are. Most of the work in Table 16-1 is of this kind.

What about tabulation of the actual precise values of the zeros? Surprisingly little of this has been done, except incidentally to the other effort (i.e., verifying the Hypothesis). So far as I am aware, the first published table of this kind to any length was by Brian Haselgrove. In 1960, working on second-generation mainframe computers at the universities of Cambridge and Manchester, in England, Haselgrove and his colleagues tabulated the first 1,600 zeros, accurate to six decimal places, and published the table.

Andrew Odlyzko told me that when he began his work on the zeta zeros in the late 1970s, Haselgrove's tables were the only ones he knew of, though he thinks that Lehman, as part of his 1966 work, might have done accurate computation of more zeros. Andrew himself has a table (on computer disk, not printed) of the first two mil-

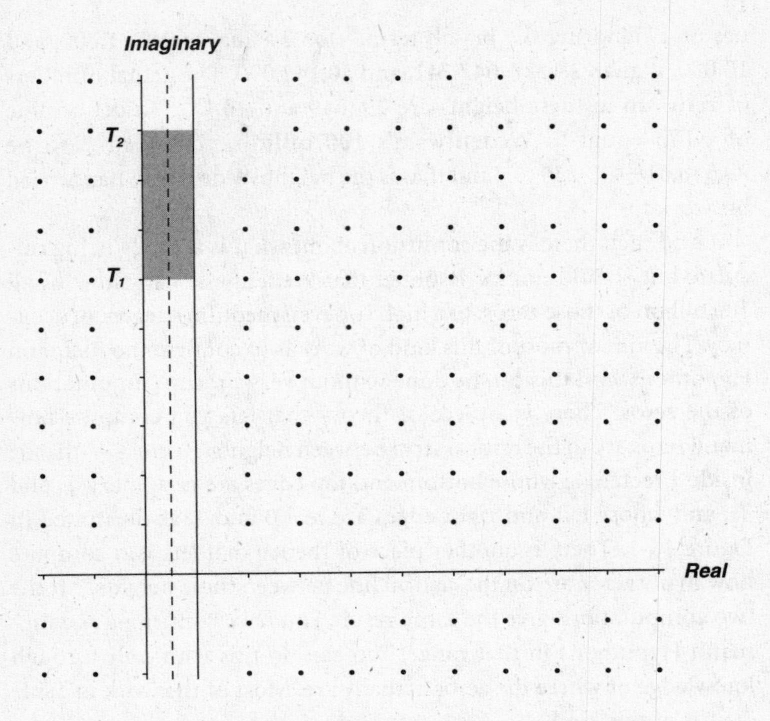

FIGURE 16-1 Heights T_1 and T_2 up the critical strip.

lion zeros, accurate to nine decimal places. At the time of writing, that is the largest known table of zeros.

All of the above work is concerned with the first *N* zeros. Andrew Odlyzko has also leapt ahead to examine small, isolated ranges very high up. He has published the highest non-trivial zero of the zeta function known to date, the 10,000,000,000,000,000,010,000-th. It is at argument $\frac{1}{2}$ + 1,370,919,909,931,995,309,568.33539*i*, to five places of decimals in the imaginary part. Andrew has also computed the first 100 zeros to 1,000 decimal places each.[93] The first zero (I mean, of course, its imaginary part) begins

14.134725141734693790457251983562470270784257115699243
3175685567460149963429809256764949010393171561012779
20297154879743676614269146988225458250536323944713
77804133812372059705496219558658602005555667258360
10773700205410982661507542780517442591306254481...

V. There are stories behind Table 16-1. That A.M. Turing, for example, is the very same Alan Turing who worked in mathematical logic, developing the idea of the Turing Test (a way of deciding whether a computer or its program is intelligent), and of the Turing machine (a very general, theoretical type of computer, a thought experiment used to tackle certain problems in mathematical logic). There is a Turing Prize for achievement in computer science, awarded annually since 1966 by the Association for Computing Machinery, equivalent to a Fields Medal[94] in mathematics, or to a Nobel Prize in other sciences.

Turing was fascinated by the Riemann Hypothesis. By 1937 (his 26th year) he had made up his mind that the Hypothesis was false and conceived the idea of constructing a mechanical computing device with which to generate a counterexample—a zero off the critical line. He applied to the Royal Society for a grant to cover the cost of construction and actually cut some of the gear wheels himself, at the engineering department of King's College, Cambridge, where he was lecturing.

Turing's work on the "zeta function machine" stopped abruptly in 1939, when World War II broke out. He joined the Government Code and Cypher School at Bletchley Park and spent the war years breaking enemy codes. Some of the gear wheels survived, however, and were found among his effects when he died, probably from suicide, on June 7, 1954.

As sad and strange as Turing's death was—he ate an apple coated, by himself, with cyanide—he enjoyed posthumous good fortune in the matter of biographers. Andrew Hodges wrote a beautiful book about him (*Alan Turing: The Enigma*, 1983), and then Hugh Whitemore made a fascinating play based on the book (*Breaking the Code*, 1986).

I have no space here to go into the details of Turing's life. I refer the reader to Hodges's fine biography, from which I shall just quote the following.

> [O]n 15 March [1952] he submitted for publication his work on the calculation of the zeta function, even though the practical attempt at doing it on the prototype Manchester computer had been so unsatisfactory. It might be that he wished to get it out of the way in case he was going to prison.

Turing was to be tried on March 31 on 12 charges of "gross indecency," consensual homosexual acts being at that time criminal offenses in Britain. In the event he did not go to prison. He was found guilty but placed on probation, with the condition that he undergo medical treatment. "There was," notes Hodges, "no concept of a right to sexual expression in the Britain of 1952."

There are other stories, too. Edward Titchmarsh, who had been a student of Hardy's (as, by the way, had Turing), worked through his 1,041 zeros[95] using punched-card machines on loan from the British Admiralty, which used them for compiling tide tables. He went on to write a classic mathematical text on the zeta function.[96] All this mechanical work came to an end with the advent of electronic computers after World War II, of course.

Other stories, too ... but I have strayed too far from my course.[97] I was going to finish telling you about the Riemann-Siegel formula.

VI. The first three entries in Table 16-1—the contributions of Gram, Backlund, and Hutchinson—all consisted of work done painstakingly with paper, pencil, and books of mathematical tables. This was computational hard labor; values of the zeta function are not easy to compute. The basic technique was one named "Euler-Maclaurin summation," developed around 1740 by Leonhard Euler and the Scottish mathematician Colin Maclaurin, working independently. It involves the approximation of integrals by long and complicated sums. Though arduous, it was the best method anyone could come up with. Gram himself tried several others, over a period of years, with very little success.

The essence of Carl Siegel's discovery, from his researches into Riemann's *Nachlass* at the Göttingen library, was this: Bernhard Riemann, in the background work for his 1859 paper, had developed a much better method for working out the zeros—and had actually implemented it and computed the first three zeros for himself! None of this was revealed in the 1859 paper. It was all hidden away in the *Nachlass.*

Says Harold Edwards: "Riemann was in fact in possession of the means to compute $\zeta\left(\frac{1}{2}+it\right)$ with amazing accuracy."[98] Riemann satisfied himself with rough calculations, however, precise knowledge of the location of the zeros not being essential for his work. He got the imaginary part of the first zero (see above) as 14.1386 and confirmed that it *is* the first; he computed the second and third to within a percentage point or two of accuracy.

The discovery of Riemann's formula, fine-tuned and published by Siegel to become the Riemann-Siegel formula, made work on the zeros much easier. All significant research depended on it up to the mid-1980s. Andrew Odlyzko's classic 1987 paper, "On the Distribution of Spacings Between Zeros of the Zeta Function," for example, which I shall have more to say about in Chapter 18.v, used Riemann-Siegel. Stimulated by this work, Odlyzko and Arnold Schönhage then developed and implemented some improved algorithms, but everything is built on Riemann-Siegel.

Carl Siegel, by the way, was not Jewish and was not directly affected by the restrictive laws of the early Nazi period. He detested the Nazis, though, and left Germany in 1940 to work at the Institute for Advanced Study in Princeton. He returned to Germany in 1951, finishing his career as a professor at that same Göttingen where, 20 years before, the archives had yielded up to him a glimpse of the astonishing powers of mind that dwelt behind Bernard Riemann's quiet diffidence.

17 A Little Algebra

I.

There should actually be a *lot* of algebra in this book, much more than I have been able to present. My focus has been on Bernhard Riemann and his work on prime numbers and the zeta function. That work was in number theory and analysis, and so my narrative has been dominated by those topics. However, modern math is, as I have already noted, very algebraic. This chapter fills in the algebraic background you need to understand two important approaches to the Riemann Hypothesis.

Like Chapters 7 and 15, this is a twofer chapter. Sections II and III give the basics of field theory; the remainder of the chapter discusses operator theory. Field theory is important because it has allowed something very much like the Riemann Hypothesis to actually be proved. Many researchers believe that field theory offers the most promising line of attack on the original, classical Riemann Hypothesis.[99] Operator theory became important following the remarkable and rather romantic developments I shall describe in the next chapter. First, though, field theory.

II. "Field" has a very particular meaning to mathematicians. A set of elements forms a field if the elements can be added, subtracted, multiplied, and divided, according to the ordinary rules of arithmetic—for example, the rule that $a \times (b + c) = ab + ac$. The results of all these operations must stay within the field.

For example, $\mathbb{N}$ is not a field. If you try to subtract 12 from 7, you get a result that is not in $\mathbb{N}$. Similarly with $\mathbb{Z}$. If you divide 12 by 7, the answer is not in $\mathbb{Z}$. These are not fields.

However, $\mathbb{Q}$, $\mathbb{R}$, and $\mathbb{C}$ are all fields. If you add, subtract, multiply, or divide two rational numbers, you get another rational number. Likewise with real and complex numbers. These are three examples of fields. Each has an infinite number of elements, of course.

Other infinite fields are easy to construct. Consider the family of all numbers with the form $a + b\sqrt{2}$, where a and b are rational numbers. Either b is zero, or it isn't. If b is not zero, since $\sqrt{2}$ is not a rational number, $a + b\sqrt{2}$ is not a rational number either. This family therefore includes all the rationals (b zero) as well as a host of very particular irrationals. It is a field. Adding $a + b\sqrt{2}$ to $c + d\sqrt{2}$ gives $(a + c) + (b + d)\sqrt{2}$, subtracting gives $(a - c) + (b - d)\sqrt{2}$, multiplying gives $(ac + 2bd) + (ad + bc)\sqrt{2}$ and dividing, using a trick similar to the one for complex numbers, gives $(ac - 2bd)/(c^2 - 2d^2) + ((bc - ad)/(c^2 - 2d^2))\sqrt{2}$. Since a and b can be any rational numbers at all, the field has infinitely many members.

Fields do not have to be infinite. Here is the simplest of all fields, with only two elements, 0 and 1. The rules for addition are: $0 + 0 = 0$, $0 + 1 = 1$, $1 + 0 = 1$, $1 + 1 = 0$. The rules for subtraction are: $0 - 0 = 0$, $0 - 1 = 1$, $1 - 0 = 1$, $1 - 1 = 0$. (Notice that these results are the same as for addition. In this field, any minus sign can be freely replaced by a plus sign!) The rules for multiplication are: $0 \times 0 = 0$, $0 \times 1 = 0$, $1 \times 0 = 0$, $1 \times 1 = 1$. The rules for division are: $0 \div 1 = 0$, $1 \div 1 = 1$, and division by zero is not allowed. (Division by zero is never allowed.) That is a perfectly sound field, and by no means trivial; I shall make good use of it in a moment. Mathematicians call it "F_2."

You can, in fact, construct a finite field for any prime number p, and even for any power of any prime number. If p is a prime number, there is a finite field with p members, one with p^2 members, one with p^3 members, and so on. Furthermore, these are all the possible finite fields. You can list them: $F_2, F_4, F_8, \ldots, F_3, F_9, F_{27}, \ldots, F_5, F_{25}, F_{125}, \ldots, \ldots$; and when you have done so, you have listed all possibilities for finite fields.

It is a mistake to think, as beginners often do, that finite fields are just a restatement of the clock arithmetic I described in Chapter 6.viii. This is true only for fields with a prime number of members. For other finite fields, the arithmetic is more subtle. Figure 17-1, for example, shows the clock arithmetic—addition and multiplication—for a clock with four hours marked (i.e., 0, 1, 2, and 3).

+	*0*	*1*	*2*	*3*
0	0	1	2	3
1	1	2	3	0
2	2	3	0	1
3	3	0	1	2

×	*0*	*1*	*2*	*3*
0	0	0	0	0
1	0	1	2	3
2	0	2	0	2
3	0	3	2	1

FIGURE 17-1 Addition and multiplication on a 4-hour clock. (That is, carrying out addition and multiplication in the usual way, then taking remainders after division by 4.)

This system of numbers and rules is interesting and useful, but it is not a field, because you can't divide 1 or 3 by 2. (If you could divide 1 by 2, then the equation $1 = 2 \times x$ would have a solution. It doesn't.) Mathematicians call it a *ring*—not unreasonably, since we are talking about clocks. In a ring, you can add, subtract, and multiply, but not necessarily divide.

The particular ring shown in Figure 17-1 has the official symbol $\mathbb{Z}/4\mathbb{Z}$. I confess I have never liked this style of symbolism though, so I am going to exercise author's privilege and create a symbol of my own for it: $\mathbb{CLOCK}_4$. Plainly, you can create such a ring for any natural number N. In my symbols, that ring would be called $\mathbb{CLOCK}_N$.

You can't create a field F_N for every number N, only for primes and prime powers. For a pure prime p, F_p looks just like $\mathbb{CLOCK}_p$—same addition table, same multiplication table. For a power of a prime, however, things get trickier. Figure 17-2 shows addition and multiplication (from which, of course, you can deduce subtraction and division) in F_4. Notice that F_4 is different from $\mathbb{CLOCK}_4$.

+	0	1	2	3
0	0	1	2	3
1	1	0	3	2
2	2	3	0	1
3	3	2	1	0

×	0	1	2	3
0	0	0	0	0
1	0	1	2	3
2	0	2	3	1
3	0	3	1	2

FIGURE 17-2 Addition and multiplication in the finite field F_4.

Every field, finite or infinite, has an important property called the *characteristic*. The characteristic of a field tells you how many times you have to add 1 to itself to get zero. If $1 + 1 + 1 + 1 + \ldots$ (N times) $= 0$, then the characteristic is N. Obviously the characteristic of F_2 is 2. Less obviously, but you will see it if you check the addition table in Figure 17-2, the characteristic of F_4 is also 2. Fields like $\mathbb{Q}$, $\mathbb{R}$, and $\mathbb{C}$, in which no amount of adding 1 to itself will ever produce zero, are said to have characteristic zero. (You might think that characteristic infinity would be more logical, and you might be right, but there are good reasons for choosing zero instead.) It can be proved that every field has characteristic either zero or a prime number.

Since this is algebra, the elements of a field need not be numbers. Algebra can deal with any kind of mathematical object. Consider all polynomials of any degree, that is, all expressions like $ax^n + bx^{n-1} + cx^{n-2} + \ldots$, where a, b, c, and so on are integers. Now form the set of all rational functions—that is, any function that is the ratio of two polynomials. That is a field. Here is an example of an addition sum in the field.

$$\frac{x}{2x^2+5x-3}+\frac{20x^2-19x+3}{x^4+3x^3}=\frac{x^4+40x^3-58x^2+25x-3}{2x^5+5x^4-3x^3}$$

(This is the kind of thing that high school algebra classes used to be devoted to.)

The coefficients of the polynomials in that field don't have to be integers. You can have some fun, in fact, by making them members of a finite field, like the F_2 I defined above. Here is an example of the kind of addition sum you would then get

$$\frac{x+1}{x}+\frac{x^3+x^2+x+1}{x^2+x+1}=\frac{x^4+x^2+x+1}{x^3+x^2+x}$$

(Remember if you check this out, that in the F_2 field, $1 + 1 = 0$; therefore, $x + x = 0$, $x^2 + x^2 = 0$, and so on.) That field would be called "the field of rational functions over F_2." It has, of course, infinitely many members; only the *coefficients* are restricted to a finite field. Thus, you can use a finite field to build up an infinite field. Notice also that since $1 + 1 = 0$, this field has characteristic 2. Thus an infinite field can have a finite characteristic.

It is not very helpful to ask what x represents in these last two examples. It is a symbol, for the manipulation of which we have watertight rules. From the algebraic point of view, that is the main thing. As a matter of fact, the answer to the question is almost certainly "x represents a number." However, the algebraist is much more interested in *what kind* of number—to what families, what groups, what fields the number belongs, and what rules of manipulation it obeys. To the analyst, my number $a+b\sqrt{2}$ is not very interesting. "It is just a real number," the analyst will say—"all right, an *algebraic* number" (Chapter 11.ii), if pressed. To the algebraist, though, it is exquisitely interesting, because it represents a field. For the most part, analysts and algebraists are not really discussing different things; they are just interested in different aspects of those things.[100]

III. That glimpse of the scope, power, and beauty of algebraic field theory is all I have space for here, though I shall revisit these topics briefly, from a different angle, in Chapter 20.v. I have given some coverage of these ideas because in 1921 the Austrian mathematician Emil Artin, in his Ph.D. thesis at the University of Leipzig, used field theory to open up a new approach to the Riemann Hypothesis. The math here is deep, and I can give only a sketch.

I mentioned in the previous section that there is a finite field for any prime power p^N. I also showed how a finite field can be used as a basis for building other fields, including even infinite ones. It turns out that if you start from a finite field, there is a way to construct one of these "extension" fields in such a way that a zeta function can be associated with it. By "a zeta function," I mean a function of a complex argument, defined over the field of complex numbers, that bears an uncanny resemblance, in its broad properties, to Riemann's zeta function. For example, these analogues to the Riemann zeta function come with their own Golden Keys, their own Euler products (see Note 36), and their own Riemann Hypotheses.

In 1933 Helmut Hasse, working at the University of Marburg in Germany, was actually able to prove a result analogous to the Riemann Hypothesis for a certain category of those base fields. In 1942 André Weil[101] extended this proof to a much wider class of objects and conjectured that similar results would apply to yet wider classes—these were the famous three "Weil Conjectures." In 1973 the Belgian mathematician Pierre Deligne, in a sensational achievement that won him a Fields Medal, proved the Weil Conjectures, essentially completing the program initiated by Artin.

Whether the techniques developed to prove these analogues of the Riemann Hypothesis for these very abstruse fields can be used to solve the classical Hypothesis is not known. A great many people think they can, and this remains a very active area of Riemann Hypothesis studies.

Are these researchers on to something? It is not clear—not, at any rate, to me. For the crux of the matter, go back to the second paragraph in this section, where I said that one of these analogue zeta functions is associated with a certain kind of field. For the classical zeta function, the one to which the original Riemann Hypothesis applies—the one this book is mainly about—the equivalent associated field is $\mathbb{Q}$, the field of ordinary rational numbers. As investigations have proceeded through these past few decades, it has become apparent that this elementary rational number field $\mathbb{Q}$ is in some sense deeper and more intractable than the subtle, artificial fields to which the results of Artin, Weil, and Deligne apply. On the other side of the argument, though, the techniques developed for the manipulation of those artificial fields have considerable power—Andrew Wiles used them to prove Fermat's Last Theorem!

IV. The physical thread of Riemann Hypothesis studies, whose genesis I shall describe in Section VI, and which has opened up wide new territories for exploration, depends on some understanding of a different algebraic topic, *operator theory*. I have, therefore, given over this section and the next to an account of operators, approaching them by way of the related theory of matrices.

Matrices are all over the place in modern math and physics, and the ability to manipulate them is a fundamental modern mathematical skill. Since my space is limited, I am going to simplify the whole business, giving just the bare essentials. In particular, I am going to ignore the entire issue of singular matrices, as if no such things existed. This is perhaps the most brazen act of simplification in the book, and I apologize to mathematically fastidious readers.

A matrix is a square array of numbers, like this, $\begin{pmatrix} 5 & 1 \\ 2 & 6 \end{pmatrix}$. I am using just whole numbers for simplicity. The numbers in a matrix can be rational, real, or even complex. That particular matrix is a 2×2. A matrix can be any size, though: 3×3, 4×4, 120×120, and so on. It can even be infinite in size, though the rules change slightly for infinite matrices. An important part of every matrix is the *lead diagonal*, the one that runs from top left to bottom right. In my example, the lead diagonal has elements 5 and 6.

Given two matrices of the same size, you can add, subtract, multiply, and divide them. The rules for doing this are not straightforward; for example, if A and B are matrices of the same size, it is not generally true that $A \times B = B \times A$. You can find the rules for manipulating matrices in any decent algebra textbook, and I don't need to go into them here. Suffice it to say that they exist and that there is an arithmetic of matrices, rather like the arithmetic of ordinary numbers, only trickier.

The important thing about matrices for us here is this. From any $N \times N$ matrix you can extract a polynomial of the N-th degree—that is, a function made up of various powers of x, up to the N-th power. This polynomial is called the *characteristic polynomial* of the matrix. I'm afraid I can't explain just how you find the characteristic polynomial of a given matrix. You must trust me, it's there, and there is a way to find it.

The characteristic polynomial of my example 2×2 matrix is $x^2 - 11x + 28$. For what values of x is this polynomial equal to zero? This is the same as asking, what are the solutions of the quadratic equation $x^2 - 11x + 28 = 0$? By the well-known formula (or, as my own schoolmaster used to say optimistically, "by inspection") the solutions are 4 and 7. And sure enough, if you substitute 4 for x, the polynomial has value $16 - 44 + 28$, which is indeed zero. Same with 7: $49 - 77 + 28$ is zero, too.

All this illustrates a general truth. Any $N \times N$ matrix has a characteristic polynomial of degree N, and that polynomial has N zeros.[102]

The zeros of a matrix's characteristic polynomial are tremendously important. They are called the *eigenvalues* of the matrix. Notice another thing. If you add up the numbers in the lead diagonal of my sample 2×2 matrix, you get 11 (because $5 + 6 = 11$). That is also the sum of the eigenvalues ($7 + 4 = 11$); and it is the negative of the first number that appears in the characteristic polynomial (the negative of -11 is 11). This is a very important number, called the *trace* of the matrix.

Characteristic polynomial; eigenvalues; trace—what's this all about? You see, the importance of matrices is not in themselves, but in what they represent. Matrix arithmetic, once you get the hang of it, is a merely mechanical skill, like ordinary arithmetic. But just as ordinary numbers can be used to represent much deeper, more fundamental things, so can matrices. It takes me 12 minutes to walk from my house to Huntington village; the distance is about 0.8 miles. If, starting tomorrow, the United States were to switch to the metric system, I should have to say "about 1.3 kilometers" instead of "about 0.8 miles." The distance, however, would not have changed; *only the numbers used to represent it would have changed.* It would still take me 12 minutes to walk it (until we switch to a metric clock).

To take another example, the calendar on my wall is a way of representing, in numbers, the motions of the sun and moon. Mainly of the sun, since we Americans have a solar calendar, whose months are out of sync with the motions of the moon. My calendar, however, was given to us by a local Chinese restaurant. If I look at it closely, I can see the months and days of the traditional Chinese lunar calendar marked, each month beginning with a new moon. The numbers are all different from the solar numbers, but they represent the same celestial events, the same passage of time, the same actual days.

Just so with matrices. The great importance of matrices is that they can be used to represent, to quantify, certain deeper and more fundamental things. What are those things? They are *operators.* The notion of an operator is one of the most important in twentieth-century math, and also in physics. I don't want to go into detail about what operators are, not until Chapter 20 at any rate. The important thing to grasp is that they are lurking underneath all this business of matrices, and it is their properties that matrices allow us to measure and study numerically.

That is why the characteristic polynomial, the eigenvalues, and the trace are such key concepts. They are properties of the underlying operator, not just of the matrix that represents it. An operator can, in fact, be represented by many matrices, all having the same eigenvalues, and so on. My sample 2×2 matrix represents a certain operator, the matrix $\begin{pmatrix} 3 & 2 \\ -2 & 8 \end{pmatrix}$ represents the same operator. So does $\begin{pmatrix} -1 & 8 \\ -5 & 12 \end{pmatrix}$, so does $\begin{pmatrix} 1000000 & 666662 \\ -1499994 & -999989 \end{pmatrix}$. All these matrices—and an infinity of others, too—have the characteristic polynomial $x^2 - 11x + 28$, the eigenvalues 4 and 7, and trace 11. That is because the underlying operator has those properties.

All of this applies to matrices of any size. Here is a 4×4 matrix.

$$\begin{pmatrix} 2 & 1 & 5 & 1 \\ 1 & 3 & 7 & 0 \\ 0 & 0 & 2 & 1 \\ 2 & 4 & 1 & 4 \end{pmatrix}$$

Its characteristic polynomial is $x^4 - 11x^3 + 40x^2 - 97x + 83$. (Notice that this matrix, like the other one, has trace 11. This is just coincidence; they are otherwise unrelated.) That polynomial has a full set of four zeros. To five decimal places, they are: 1.38087, 7.03608, $1.29152 - 2.62195i$, and $1.29152 + 2.62195i$. Those are, of course, the eigenvalues of the matrix. Two of them, as you can see, are complex numbers. (And complex conjugates of each other—which is always the case for a polynomial with real coefficients.) That is quite normal, even when, as here, all the numbers in the home matrix are real numbers. If you add up the four eigenvalues, you get 11; the imaginary components cancel out on addition.

V. By the time mathematicians had studied matrices for a few decades, they had classified them into different types. They had developed, so to speak, a taxonomy of matrices, in which the entire family of $N \times N$ matrices—referred to by mathematicians as "the general linear group for N," and symbolized by "GL_N"—was organized into species and genera.

I am going to pluck just one species out of that great matrix zoo, the *Hermitian* matrix, named after the great French mathematician Charles Hermite, whom we met briefly in Chapter 10.v. The numbers in a Hermitian matrix are complex numbers and they have the following pattern: if the number in the m-th row of the n-th column is $a + bi$, then the number in the n-th row of the m-th column is $a - bi$. In other words, every element of the matrix is the complex conjugate (Chapter 11.v) of its reflection in the lead diagonal. An example will, I hope, make this clear. Here is a 4×4 Hermitian matrix.

$$\begin{pmatrix} -2 & 8-3i & 4+7i & -3+2i \\ 8+3i & 4 & 1-i & -1-5i \\ 4-7i & 1+i & -5 & -6i \\ -3-2i & -1+5i & 6i & 1 \end{pmatrix}$$

You see how the element in the third row, first column is the complex conjugate of the one in the first row, third column? That's a Hermitian matrix. Note that it follows from the definition that all the numbers in the lead diagonal must be real, because the definition requires each number in the diagonal to be its own complex conjugate, and only a real number can be its own complex conjugate: $a + bi = a - bi$ if and only if b is zero.

Now, there is a famous theorem about Hermitian matrices, which says that *all the eigenvalues of a Hermitian matrix are real.* This is pretty surprising, when you think about it. Even if all the elements of a matrix are real, the eigenvalues can still be complex, as I showed with my first 4×4 matrix example. That a matrix with complex elements should nonetheless have real eigenvalues is remarkable. Well, if the matrix is Hermitian, it does. The eigenvalues of that matrix I just showed are (approximately): 4.8573, 12.9535, −16.553, and −3.2578. All real (and adding up to −2, the trace).

This theorem, by the way, implies that *all the coefficients of the characteristic polynomial of a Hermitian matrix are real.* This follows from the fact that the eigenvalues of any matrix are, by definition, the zeros of the matrix's characteristic polynomial. If a polynomial has zeros a, b, c, ..., then it can be factorized as $(x-a)(x-b)(x-c)\ldots$. You can just multiply out all the parentheses to get back to the usual form of the polynomial. Well, if a, b, c, ... are all real numbers, then multiplying out those parentheses gives you an expression with real-number coefficients. Since I have already stated the eigenvalues of the 4×4 Hermitian matrix above, we know that the characteristic polynomial is $(x-4.8573)(x-12.9535)(x+16.553)(x+3.2578)$. If you multiply out all the parentheses, you get the following as the characteristic polynomial: $x^4 + 2x^3 - 236x^2 + 286x + 3393$.

VI. This was all known 100 years ago ... at the time, that is, when David Hilbert was just embarking on his investigation of integral equations, in which the study of operators played a key role. Other mathematicians—some independently, some inspired by Hilbert's work—also spent the early years of the twentieth century absorbed in the study of operators. It was in the air. The Riemann Hypothesis was not nearly so much in the air at this point; although following Hilbert's 1900 address and the publication of Landau's book in 1909, a lot of the best minds were beginning to think hard about it.

It is, therefore, not altogether surprising that the two things came together in two of the most brilliant and wide-ranging intellects of the time. One of those intellects was Hilbert's, the other was George Pólya's, and they seem to have reached the same insight independently. Their thought processes probably went something like this.

> Here is a mathematical object, the Hermitian matrix, which is built up of complex numbers; yet its most intimate and essential property—the list of its eigenvalues—consists entirely and unexpectedly of real numbers. And now here is a function, the Riemann zeta function, which is built up of complex numbers; and its most intimate and essential property is the list of its non-trivial zeros. (Let's ignore the other zeros for this argument.) Every one of these zeros is in the critical strip. They are symmetrical about the critical line, whose real part is $\frac{1}{2}$. Let's say a typical zero is $\frac{1}{2} + zi$, for some number z. Then the Riemann Hypothesis says that all the z's are real.

Mathematicians of the 1910s would actually have said "operator," not "matrix." Matrices, though they had been lying around since Arthur Cayley invented them in 1856, did not become common currency until quantum mechanics took off around 1925. Still, you can see the rough analogy here. With both the eigenvalues of a Hermitian matrix and the non-trivial zeta zeros, we have a list of unexpectedly real numbers emerging from the key property of an essentially complex object. Hence,

The Hilbert-Pólya Conjecture

The non-trivial zeros of the Riemann zeta function correspond to the eigenvalues of some Hermitian operator.

The origins of the Conjecture are rather murky. Both Hilbert and Pólya are supposed to have mentioned the possibility of some such equivalence in lectures or conversations, at some time in the years 1910–1920. However, neither, so far as I have been able to discover, committed the thought to paper for publication. So far as I know—

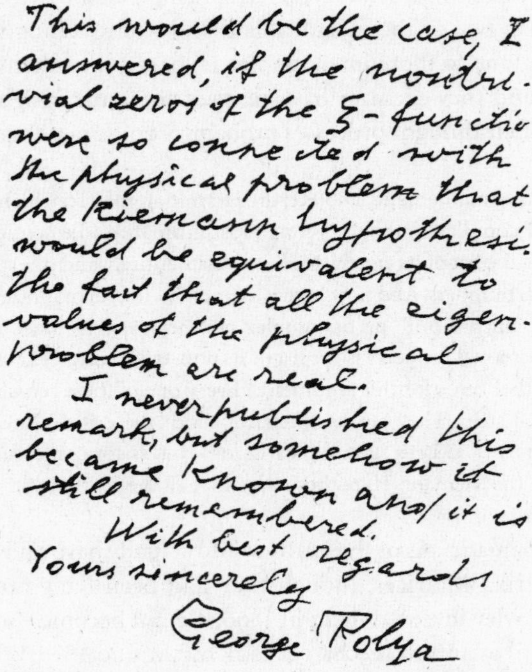

This would be the case, I
answered, if the nontri-
vial zeros of the ξ-functio
were so connected with
the physical problem that
the Riemann hypothesis
would be equivalent to
the fact that all the eigenvalues of the physical
problem are real.
I never published this
remark, but somehow it
became known and it is
still remembered.
With best regards
Yours sincerely
George Pólya

FIGURE 17-3 Part of George Pólya's letter to Andrew Odlyzko.

and, says Peter Sarnak, so far as he knows—the only written evidence for the Hilbert-Pólya Conjecture having been conjectured consists of a letter Pólya wrote to Andrew Odlyzko 60 years later, part of which is shown in Figure 17-3. In it, Pólya said that he had been asked the following question by Edmund Landau: "Can you think of any *physical* reason why the Riemann Hypothesis might be true?" Of Hilbert's own conjecturing, there is no material evidence at all that I am aware of.

It must be remembered, though, that Hilbert was a figure of towering stature in early twentieth-century mathematics; and also that he lived and worked in the atmosphere of German academia, where university professors were looked up to by their students and subordinates as remote and omniscient gods, to be approached only with the utmost deference. Not only was a professor not ever to be addressed as anything less elevated than "Herr Professor," even his wife became "Frau Professor." For the very grandest of these Olympians, indeed, even "Herr Professor" was inadequate. The most exceptional individuals were awarded the title "Geheimrat" by the German government—a rank roughly equivalent to a British knighthood. The correct form of address was then "Herr Geheimrat," though Hilbert himself did not care for this level of formality.

Given all of this, it is not surprising that if by good fortune you got sufficiently close to one of these deities to hear him speak, you would not soon forget his words. It is also the case, to be sure, that such giants caused a certain amount of unverifiable apocrypha to be generated about them. Still, I think the balance of the evidence, circumstantial though it be, leads one to believe that Hilbert did, indeed, at some point utter the Hilbert-Pólya Conjecture, or something equivalent to it. (To simply say "the Pólya Conjecture" would be confusing, by the way, as there is another, different conjecture known by that name.)

18 Number Theory Meets Quantum Mechanics

I. In the last chapter I gave the mathematical background and a little historical background to the Hilbert-Pólya Conjecture. The Conjecture was far ahead of its time and lay there untroubled for half a century.

That was, however, a very eventful half-century in physics, the most eventful ever. In 1917, just around the time of the Conjecture, Ernest Rutherford observed the splitting of the atom; 15 years later, Cockroft and Walton split the atom by artificial means. This led in turn to Enrico Fermi's work, to the first controlled chain reaction in 1942, and to the first nuclear explosion on July 16, 1945.

"Splitting the atom" is, as all high-school physics teachers tell their classes, a misnomer. You split atoms every time you strike a match. What we are really talking about here is the splitting of the atomic nucleus, the heart of the atom. To get a nuclear reaction—controlled or otherwise—going, you must fire a subatomic particle into the nucleus of a very heavy element. If you do that in a certain way, the nucleus splits, firing off new subatomic particles as it does. These particles penetrate the nuclei of neighboring atoms ... and so on, leading to a chain reaction.

Now, the nucleus of a heavy element is a very peculiar beast. You can visualize it as a seething, wobbling blob of protons and neutrons, welded together in such a way that it's hard to say where one particle starts and another ends. In the case of seriously heavy elements like uranium, the whole blob is teetering on the edge of instability. It might, in fact, depending on the precise mix of protons and neutrons, be actually unstable, liable to fly apart of its own volition.

As nuclear physics developed through the middle decades of the twentieth century, it became very important to understand the behavior of this strange beast and, in particular, to understand what happens if you fire a particle into it. Now, this nucleus, this wobbling blob, can exist in a number of states, some having high energy (imagine really energetic wobbling), some having low energy (a dull, languorous kind of wobbling). If a particle is fired into it so that the nucleus absorbs the particle instead of flying apart, then—since the energy of the particle must go somewhere—the nucleus moves from a lower state of energy to a higher. Later, tired of all the excitement, it might eject an equivalent particle, or perhaps a different type of particle altogether, and resume a lower energy state.

How many possible energy levels are there? When does a nucleus pass from level *a* to level *b*? How are the energy levels spaced relative to each other, and why are they spaced like that? Questions like these placed the study of the nucleus in a larger class of problems, problems about *dynamical systems*, collections of particles each of which has, at any point in time, a certain position and a certain velocity. As investigations proceeded through the 1950s, it became apparent that some of the most interesting dynamical systems in the quantum realm, including the heavy nucleus, were too complicated to yield to exact mathematical analysis. The number of energy levels was too large, the possible configurations too numerous. The whole thing was a nightmare version of the "many-body" problem of classical (that is, pre-quantum) mechanics, where several objects—the planets of the solar system, for example—are all acting on each other through gravity.

Precise mathematics has trouble coping with this level of complexity, so investigators fall back on statistics. If we can't discover exactly what's going to happen, perhaps we can discover what, on average, is most likely to happen. Such statistical approaches had been extensively developed in classical mechanics, beginning as far back as the 1850s, long before quantum theory appeared. Things go somewhat differently in the quantum world, but at least there was a good body of classical theory to provide inspiration. The necessary work was done, the necessary statistical tools for complex quantum dynamical systems like heavy-element nuclei were developed in the late 1950s and early 1960s, key players being the nuclear physicists Eugene Wigner and Freeman Dyson. One central concept was that of a *random matrix.*

II. A random matrix is just what its name suggests, a matrix made up of numbers chosen at random. Not quite at random, actually. Let me offer an illustration. Here is a random 4×4 matrix, of a rather particular type whose relevance I shall explain later. I round everything to four places of decimals in what follows, to save space.

$$\begin{pmatrix} 1.9558 & 0.0104-0.4043i & 1.8694-1.2410i & 0.8443-0.4180i \\ 0.0104+0.4043i & 1.8675 & 0.7520+1.1290i & 0.2270+0.1323i \\ 1.8694+1.2410i & 0.7520-1.1290i & 0.0781 & -1.6122+0.8667i \\ 0.8443+0.4180i & 0.2270-0.1323i & -1.6122-0.8667i & -2.0378 \end{pmatrix}$$

The first thing you may notice about this contraption is that it is Hermitian—it has that not-quite-symmetry about the lead diagonal that I described in Chapter 17.v. Recall the following facts from that chapter.

- Associated with every $N \times N$ matrix is a polynomial of degree N, called the *characteristic polynomial.*

- The zeros of the characteristic polynomial are called the *eigenvalues* of the matrix.
- The sum of the eigenvalues is called the *trace* of the matrix (and is equal to the sum of the lead-diagonal elements).
- In the particular case of a Hermitian matrix, the eigenvalues are all real and so, therefore, are the coefficients of the characteristic polynomial, and also the trace.

For the sample matrix I have shown here, the characteristic polynomial is

$$x^4 - 1.8636x^3 - 15.3446x^2 + 26.0868x - 2.0484$$

The eigenvalues are: –3.8729, 0.0826, 1.5675, and 4.0864. The trace is 1.8636.

Now turn your attention to the actual numbers that make up that sample matrix. The numbers you see there—the real numbers that make up the lead diagonal (with a slight qualification, see below), and all the real parts and imaginary parts of the off-diagonal complex numbers—are random in a certain special sense. They are plucked at random from a Gaussian-normal distribution—the famous "bell curve" that crops up all over the place in statistics.

Imagine the standard bell curve drawn on a sheet of fine-ruled graph paper, so that there are hundreds of graph-paper squares under the curve (Figure 18-1). Pick one of those squares at random; its horizontal distance from the peak center-line is a Gaussian-normal random number. There are many more of those squares clustered round the peak than there are out in the tails of the curve so you are much more likely to get a number between –1 and +1 than you are to get a number to the right of +2, or to the left of –2. So you see if you look at the numbers in the sample matrix shown at the beginning of this section. (Though the lead-diagonal elements are actually Gaussian-normal random numbers multiplied by $\sqrt{2}$ for technical reasons, and are, therefore, a bit bigger than you would expect.)

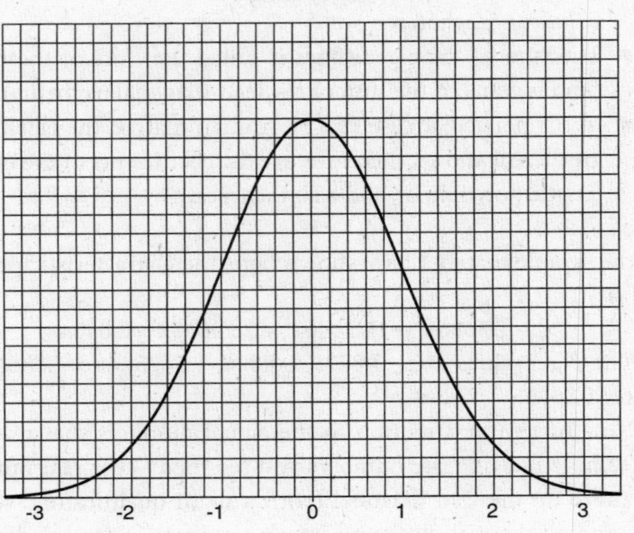

FIGURE 18-1 The Gaussian-normal distribution.

Gaussian-random Hermitian matrices like that one, though much, much bigger, proved to be just the ticket for modeling the behavior of certain quantum-dynamical systems. In particular, their eigenvalues turned out to provide an excellent fit for the energy levels observed in experiments. Therefore these eigenvalues, the eigenvalues of random Hermitian matrices, became the subject of intensive study through the 1960s. Their spacing in particular turned out to be very interesting. They were not spaced at random. It was, for example, much more unusual than you would expect, on a random basis, for two levels to be close to each other. This is the phenomenon called "repulsion"—energy levels trying to get as far as possible from each other, like a long standing line of antisocial people.

To give you a visual aid to what I am describing here, I asked my math software package, *Mathematica 4*, to generate a random 269×269 Hermitian matrix and compute its eigenvalues (see Figure

18-2). The reason for using the number 269 here will become clear very shortly. *Mathematica*, which never ceases to amaze me, did this in a trice. The 269 eigenvalues ranged from −46.207887 to 46.3253478. My idea was to string them out as blobs on a line going from −50 to +50, like raindrops on a fence wire, to show you the pattern of spacings. There was no way I could fit this neatly on a book page, though; so I chopped up the line into 10 equal segments (−50 to −40, −40 to −30, and so on) and just stacked the segments on top of each other to make Figure 18-2.

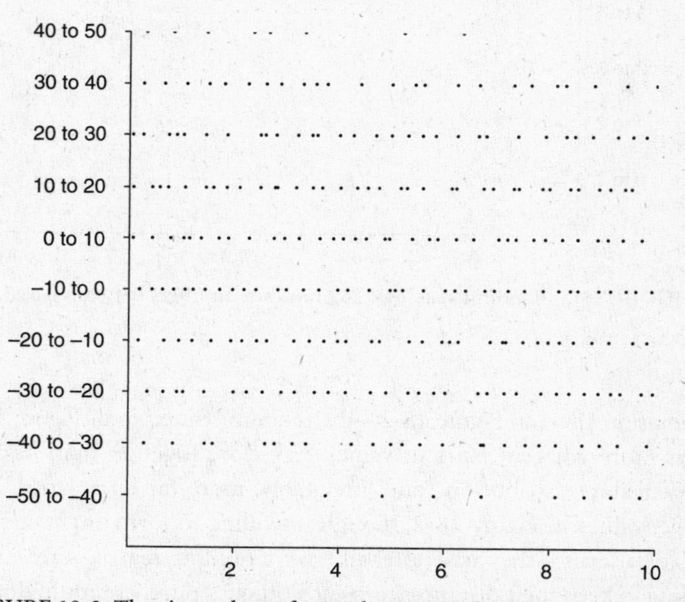

FIGURE 18-2 The eigenvalues of a 269-by-269 random Hermitian matrix.

There is no obvious pattern to the spacing. You might say that it is random. Not at all! Figure 18-3 shows 269 numbers picked entirely at random in the range 0–10 and plotted in the same way. Comparing Figures 18-2 and 18-3, you can see that the eigenvalues of a random matrix are not randomly scattered across their range. You can see the

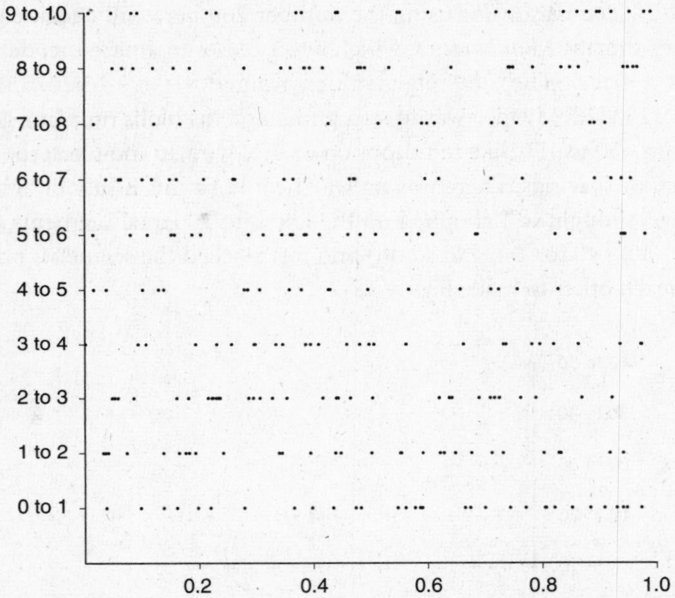

FIGURE 18-3 Random spacings: 269 random numbers between 0 and 10.

repulsion effect in Figure 18-2—the random scattering of Figure 18-3 has more adjacent pairs of values very close together than has the eigenvalue distribution (and, inevitably, more far apart, too). The eigenvalues in Figure 18-2, though unwilling to form any recognizable patterns—they arise, after all, from a random matrix—are struggling to keep their distance from each other. A purely random dot, by contrast, doesn't seem to mind at all if it finds itself jammed up against another random dot.

Permit me to introduce three terms of art here. The set of random (that is, Gaussian-random) Hermitian matrices[103] of the type I have been describing is called, in its totality, the "Gaussian Unitary Ensemble," or GUE (pronounced "goo"). The precise statistical properties of the spacings between a long non-uniform string of numbers like those I have illustrated are encapsulated in a creature called the "pair correlation function." And a certain ratio associated with this function, and highly characteristic of it, is called its "form factor."

Now I am in a position to tell you about a remarkable meeting, one that opened up very strange and mysterious questions about the Riemann Hypothesis and launched a thousand research projects.

III. The meeting, a chance encounter between a number theorist and a physicist, occurred at Princeton's Institute for Advanced Study in the spring of 1972. The number theorist was Hugh Montgomery, a young American doing graduate work at Trinity College, Cambridge—G.H. Hardy's old college. The physicist was Freeman Dyson, who held a professorship at the Institute. Dyson, whom I mentioned earlier, was a renowned physicist. He had not yet embarked on his second career as an author of thought-provoking bestsellers about the origins of life and the future of the human race.

Hugh Montgomery's most recent work had been an investigation of the spacing between non-trivial zeros of the zeta function. This was not part of any attempt to prove the Riemann Hypothesis. It just happens that a result about the nature of that spacing has consequences in the theory of number fields, fields somewhat like the $a + b\sqrt{2}$ field that I showed you in Chapter 17.ii.[104] This was Montgomery's area of interest. Here is the story as he tells it.

> I was still a graduate student when I did this work. I had written my thesis but not yet defended it. When I first did the work, I didn't understand what it meant. I felt that there should be something this was telling me, but I didn't know what, and I was troubled by that.
>
> That spring, the spring of '72, Harold Diamond[105] organized an analytic number theory conference in St. Louis. I went and lectured at that, then I flew to Ann Arbor. I'd accepted a job at Ann Arbor and I wanted to buy a house. Well, I bought a house. Then I

stopped off in Princeton on my way back to England, specifically to talk to Atle [Selberg] about this. I was a little worried that when I showed him my results he'd say: "This is all very nice, Hugh, but I proved it many years ago." I heaved a big sigh of relief when he didn't say that. He seemed interested but rather noncommittal.

I took afternoon tea that day in Fuld Hall with Chowla.[106] Freeman Dyson was standing across the room. I had spent the previous year at the Institute and I knew him perfectly well by sight, but I had never spoken to him. Chowla said: "Have you met Dyson?" I said no, I hadn't. He said: "I'll introduce you." I said no, I didn't feel I had to meet Dyson. Chowla insisted, and so I was dragged reluctantly across the room to meet Dyson. He was very polite, and asked me what I was working on. I told him I was working on the differences between the non-trivial zeros of Riemann's zeta function, and that I had developed a conjecture that the distribution function for those differences had integrand $1 - (\sin \pi u / \pi u)^2$. He got very excited. He said: "That's the form factor for the pair correlation of eigenvalues of random Hermitian matrices!"

I'd never heard the term "pair correlation." It really made the connection. The next day Atle had a note Dyson had written to me giving references to Mehta's book,[107] places I should look, and so on. To this day I've had one conversation with Dyson and one letter from him. It was very fruitful. I suppose by this time the connection would have been made, but it was certainly fortuitous that the connection came so quickly, because then when I wrote the paper for the proceedings of the conference, I was able to use the appropriate terminology and give the references and give the interpretation. I was amused when, a few years later, Dyson published a paper called "Missed Opportunities." I'm sure there are lots of missed opportunities, but this was a counterexample. It was real serendipity that I was able to encounter him at this crucial juncture.

You can understand why Freeman Dyson was so excited. The expression that Hugh Montgomery mentioned, the expression that had emerged from his inquiries into the Riemann zeta function's non-trivial zeros, was precisely the form factor associated with a random

Hermitian matrix—the kind of thing Dyson had been involved with for several years in his researches into quantum dynamical systems. (And Montgomery even understated the degree of serendipity involved in this meeting. Though he made his name as a physicist, Dyson's first degree was in math, and his first area of interest was number theory. If this had not been so, he might not have grasped what Montgomery was talking about.[108])

To illustrate the point, I am going to take all the non-trivial zeros of the Riemann zeta function up to the height 500*i*—that is, on the critical line (they all *are* on the critical line; the Riemann Hypothesis is certainly true down at these low levels)—from $\frac{1}{2}$ to $\frac{1}{2} + 500i$. There are 269 zeros in this range. (That is why I picked the number 269 for Figures 18-2 and 18-3.) Figure 18-4 shows them, their range chopped into 10 segments and stacked up just as before. If you compare this with Figures 18-2 and 18-3, you see that it resembles Figure 18-2, not Figure 18-3.

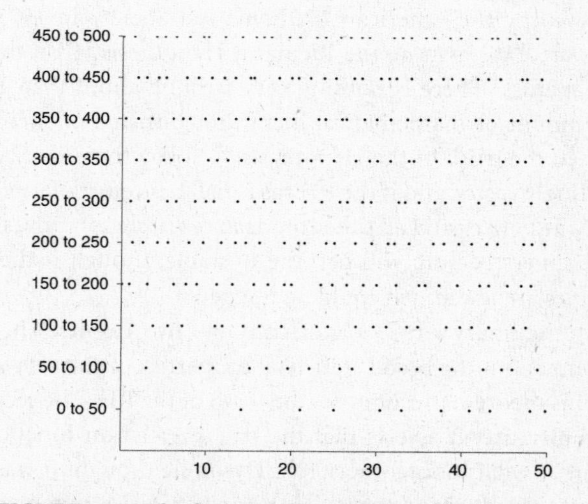

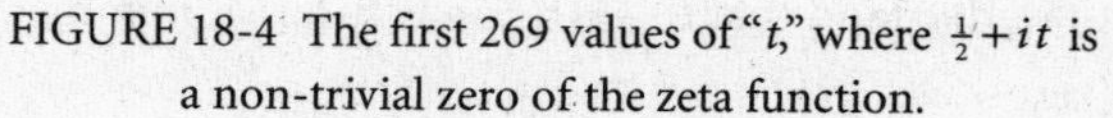
FIGURE 18-4 The first 269 values of "*t*," where $\frac{1}{2} + it$ is a non-trivial zero of the zeta function.

You should make some modest allowances when comparing these figures. The zeta zeros in Figure 18-4 take a while to get started and pack closer further up the critical line, according to the principle I stated in Chapter 13.viii. Also, the eigenvalues in Figure 18-2 are stretched out somewhat at the beginning and end of their range and correspondingly squished in the middle. Both effects can be reduced by taking more zeros and bigger matrices and by normalizing (see below). Even allowing for these distortions, the following points look pretty plausible on the basis of these diagrams.

- Neither the zeta zeros nor the eigenvalues look much like randomly scattered points.
- They resemble each other.
- In particular, they both show the repulsion effect.

IV. Hugh Montgomery's paper on the spacing of zeta zeros was published by the American Mathematical Society in 1973. Its first words are, "We assume the Riemann Hypothesis (RH) throughout this paper...." There is nothing very striking about that. By 1973 a vast amount of mathematical literature consisted of theorems that assumed the truth of the Hypothesis.[109] Today the quantity is correspondingly vaster, and if the RH (as I shall henceforth term it, following Montgomery and all other modern researchers) proves false, this entire superstructure will become unstable; though if the counterexamples are few, much could be rescued.

Montgomery's 1973 paper contains two results. The first is a theorem about the broad statistical properties of the zeta-zero spacing. This theorem presupposes the truth of the RH. The second result is a conjecture. It asserts that the pair correlation function for the spacing is what Montgomery told Dyson he thought it was. It is important to understand that this is a conjecture. Montgomery was not able to prove it, not even on the assumption that the RH is true. Nobody else has been able to prove it, either.

Most of the features of the Riemann zeros that you will see reported and discussed, most of the ideas that have come up in the last 30 years, are likewise conjectural. Hard proofs are in desperately short supply in this area. In part this is because, following the link established by Montgomery, so much of the recent thinking about the RH has been done by physicists and applied mathematicians. Sir Michael Berry[110] likes to quote the Nobel Prize-winning physicist Richard Feynman in this context, "A great deal more is known than has been proved." In part it is also because the RH is a very, very tough problem. There is such an immense quantity of literature on the RH now that you have to keep reminding yourself of the truth that very little is known for certain about the zeros of the zeta function, and even with all the rising interest during the past few years, results that are mathematically watertight still come only occasionally, at long intervals.

V. The Institute for Advanced Study in Princeton, New Jersey, is only 32 miles from AT&T's Bell Labs research center in Murray Hill. Hugh Montgomery lectured on what was, by that time, the "Montgomery pair correlation conjecture" at Princeton in 1978. Among those present was Andrew Odlyzko, a young researcher from the AT&T facility. At just about this time, his laboratory acquired a Cray-1 supercomputer. Researchers were encouraged to run projects on the Cray, to familiarize themselves with the kinds of algorithms appropriate to its architecture.

Ruminating on Montgomery's lecture, Odlyzko reasoned as follows. The Montgomery conjecture asserts that the spacing of zeta zeros follows such-and-such a statistical law. This law also appears in the investigation of a certain family of quantum dynamical systems that conform to the GUE model. The statistical properties of that family have been the subject of intensive analysis over several years. The statistical properties of the zeta zeros, however, have been very little investigated. Useful work could be done, the balance redressed, by undertaking a statistical study of the zeta zeros.

That is what Andrew Odlyzko proceeded to do. Using spare time in 5-hour segments on the Cray computers[111] at Bell Labs, he generated the first 100,000 non-trivial zeros of the Riemann zeta function to high accuracy (around 8 decimal places), using the Riemann-Siegel formula. Then, to get a snapshot of the situation much higher up the critical line, he generated another 100,000 zeros starting with the 1,000,000,000,001st. He then ran these two sets of zeros through various statistical tests, to see how they compared with the eigenvalues of matrices for GUE operators. The results of all this work were published in a landmark paper in 1987, under the title, "On the Distribution of Spacings Between Zeros of the Zeta Function."

The results were not quite conclusive. As Odlyzko put it very delicately in the paper, "The data presented so far are fairly consistent with the GUE predictions." There were slightly more small spacings than the GUE model predicted. Odlyzko's results were sufficiently impressive, though, to capture the attention of a wide range of researchers. Further work cleared up the discrepancies noted in the 1987 paper, and the Montgomery Pair Correlation Conjecture became the Montgomery-Odlyzko Law.[112]

The Montgomery-Odlyzko Law

The distribution of the spacings between successive non-trivial zeros of the Riemann zeta function (suitably normalized) is statistically identical with the distribution of eigenvalue spacings in a GUE operator.

VI. I can give only a brief sketch of the nature of Odlyzko's results. To do so, I duplicated them on my own PC, using a list of the zeros that Odlyzko has helpfully posted on his web site. To avoid any start-up anomalies, I took the 90,001st to the 100,000th zeros, counting up the critical line from $z = \frac{1}{2}$. That's 10,000 zeros—quite sufficient to make some statistical sense of. The 90,001st zero is at $\frac{1}{2}$ + 68194.3528*i*; the 100,000th is at $\frac{1}{2}$ + 74920.8275*i* (rounding to 4 decimal places). I am thus going to investigate the statistical properties of a sequence of 10,000 real numbers, a sequence that starts with 68194.3528 and ends with 74920.8275.

Since, as I pointed out in Chapter 13.viii, the zeros get closer together, on average, as you go up the critical line, I must make an adjustment to stretch out the higher end of this range. This I can do quite simply by multiplying every number by its log. Bigger numbers have bigger logs, and this is just what I need to even out the average spacing. This is the meaning of the word "normalized" in the statement of the Montgomery-Odlyzko Law given above. My sequence now begins with 759011.1279 and ends with 840925.3931.

Furthermore, I am interested in the *relative* spacing of the zeros; so I can subtract 759011.1279 from every number in the sequence without affecting my result. The sequence now goes from zero to 81914.2653. Finally, just to make the numbers neater, I am going to switch to a different scale, dividing every number in my sequence by 8.19142653. Again, this doesn't affect the relative spacing; I have merely switched rulers. This final form of my sequence starts like this: 0, 1.2473, 2.5840, ..., and ends like this: 9997.3850, 9999.1528, 10,000.

If you include the end points, I now have 10,000 numbers laid out for study, ranging from 0 to 10,000. Since there are 9,999 spaces between consecutive numbers, the average spacing is 10,000 ÷ 9,999, which is just a shade greater than 1.

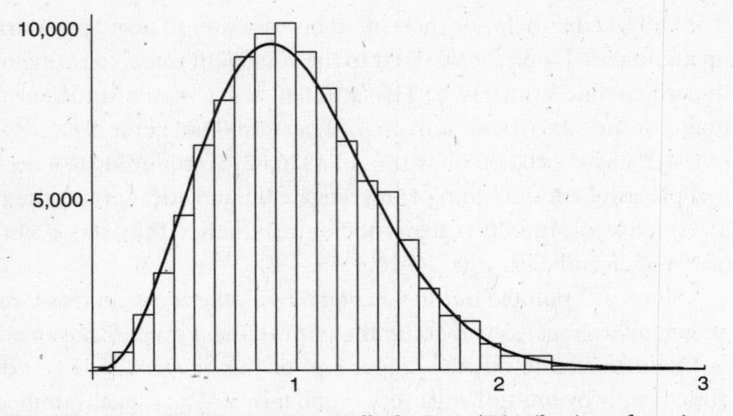

FIGURE 18-5 The Montgomery-Odlyzko Law. (Distribution of spacings for the 90,001st to 100,000th zeta-function zeros.)

Now I can ask statistical questions. Sample question: How do the spacings depart from that average? How many have a length less than one?[113] The answer is 5,349. How many have a length of more than 3? None. Now, this is at total variance with the counts you get from a perfectly random scattering, which are 6,321 and 489, respectively.[114] That confirms the lesson of Figures 18-2 and 18-3. The zeros are not randomly scattered. They are more bunched around the average spacing (a tad more than 1), with a dearth of small spacings and a dearth of large ones.

Tallying the number of spacings with length between 0 and 0.1, between 0.1 and 0.2, and so on, and making a histogram of the tallies, scaled so that the whole area is 9,999, I get Figure 18.5. This shows the spacings for my 10,000 zeros against the curve predicted by GUE theory. It's not a sensationally good fit, but then my sample isn't very big, or very high up the critical line. The fit is good enough, well within the variation allowed by chance; and the fits in Andrew Odlyzko's paper are of course much better.[115]

VII. So yes, it seems that the non-trivial zeros of the zeta function and the eigenvalues of random Hermitian matrices are related in some way. This raises a rather large question, a question that has been hanging in the air ever since that encounter in Fuld Hall in 1972.

The non-trivial zeros of Riemann's zeta function arise from inquiries into the distribution of prime numbers. The eigenvalues of a random Hermitian matrix arise from inquiries into the behavior of systems of subatomic particles under the laws of quantum mechanics. *What on earth does the distribution of prime numbers have to do with the behavior of subatomic particles?*

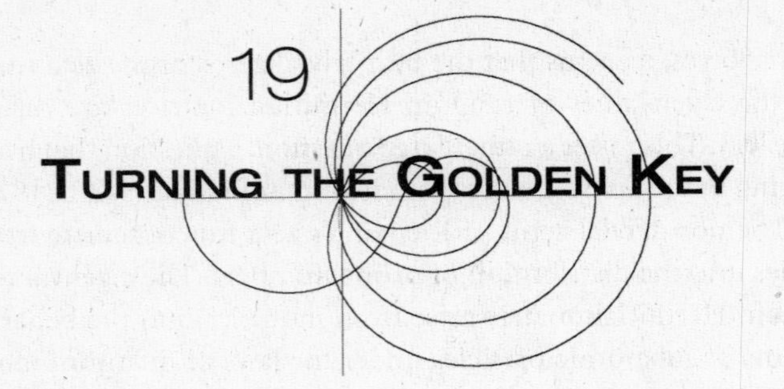

19
TURNING THE GOLDEN KEY

I. Now I am going to try to make a run right for the heart of Riemann's 1859 paper. This involves taking in some of Riemann's math, which is very advanced. I shall skip nimbly over the really difficult parts, presenting them as *faits accomplis*, and just try to give the logical steps in Riemann's argument by saying things like, "Mathematicians have a way to get from here to here," without explaining what that way is, or why it works.

My hope is that you will end up with at least an outline of the main logical steps Riemann followed. I can't do even that much, though, without a very small amount of calculus, the essential points of which I have already laid out in Chapter 7.vi-vii. You might find the following few sections challenging. The reward will be a result of great beauty and power, from which flows everything—the Hypothesis, its importance, and its relevance to the distribution of prime numbers.

II. To begin, I'm going to contradict something I said way back in Chapter 3.iv. Well, sort of contradict it. I said there wasn't much point trying to draw a graph of the Prime Counting Function $\pi(N)$. At that point in the book, there wasn't. Well, now there is.

However, I am first going to make some slight adjustments. Instead of writing $\pi(N)$, which, to the mathematical eye, translates as "the number of primes up to and including the natural number *N*," I'm going to write $\pi(x)$, which should be taken to mean "the number of primes up to and including the real number *x*." This is not a big deal. Obviously the number of primes up to 37.51904283 inclusive is just the number of primes up to 37 inclusive, which is twelve: 2, 3, 5, 7, 11, 13, 17, 19, 23, 29, 31, 37. But we are heading for some calculus, so we want to be in the realm of numbers at large, not just whole numbers.

And one more adjustment. As I advance the argument *x* smoothly through a range of values, $\pi(x)$ is going to make sudden jumps. Suppose *x* moves smoothly from 10 to 12, for example. The number of primes less than 10 is 4 (2, 3, 5, and 7), so the function value is 4 when $x = 10$, and also of course when $x = 10.1$, 10.2, 10.3, and so on. At argument 11, however, it will suddenly jump to 5; and for 11.1, 11.2, 11.3, ... the value will hold at a steady 5. This is what mathematicians call a "step function." And here is an adjustment rather commonly made with step functions. At precisely the point where $\pi(x)$ jumps, I am going to give it a value half-way up the jump. So at argument 10.9, or 10.99, or 10.999999, the function value is 4; at argument 11.1, or 11.01, or 11.000001, the function value is 5; but at argument 11 it is 4.5. I am sorry if this seems a little wacky, but it is essential for what follows. If I do this, then all the arguments in this chapter and Chapter 21 work; if I don't, they don't.

Now I can show a graph of $\pi(x)$ (see Figure 19-1). Step functions are a little hard to get used to at first, but from a mathematical point of view they are perfectly sound. The domain here is all non-negative numbers. In that domain, every argument has a single

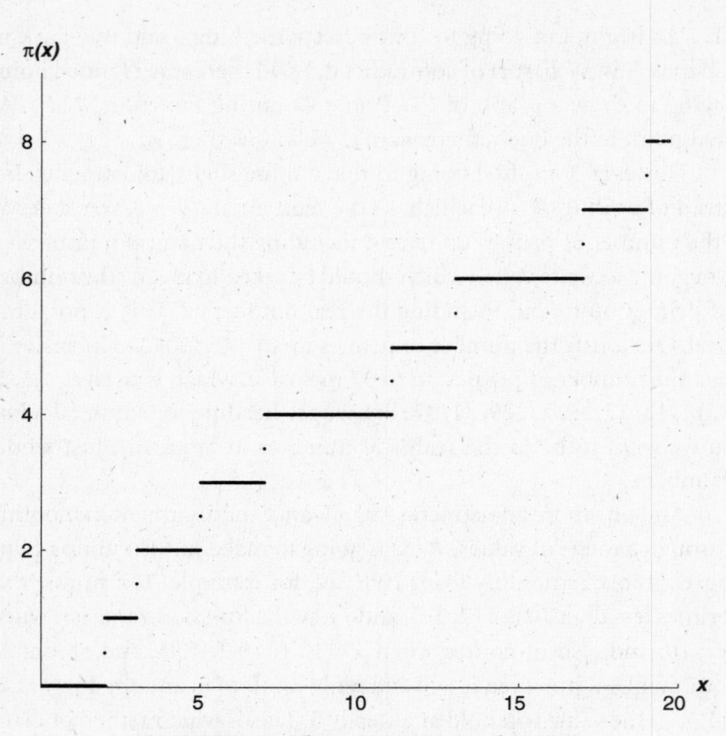

FIGURE 19-1 The prime counting function.

unique value. Give me an argument, I'll give you a value. Math has much stranger functions than this.

III. Now I am going to introduce another function, also a step function, just a little bit odder than $\pi(x)$. Riemann, in his 1859 paper, calls it the "*f*" function, but I'm going to follow Harold Edwards and refer to it as the "*J*" function. Since Riemann's time, mathematicians have got into the habit of using "*f*" to represent a generic function, "Let *f* be any function…," so it's a bit difficult for them to see *f* as referring to a particular function.

OK, here goes with the definition of the *J* function. For any non-negative number *x*, the *J* function has the value shown in Expression 19-1.

$$J(x)=\pi(x)+\frac{1}{2}\pi\left(\sqrt{x}\right)+\frac{1}{3}\pi\left(\sqrt[3]{x}\right)+\frac{1}{4}\pi\left(\sqrt[4]{x}\right)+\frac{1}{5}\pi\left(\sqrt[5]{x}\right)+\cdots$$

Expression 19-1

The symbol "π" here is the Prime Counting Function as defined up above, for any real number *x*.

Notice that this is *not* an infinite sum. To see why not, consider some definite number *x*, say, $x = 100$. The square root of 100 is 10; the cube root is 4.641588…; the fourth root is 3.162277…; the fifth root is 2.511886…; the sixth root is 2.154434…; the seventh root is 1.930697…; the eighth root is 1.778279…; the ninth root is 1.668100…, and the tenth root is 1.584893…. I could, of course, go on working out the eleventh, twelfth, thirteenth roots, and so on for as far as you please. I don't need to, though, because the prime counting function has a very nice property; if *x* is less than 2, $\pi(x)$ is zero, because there aren't any primes less than 2! In fact I could have stopped calculating roots of 100 after the seventh. What I have in this case is

$$J(100)=\pi(100)+\frac{1}{2}\pi(10)+\frac{1}{3}\pi(4.64...)+\frac{1}{4}\pi(3.16...)+$$
$$\frac{1}{5}\pi(2.51...)+\frac{1}{6}\pi(2.15...)+0+0\cdots$$

which, if you count up primes, means

$$J(100)=25+\left(\frac{1}{2}\times 4\right)+\left(\frac{1}{3}\times 2\right)+\left(\frac{1}{4}\times 2\right)+\left(\frac{1}{5}\times 1\right)+\left(\frac{1}{6}\times 1\right)$$

and that works out to $28\frac{8}{15}$, or 28.53333…. If you keep taking roots of any number, sooner or later the root drops below 2, and all the terms in the *J* function from there on are zero. So for any argument *x*, the value of $J(x)$ can be worked out from a finite sum—a great improvement over some of the functions we've been dealing with!

This J function is, as I said, a step function. Figure 19-2 shows what it looks like for arguments up to 10. You can see that the J function jumps suddenly from one value to another, holds the new value for a while, then makes another jump. What are these jumps? What's the pattern?

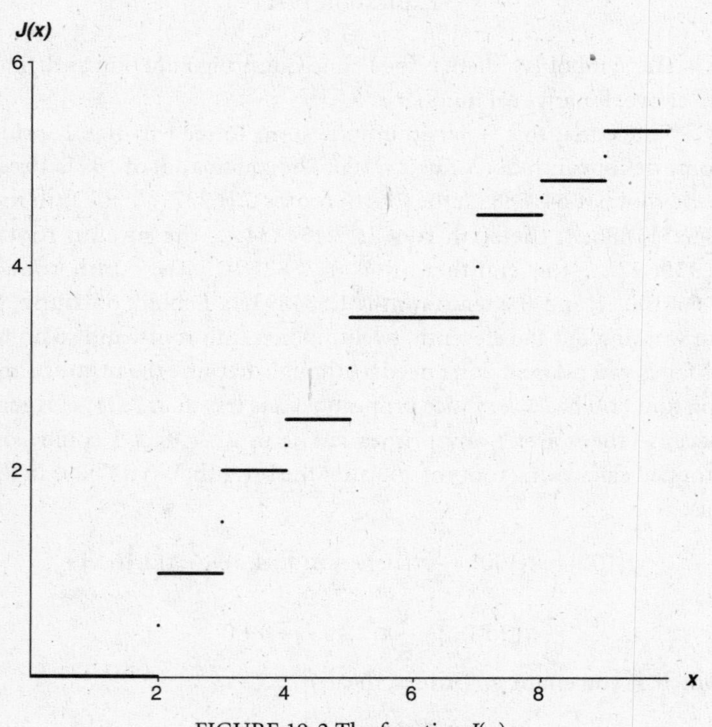

FIGURE 19-2 The function $J(x)$.

If you stare hard at Expression 19-1, you will see the following pattern. First, when x is any prime number, $J(x)$ jumps up 1, because $\pi(x)$—the number of primes up to and including x—jumps up 1. Second, when x is the exact square of a prime (e.g., $x = 9$, which is the square of 3), $J(x)$ jumps up one-half, because the square root of x is a prime, so $\pi(\sqrt{x})$ jumps up 1. Third, when x is the exact cube of a prime (e.g., $x = 8$, which is the cube of 2), $J(x)$ jumps up one-third, because the cube root of x is a prime, so $\pi(\sqrt[3]{x})$ jumps up 1. And so on.

Note, by the way, that the J function preserves the feature I introduced with $\pi(x)$. At the actual point where a jump occurs, the function value is halfway up the jump.

To give a fuller picture of the J function, Figure 19-3 is a graph of $J(x)$ for arguments up to 100. The smallest jump here is at $x = 64$, which is a sixth power ($64 = 2^6$), so the J function jumps up one-sixth at $x = 64$.

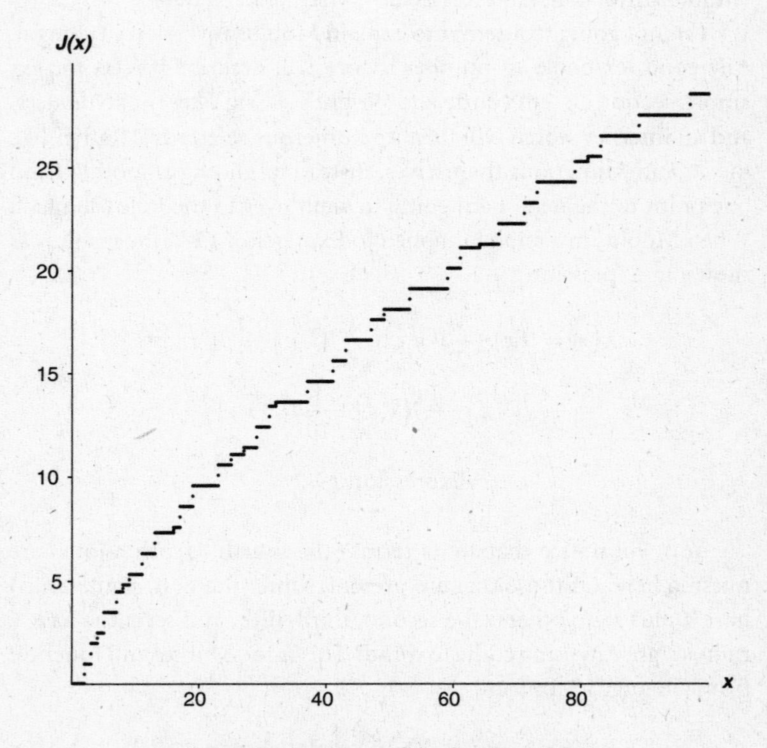

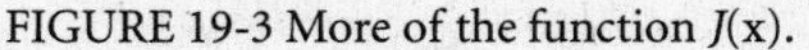
FIGURE 19-3 More of the function J(x).

What possible use is a function like this? Patience, patience. But first, I am going to take one of those leaps I warned you about at the start of the chapter.

IV. I mention yet again that mathematicians have any number of ways to invert relationships. We have an expression for *P* in terms of *Q*? OK, let's see if we can find a way to express *Q* in terms of *P*. Over the centuries, mathematics has developed a huge toolbox of inversion tricks, for use in all kinds of circumstances. One of them is called "Möbius inversion," and it is exactly what we need here.

I'm not going to attempt to explain Möbius inversion in general. Any good textbook on number theory will describe it (see, for example, section 16.4 in Hardy and Wright's classic *Theory of Numbers*) and an internet search will turn up numerous references. Rather like the π and J functions themselves, instead of gliding smoothly from one point to the next, I am going to vault over to the following fact: When Möbius inversion is applied to Expression 19-1, the result is as shown in Expression 19-2.

$$\pi(x) = J(x) - \frac{1}{2}J\left(\sqrt{x}\right) - \frac{1}{3}J\left(\sqrt[3]{x}\right) - \frac{1}{5}J\left(\sqrt[5]{x}\right) + \frac{1}{6}J\left(\sqrt[6]{x}\right) - \frac{1}{7}J\left(\sqrt[7]{x}\right) + \frac{1}{10}J\left(\sqrt[10]{x}\right) - \cdots$$

Expression 19-2

You will notice that some terms (the fourth, eighth, ninth) are missing here. Of those that are present, some (the first, sixth, tenth) have a plus sign; others (the second, third, fifth, and seventh) have a minus sign. Anything come to mind? This is the Möbius mu function from Chapter 15. In fact

$$\pi(x) = \sum_{n} \frac{\mu(n)}{n} J\left(\sqrt[n]{x}\right)$$

(where $\sqrt[1]{x}$, here and everywhere in the book, is understood to mean just x). Why did you think it was called "Möbius inversion"?

I have now written $\pi(x)$ in terms of $J(x)$. That is a wonderful thing, because Riemann found a way to express $J(x)$ in terms of $\zeta(x)$.

Before leaving Expression 19-2, I'd just like to point out that it is, like Expression 19-1, a finite sum, not an infinite one. This is because the J function, like the π function, is zero when x is less than 2 (check the graph), and if you keep taking roots of a number, the answers eventually drop below 2 and stay there. For example,

$$\begin{aligned}\pi(100) &= J(100) - \tfrac{1}{2}J(10) - \tfrac{1}{3}J(4.64...) - \tfrac{1}{5}J(2.51...) + \\ &\quad \tfrac{1}{6}J(2.15...) - 0 + 0 - \cdots \\ &= 28\tfrac{8}{15} - \left(\tfrac{1}{2}\times 5\tfrac{1}{3}\right) - \left(\tfrac{1}{3}\times 2\tfrac{1}{2}\right) - \left(\tfrac{1}{5}\times 1\right) + \left(\tfrac{1}{6}\times 1\right) \\ &= 28\tfrac{8}{15} - 2\tfrac{2}{3} - \tfrac{5}{6} - \tfrac{1}{5} + \tfrac{1}{6}\end{aligned}$$

which is precisely 25, which is indeed the number of primes less than 100. Magic.

Now let's turn the Golden Key.

V. Here is the Golden Key, the very first equation in Riemann's 1859 paper, the one I developed in Chapter 7, arguing that it is just a fancy way to write out the sieve of Eratosthenes.

$$\zeta(s) = \frac{1}{1-\frac{1}{2^s}} \times \frac{1}{1-\frac{1}{3^s}} \times \frac{1}{1-\frac{1}{5^s}} \times \frac{1}{1-\frac{1}{7^s}} \times \frac{1}{1-\frac{1}{11^s}} \times \frac{1}{1-\frac{1}{13^s}} \times \cdots$$

Remember that the numbers on the right-hand side are just primes.

I am going to take the log of both sides. If one thing is equal to another thing, of course its log must be equal to the other thing's log. From Power Rule 9, which says that $\log(a \times b) = \log a + \log b$,

$$\log\zeta(s) = \log\left(\frac{1}{1-\frac{1}{2^s}}\right) + \log\left(\frac{1}{1-\frac{1}{3^s}}\right) + \log\left(\frac{1}{1-\frac{1}{5^s}}\right) + \log\left(\frac{1}{1-\frac{1}{7^s}}\right)$$
$$+\log\left(\frac{1}{1-\frac{1}{11^s}}\right) + \cdots$$

Since $\log\frac{1}{a} = -\log a$ by Power Rule 10, this is

$$-\log\left(1-\frac{1}{2^s}\right) - \log\left(1-\frac{1}{3^s}\right) - \log\left(1-\frac{1}{5^s}\right) - \log\left(1-\frac{1}{7^s}\right)$$
$$-\log\left(1-\frac{1}{11^s}\right) - \cdots$$

Now recall Sir Isaac Newton's infinite series for log $(1 - x)$ in Chapter 9.vii. It applied to x between -1 and $+1$, which is certainly the case here, so long as s is positive. So I can expand each log as an infinite series as shown in Expression 19-3.

$$\frac{1}{2^s} + \left(\frac{1}{2}\times\frac{1}{2^{2s}}\right) + \left(\frac{1}{3}\times\frac{1}{2^{3s}}\right) + \left(\frac{1}{4}\times\frac{1}{2^{4s}}\right) + \left(\frac{1}{5}\times\frac{1}{2^{5s}}\right) + \left(\frac{1}{6}\times\frac{1}{2^{6s}}\right) + \cdots$$
$$+\frac{1}{3^s} + \left(\frac{1}{2}\times\frac{1}{3^{2s}}\right) + \left(\frac{1}{3}\times\frac{1}{3^{3s}}\right) + \left(\frac{1}{4}\times\frac{1}{3^{4s}}\right) + \left(\frac{1}{5}\times\frac{1}{3^{5s}}\right) + \left(\frac{1}{6}\times\frac{1}{3^{6s}}\right) + \cdots$$
$$+\frac{1}{5^s} + \left(\frac{1}{2}\times\frac{1}{5^{2s}}\right) + \left(\frac{1}{3}\times\frac{1}{5^{3s}}\right) + \left(\frac{1}{4}\times\frac{1}{5^{4s}}\right) + \left(\frac{1}{5}\times\frac{1}{5^{5s}}\right) + \left(\frac{1}{6}\times\frac{1}{5^{6s}}\right) + \cdots$$
$$+\frac{1}{7^s} + \left(\frac{1}{2}\times\frac{1}{7^{2s}}\right) + \left(\frac{1}{3}\times\frac{1}{7^{3s}}\right) + \left(\frac{1}{4}\times\frac{1}{7^{4s}}\right) + \left(\frac{1}{5}\times\frac{1}{7^{5s}}\right) + \left(\frac{1}{6}\times\frac{1}{7^{6s}}\right) + \cdots$$

$$+\frac{1}{11^s}+\left(\frac{1}{2}\times\frac{1}{11^{2s}}\right)+\left(\frac{1}{3}\times\frac{1}{11^{3s}}\right)+\left(\frac{1}{4}\times\frac{1}{11^{4s}}\right)+\left(\frac{1}{5}\times\frac{1}{11^{5s}}\right)$$

$$+\left(\frac{1}{6}\times\frac{1}{11^{6s}}\right)+\cdots$$

$$+\cdots$$

Expression 19-3

This is an infinite sum of infinite sums—a bit startling at first sight, I suppose, but not actually an unusual situation in math.

At this point you might think I am much worse off than when I started. From a fairly neat little infinite product, I have now got myself an infinite sum of infinite sums. The situation might seem hopeless. Ah, but that is to reckon without the power of the calculus.

VI. Let me pick on just one term in that sum of sums. I'll pick on the term $\frac{1}{2}\times\frac{1}{3^{2s}}$. Consider this function: x^{-s-1}, and assume for the time being that s is a positive number. What's the integral of x^{-s-1}? By the general rule for powers, which I gave in Chapter 7.vii, it's $x^{-s}/(-s)$, i.e., $(-1/s)\times(1/x^s)$. If I take the integral at infinity minus the integral at 3^2, what do I get? Well, if x is a very large number, $(-1/s)\times(1/x^s)$ is a very small number, so it's fair to say that when x is infinite, it's zero. From that—from zero—I'm going to subtract $(-1/s)\times(1/(3^2)^s)$. The answer to this subtraction is $(1/s)\times(1/(3^2)^s)$. The long and short of it is that the term in Expression 19-3 that I picked on can be rewritten as an integral,

$$\frac{1}{2}\times\frac{1}{3^{2s}}=\frac{1}{2}\times s\times\int_{3^2}^{\infty}x^{-s-1}dx$$

Why in Heaven's good name would I want to do that? To get back to the J function, that's why.

You see, $x = 3^2$ is where the J function takes a step up of $\frac{1}{2}$. In a mathematician's mind—certainly in the mind of a great mathematician like Riemann—that part-expression $\frac{1}{2} \times \int_{3^2}^{\infty} \ldots$ conjures up an image. The image it conjures up is the one in Figure 19-4. It's the *J* function, with a strip filled in. The strip goes from 3^2 (that is, from 9) to infinity, and it has height one-half. Plainly, the whole area under ("area under"—think "integral") the *J* function is made up of strips

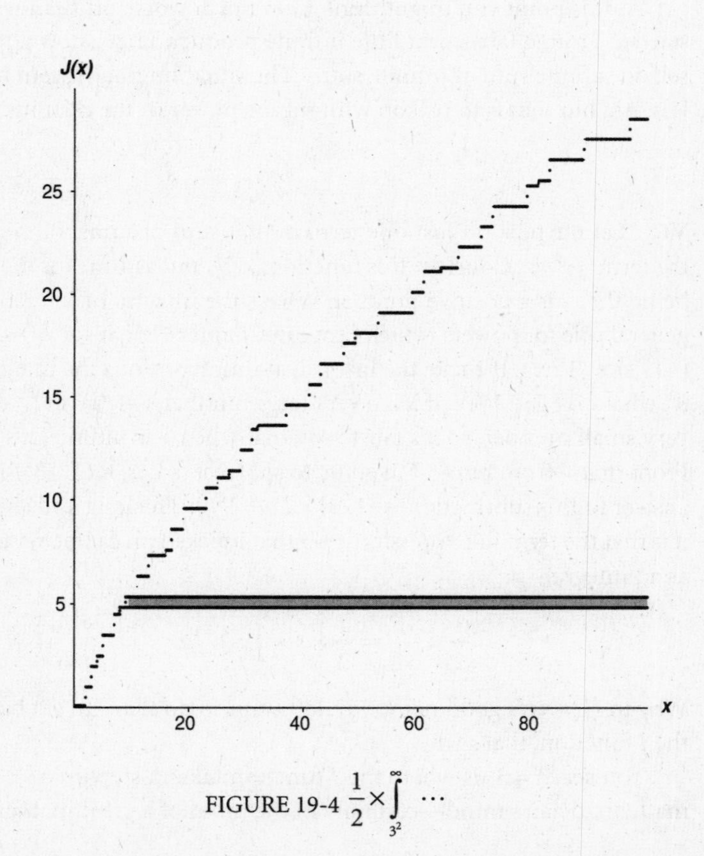

FIGURE 19-4 $\frac{1}{2} \times \int_{3^2}^{\infty} \cdots$

like that. Strips going from each prime to infinity, with height 1; strips going from each square of a prime to infinity, with height one-half; strips going from each cube of a prime to infinity with height one-third.... See how it all keys in to that infinite sum of infinite sums in Expression 19-3?

Of course, the area under the *J* function is infinite. The strip I showed has infinite area (height $\frac{1}{2}$, length infinite, $\frac{1}{2} \times \infty = \infty$). So do all the other strips. Together, they add up to an infinity. But what if I were to squish down the *J* function at the right, so that the area under it is finite? So that each one of those strips tapers away to nothing, with a finite area? How might I accomplish such a squishing-down?

That last integral suggests a way. Suppose I pick some number *s* (which I shall suppose to be greater than 1). For every argument *x*, I shall multiply $J(x)$ by x^{-s-1}. By way of illustration, take $s = 1.2$. Then x^{-s-1} means $x^{-2.2}$; or, to put it another way, $1/x^{2.2}$. Take an argument *x*; say, $x = 15$. Now $J(15)$ has the value 7.333333...; $15^{-2.2}$ has the value 0.00258582.... Multiplying them together, $J(x)x^{-s-1}$ has the value 0.018962721.... If I take a bigger argument, the squishing-down is more pronounced. For $x = 100$, $J(x)x^{-s-1}$ has the value 0.001135932....

Figure 19-5 shows a graph of the function $J(x)x^{-s-1}$, with $s = 1.2$. To emphasize the squish-down effect, I have shown the same strip I showed before, now in its squished-down form. You can see how it gets skinnier and skinnier as the arguments head off east. There is a fighting chance that its entire area will be finite, even though it's infinitely long. Supposing that were true, and supposing it were true of every other strip, what would be the entire area under this function? What, to put it mathematically, would be the value of $\int_0^\infty J(x)\,x^{-s-1}\,dx$?

Let's see. Taking the primes one by one, for the prime 2 I have, pre-squish, a strip going from 2 to infinity, height 1, then a strip go-

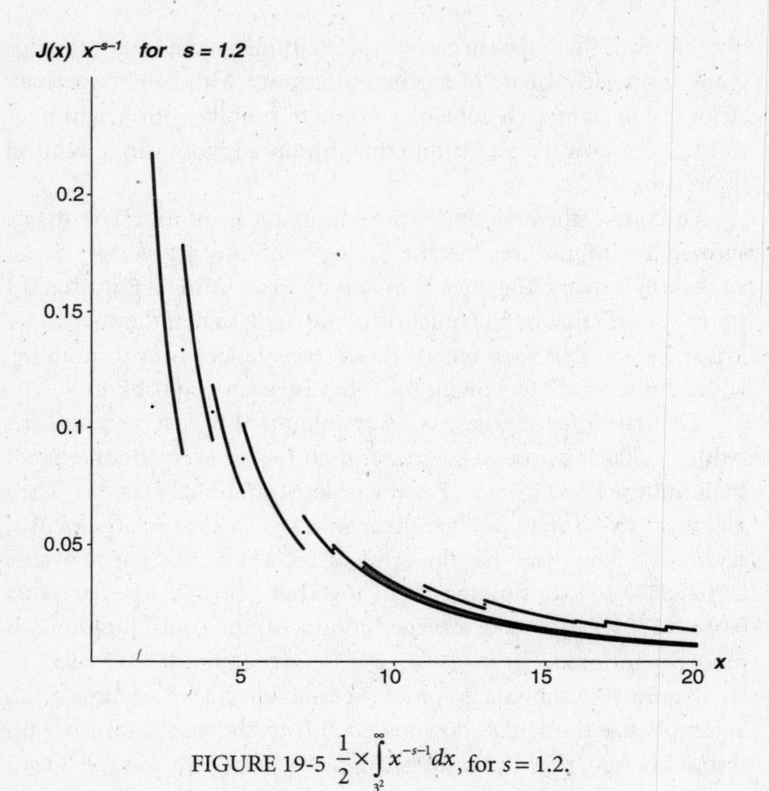

FIGURE 19-5 $\frac{1}{2}\times\int_{3^2}^{\infty} x^{-s-1}dx$, for $s = 1.2$.

ing from 2^2 to infinity, height $\frac{1}{2}$, then a strip going from 2^3 to infinity, height $\frac{1}{3}$, and so on. The sum of the squished strips, considering just the prime 2, is shown in Expression 19-4.

$$\int_{2}^{\infty} 1\times x^{-s-1}dx + \int_{2^2}^{\infty} \tfrac{1}{2}\times x^{-s-1}dx + \int_{2^3}^{\infty} \tfrac{1}{3}\times x^{-s-1}dx$$

$$+\int_{2^4}^{\infty} \tfrac{1}{4}\times x^{-s-1}dx + \int_{2^5}^{\infty} \tfrac{1}{5}\times x^{-s-1}dx + \cdots$$

Expression 19-4

Of course, that's just the 2-strips. There is a similar infinite sum of integrals for the 3-strips, shown in Expression 19-5.

$$\int_3^\infty 1\times x^{-s-1}dx+\int_{3^2}^\infty \tfrac{1}{2}\times x^{-s-1}dx+\int_{3^3}^\infty \tfrac{1}{3}\times x^{-s-1}dx$$
$$+\int_{3^4}^\infty \tfrac{1}{4}\times x^{-s-1}dx+\int_{3^5}^\infty \tfrac{1}{5}\times x^{-s-1}dx+\cdots$$

Expression 19-5

There's another for 5, another for 7, and so on for all the primes. An infinite sum of infinite sums of integrals! Worse and worse! Ah, but things always look darkest before the dawn.

That brings us back to the beginning of this section. Since integration is transparent to a multiplying factor, $\int_{3^2}^\infty \tfrac{1}{2}\times x^{-s-1}dx$ is the same as $\tfrac{1}{2}\times\int_{3^2}^\infty x^{-s-1}dx$. But I showed at the beginning of this section that my sample term from Expression 19-3, $\tfrac{1}{2}\times\tfrac{1}{3^{2s}}$, is equal to $s\times\tfrac{1}{2}\times\int_{3^2}^\infty x^{-s-1}dx$; that is, *s* times the thing I just got. So what does Expression 19-5 amount to? Why, just the second line of Expression 19-3, divided by *s*! And Expression 19-4, plus Expression 19-5, plus the similar expressions for all the other primes, add up to all of Expression 19-3, divided by *s*.

Here comes the dawn. It follows that the thing I am currently fooling with, which is $\int_0^\infty J(x)\,x^{-s-1}\,dx$, is just Expression 19-3, divided by *s*. But Expression 19-3 is equal to log ζ (*s*), from the Golden Key. Hence the result shown in Expression 19-6.

The Golden Key (calculus version)

$$\frac{1}{s}\log\zeta(s) \;=\; \int_0^\infty J(x)x^{-s-1}dx$$

Expression 19-6

I simply cannot tell you how wonderful this result is. It leads directly to the central result in Riemann's paper, a result I shall show in Chapter 21. Really, it is just a rewriting of the Golden Key in terms of calculus. This is a great and marvelous thing to do, however, as it now opens up the Golden Key to all the powerful tools of nineteenth-century calculus. That was Riemann's achievement.

One of those tools is yet another inversion method, which allows us to turn this new expression inside out to get an expression for J in terms of ζ. I'm going to hold off showing this inverted expression for the time being. The logic is clear, though.

- I can express $\pi(x)$ in terms of $J(x)$ (Section IV in this chapter).
- By inverting Expression 19-6, I can express $J(x)$ in terms of the zeta function.

Therefore,

- I can express $\pi(x)$ in terms of the zeta function.

Which is exactly what Riemann had set out to do, because then all the properties of the π function will be found encoded somehow in the properties of the ζ function.

The π function belongs to number theory; the ζ function belongs to analysis and calculus; and we have just thrown a pontoon bridge across the gap between the two, between counting and measuring. In short, we have just created a powerful result in analytic number theory. Figure 19-6 shows a graphical representation of Expression 19-6, the Golden Key in calculus form.

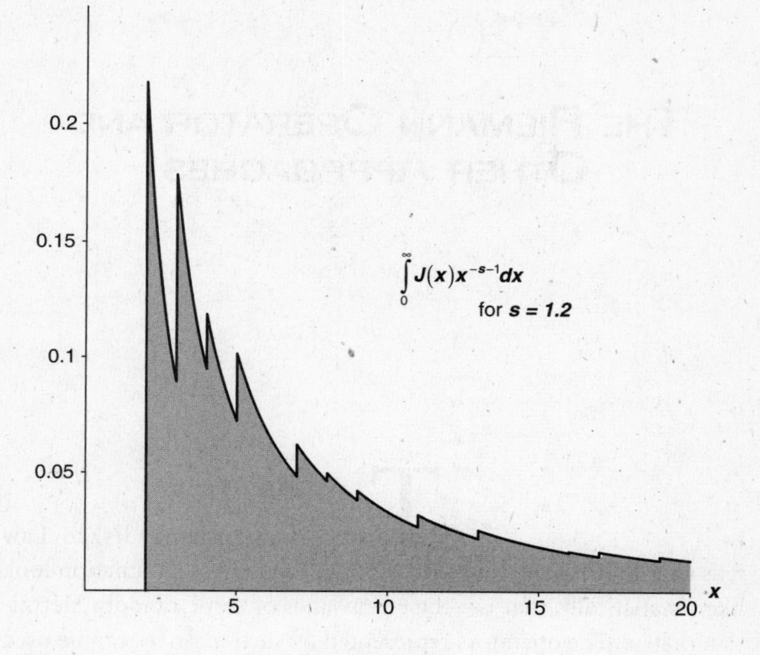

FIGURE 19-6

The area shown shaded is $\int_0^\infty J(x)\, x^{-s-1} dx$, for $s = 1.2$. Its numerical value is actually 1.434385276163…. This is equal to $\frac{1}{s}\log\zeta(s)$.

20

The Riemann Operator and Other Approaches

I. The Montgomery-Odlyzko Law tells us that the non-trivial zeros of the Riemann zeta function look like—statistically, that is—the eigenvalues of some random Hermitian matrix. The operators represented by such matrices can be used to model certain dynamical systems in quantum physics. Is there, then, a Riemann operator, an operator whose eigenvalues are precisely the zeta zeros? If there is, what dynamical system does it represent? Could that system be created in a physics lab? And if it could, would that help to prove the Hypothesis?

Research into these questions was under way even before the publication of Odlyzko's 1987 paper. The previous year, in fact, Michael Berry had published a paper titled "Riemann's Zeta Function: A Model for Quantum Chaos?" Using results that were being widely circulated and discussed at the time, including some of Odlyzko's, Berry tackled the following question. Suppose there is a Riemann operator: what kind of dynamical system would it model? His answer was: a chaotic system. To explain this, I must make a brief detour through Chaos Theory.

II. That pure number theory—ideas about the natural numbers and their relations with each other—should have relevance to subatomic physics is not all that surprising. Quantum physics has a much stronger arithmetical component than classical physics, since it depends on the idea that matter and energy are not infinitely divisible. Energy comes in 1, 2, 3, or 4 quanta, but not in $1\frac{1}{2}$, $2\frac{17}{32}$, $\sqrt{2}$, or π quanta. That is by no means the whole story, and quantum mechanics could not have been developed without the most powerful tools of modern analysis. Schrödinger's famous wave equation, for example, is written in the language of traditional calculus. Still, the arithmetical component is there in quantum mechanics, whereas in classical mechanics it is almost entirely absent.

The foundations of classical physics—the physics of Newton and Einstein—are quintessentially analytical, in the mathematical sense. They rest on mathematical analysis, on the notions of infinite divisibility, of smoothness and continuity, of limit and derivative, of real numbers. Newton invented the calculus, too, remember—the ultimate application of the concept "limit"—that eventually took over most of analysis.

Take the classical problem of one body in an elliptical orbit around another, under mutual gravitational attraction. At a certain distance from the parent body (measured by r, a real number) the satellite body has some precise velocity (measured by v, another real number). The relationship between v and r has a precise mathematical expression; v is in fact a function of r, expressed by the so-called *vis viva* equation familiar to all students of elementary celestial mechanics,

$$v = \sqrt{M\left(\frac{2}{r} - \frac{1}{a}\right)},$$

where M and a are some fixed numbers determined by the components and initial conditions of the system under observation—on the masses of the two bodies, and so on.

Now, of course, in practice we cannot attain the infinite precision needed to assign actual real numbers to r and v. We might be able to measure r to 10 decimal places, or even 20; but to pin down a real number you need infinitely many decimal places, and that we cannot get. In the case of any actual orbit, therefore, there will be some modest error in assigning a real number to r, and a corresponding error in the computed value of v. This doesn't matter much. Kepler's laws assure us that we will still get a regular ellipse, and the mathematics of the *vis viva* equation tell us that a 1 percent error in r typically turns up only a 0.5 percent error in v. The situation is manageable, predictable. It is, as mathematicians say, "integrable."

That, however, is an extremely simple problem. Almost all actual physical problems are more complex than that. Take the case of three bodies under mutual gravitational attraction, for instance—the famous "three-body problem." Can we solve it with closed-form solutions like the *vis viva* equation? Is it integrable? By the end of the nineteenth century it was apparent that the answers are: No, we cannot, and it is not. The only way to get solutions is by extensive numerical calculation, leading to approximations.

In 1890, in fact, Henri Poincaré published a definitive paper on the three-body problem, making it clear not only that the problem has no closed-form solutions, but that it has another, even more disturbing quality: Its solutions are sometimes chaotic. That is, if you vary the initial conditions of the problem—the numbers equivalent to M and a in my two-body example—very slightly, the resulting orbits change drastically, beyond all recognition. Poincaré himself commented that one set of conditions produced "orbits so tangled that I cannot even begin to draw them."

Poincaré's paper is generally taken to mark the birth of modern chaos theory. Nothing much happened in chaos theory for several decades, mainly because mathematicians had no way to do number-crunching on the scale required to analyze chaotic results. That changed when computers became available, and chaos theory was reborn with the work of meteorologist Ed Lorentz at M.I.T. in the

1960s.[116] Chaos theory is now a vast subject embracing many different subdisciplines within physics, mathematics, and computer science.

It is important to grasp that a chaotic system, like a solution to the three-body problem, need not (and, in general, does not) consist of random motions. The beauty of chaos theory is that there are patterns embedded in chaotic systems. While in general a chaotic system never retraces its steps, it does exhibit these recurring patterns; and underlying these patterns are certain regular, but unstable, periodic orbits into which, in theory, if infinite precision were available to the nudger, a chaotic system could be nudged.

III. When modern chaos theory first came up, physicists took it to be entirely a classical matter, with no relevance for quantum theory. Chaos arises from issues like the three-body problem because the numbers defining the initial conditions are real numbers, measuring numbers, infinitely divisible; they can be varied by 1 percent, or by 0.1 percent, or by 0.001 percent.... Since the conditions are infinitely variable, an infinity of outcomes presents itself. In quantum theory, by contrast, you can vary those initial conditions by 1, 2, or 3 units, but not by $1\frac{1}{2}$ or 2.749. There should be "no room" for chaos in quantum theory. It is true that there is a degree of uncertainty in quantum mechanics, but the controlling equations are nonetheless linear. Small perturbations lead to small consequences, as with the classical *vis viva* equation for two-body motion.

Yet in fact, a certain level of chaos can be observed in quantum-scale dynamical systems. The orderly energy-level structure of the electrons in orbit around the nucleus of an atom, for example, can be scrambled into an irregular pattern by the application of a sufficiently strong magnetic field. (This is, in fact, one of the dynamical systems modeled by GUE operators.) The atom's subsequent behavior is chaotic—wildly different for only slightly different initial conditions.

If such quantum-chaotic systems persist for a period, however, the laws of quantum mechanics eventually impose order on them, draining away the chaos. The number of permitted states dwindles; the number of forbidden states swells. The bigger and more complex the system, the longer it takes for the quantum rules to assert order, and the larger the number of permitted states...until, on the scale of the everyday world, it would take trillions of years for the quantum order to assert itself, and the number of permitted states is large enough to be taken as infinite. That is why we have chaos in classical physics.

Back in 1971, physicist Martin Gutzwiller found a way to relate chaotic systems on the classical scale with analogous systems down in the quantum world, by allowing the quantum factor, Planck's constant, in the quantum-mechanical equations to tend to zero, and taking limits. The periodic orbits that underlie a classical-chaotic system correspond to the eigenvalues of the operator defining this "semiclassical" system.

Michael Berry argued that if there is a Riemann operator, it models one of these semiclassical chaotic systems, and its eigenvalues, the imaginary parts of the zeta zeros, are the energy levels of that system. The periodic orbits in the analogous classical-chaotic system would correspond to ... the prime numbers! (To their logs, to be precise.) He further argued that this semiclassical system would not have the quality of "time reversal symmetry"—that is, if all the velocities of all the particles in the system were to be instantly and simultaneously reversed, the system would not return to its initial state. (Chaotic systems can be time-reversible or not. The ones that are time-reversible are modeled not by operators of the GUE type, but by another kind belonging to a different ensemble, the GOE—Gaussian Orthogonal Ensemble.)

Berry's work (much of it in collaboration with a Bristol colleague, Jonathan Keating) is subtle and deep. He has, for example, analyzed the Riemann-Siegel formula in great detail in search of insights into the zeros, and their effects on each other at different ranges. At the time of writing, he has not identified any dynamical system that corresponds to the Riemann operator, but thanks to his work, if such an operator exists, we shall know it at once when we see it.

IV. Another researcher, Alain Connes, Professor of Mathematics at the Collège de France in Paris, has taken an alternative approach. Instead of seeking to pin down the kind of operator the zeta zeros might be eigenvalues of, he has actually constructed such an operator.

That was no mean feat. An operator must have something to operate on. The kind of operators I have been speaking about operate on *spaces.* A flat two-dimensional space will do to illustrate the general principle, with a sheet of graph paper for purposes of visualization, though you must imagine the paper extending to infinity in all directions. Suppose that I rotate that space by 30 degrees counterclockwise, sending every point of the space to some other point thereby (except the point about which I am rotating—that stays put). This rotation is an instance of an *operator.* The characteristic polynomial of this particular operator is $x^2 - \sqrt{3}\,x + 1$. The eigenvalues are $\frac{1}{2}\sqrt{3}+\frac{1}{2}i$ and $\frac{1}{2}\sqrt{3}-\frac{1}{2}i$; the trace is $\sqrt{3}$.

If you wanted, you could set up a coordinate system to describe all the points of the space, drawing a horizontal x-axis and a vertical y-axis to meet at the rotation point, and marking off distances along them in inches or centimeters, in the usual way. You might then notice that my rotation operator sends the point (x, y) to a new point with different coordinates—actually, to $(\frac{1}{2}\sqrt{3}x-\frac{1}{2}y,\ \frac{1}{2}x+\frac{1}{2}\sqrt{3}y)$. That is incidental to the nature of the operator, though, which exists, and which moves the points of the space to new points, independent of any coordinate system. A rotation is a rotation, even if you forgot to draw in a pair of axes.

The operators used in mathematical physics operate on much more complicated spaces than that, of course. Their spaces are not merely two-dimensional, nor just three-dimensional like the space we live out our everyday lives in. Nor are they even four-dimensional, like the one required by Relativity Theory. They are abstract mathematical spaces with *infinitely many* dimensions. Each point of such a space is a function. An operator transforms one function into another function, that is, in the language of spaces and points, it sends one point to another point.

To get a very elementary idea of how a function might be identified with a point in a space, consider one simple class of functions, the quadratic polynomials $p + qx + rx^2$. The family of all such polynomials could be represented by a three-dimensional space, the point with coordinates (p, q, r) standing for the polynomial $p + qx + rx^2$. A four-dimensional space could model cubic polynomials; a five-dimensional space could model quartics ... and so on. Now, since some functions can be written as series, and a series looks like an infinite polynomial (e^x, for example, as $1 + x + \frac{1}{2}x^2 + \frac{1}{6}x^3 + \frac{1}{24}x^4 + \ldots$), you can see how a space of infinitely many dimensions might be useful for modeling functions. Then e^x would be the point in that space located by the infinity of coordinates $\left(1, 1, \frac{1}{2}, \frac{1}{6}, \frac{1}{24}, \ldots\right)$.

In quantum mechanics, the functions are wave functions, defining the probability that the particles of a system are at certain places, with certain velocities, at a given moment in time. Each point of the space, in other words, represents a state of the system. The operators used in quantum mechanics encode observable features of the system—most famously, the Hamiltonian operator, which encodes the system's energy. The eigenvalues of the Hamiltonian operator are the fundamental energy levels of the system. Each eigenvalue is particularly associated with a key point—function—of the space, called an eigenfunction, representing the state of the system at that energy level. These eigenfunctions are essential and fundamental states of the system. Every possible state of the system, every physical manifestation, is some linear combination of the eigenfunctions, just as every point in a three-dimensional space can be written as (x, y, z), a linear combination of the points (1,0,0), (0,1,0), and (0,0,1).

Alain Connes has constructed a very peculiar space for his Riemann operator to operate on. The prime numbers are built in to this space in a way derived from concepts in algebraic number theory. Here is a sketch of Connes's work.

V. Classical physics is built around real numbers like this, 22.45915771836..., for which—absent a closed form—an infinite number of digits is required to give full theoretical accuracy. Actual physical measurements, though, are approximate, like this: 22.459. That is a rational number, $\frac{22459}{1000}$. The entire world of physical experiment can therefore be written down in rational numbers, members of $\mathbb{Q}$. To pass from the experimental world to the theoretical, we have to *complete* $\mathbb{Q}$ (see Chapter 11.v). That is, we have to enlarge it, so that if an infinite sequence of numbers in $\mathbb{Q}$ has a limit, that limit is either in $\mathbb{Q}$ itself, or in the enlarged field. The normal and natural way to do this is with $\mathbb{R}$, the real numbers, or $\mathbb{C}$, the complex numbers.

Algebraic number theory, however, has other ways to complete $\mathbb{Q}$. In 1897 the Prussian mathematician Kurt Hensel[117] devised an entire new family of objects to deal with certain problems in the theory of algebraic fields, like that $a+b\sqrt{2}$ field that I discussed in Chapter 17.ii. These objects are called "*p*-adic numbers." There is one field of these exotic creatures, with infinitely many members in it, for any prime number *p*. The building blocks of this field are the clock rings of size p, p^2, p^3, p^4, and so on, that I discussed in Chapter 17.ii. In the symbols I introduced there, they are $\mathbb{CLOCK}_p$, $\mathbb{CLOCK}_{p^2}$, $\mathbb{CLOCK}_{p^3}$.... The field of 7-adic numbers, for example, is built up from the rings $\mathbb{CLOCK}_7$, $\mathbb{CLOCK}_{49}$, $\mathbb{CLOCK}_{343}$, $\mathbb{CLOCK}_{2401}$.... Recall my illustration of how a finite field can be used to help build an infinite field? Well, here we are using an infinity of finite rings to build a new infinite field!

The field of *p*-adic numbers goes by the symbol "$\mathbb{Q}_p$." So there is a field $\mathbb{Q}_2$, a field $\mathbb{Q}_3$, a field $\mathbb{Q}_5$, a field $\mathbb{Q}_7$, a field $\mathbb{Q}_{11}$, and so on. Each is a complete field: $\mathbb{Q}_2$ the field of 2-adic numbers, $\mathbb{Q}_3$ the field of 3-adic numbers, and so on.

As the symbol suggests, the p-adic numbers bear a certain resemblance to ordinary rational numbers. However, $\mathbb{Q}_p$ is richer and more complicated than $\mathbb{Q}$ and in some respects is more like $\mathbb{R}$, the field of real numbers. In particular, $\mathbb{Q}_p$ can, like $\mathbb{R}$, be used to complete $\mathbb{Q}$.

You might at this point be wondering, "All well and good; but you say there is a field $\mathbb{Q}_p$ of these strange new objects, these p-adic numbers, for any prime number p, and that any old $\mathbb{Q}_p$ can be used to complete $\mathbb{Q}$. So … which one is best, $\mathbb{Q}_2$? $\mathbb{Q}_3$? $\mathbb{Q}_{11}$? $\mathbb{Q}_{45827}$? Which prime should Professor Connes use to carry out this stunt, to throw a bridge from the prime numbers to the physics of dynamical systems?"

The answer is, all of them! You see, there is an algebraic concept called an *adele* that embraces within its broad arms all the $\mathbb{Q}_p$, for all the prime numbers 2, 3, 5, 7, 11, …. In fact, it embraces real numbers, too! Adeles are built up from $\mathbb{Q}_2$, $\mathbb{Q}_3$, $\mathbb{Q}_5$, $\mathbb{Q}_7$, …, and $\mathbb{R}$, in much the same way that p-adic numbers are built up from CLOCK_p, CLOCK_{p^2}, CLOCK_{p^3}, …. Adeles are, if you like, one further level of abstraction up from p-adic numbers, which are themselves one level of abstraction up from ordinary rational numbers.

If all this has your head spinning, just suffice it to say that we have a class of super-numbers that are simultaneously 2-adic, 3-adic, 5-adic, … and also real. *Every one of these super-numbers has all the primes imbedded in it.*

The adele is certainly a very abstruse concept. Nothing is so abstruse that it doesn't find its way into physics eventually, though. In the 1990s mathematical physicists set about constructing adelic quantum mechanics, in which the actual rational-number measurements that show up in experiments were taken to be manifestations of these bizarre creatures hauled up from the lightless depths of the mathematical abyss.

This is the kind of space Alain Connes built for his Riemann operator to play in, an adelic space. Being adelic, it has the prime numbers built in, so to speak. Operators that act on this space are perforce prime-based. You can now, I hope, see how it is possible to build a Riemann operator whose eigenvalues are precisely the non-trivial zeros of the zeta function, and whose space—the space on which it operates—has the primes built in, in the way I have attempted to describe, while yet being relevant to actual physical systems, actual assemblies of subatomic particles.

The Riemann Hypothesis (RH) is then reduced to the matter of proving a certain trace formula—that is, a formula like Gutzwiller's, relating the eigenvalues of an operator on Connes's adelic space to the periodic orbits in some analogous classical system. Having the prime numbers already built in to one side of the formula ought to make everything easy. In a way it does, and Connes's construction is brilliant, and extremely elegant, with energy levels that are precisely zeta zeros on the critical line. Unfortunately, it has so far offered no clue as to why there might not be zeta zeros *off* the critical line!

Opinions as to the value of Connes's work vary widely. Not at all sure that I understood it myself, I canvassed some real mathematicians working in the field. I shall tread carefully here. For all I know, Alain Connes might announce a proof of the RH the day this book comes out, and I don't want to make anyone look foolish. Here are two quotes from professionals.

Mathematician X: "Tremendously important work! Connes will not only prove the RH, he will give us a Unified Field Theory, too!"

Mathematician Y: "What Connes has done, basically, is to take an intractable problem and replace it with a different problem that is equally intractable."

I do not feel qualified to tell you which opinion is correct. Given the stature and abilities of X and Y, though, I feel pretty sure that one of them is....[118]

VI. Other approaches to the RH are still active, of course. The algebraic approach through finite fields that I mentioned in Chapter 17 is very much alive. And, as we glimpsed in Section V above, that approach has interesting connections with the physical lines of attack. Analytic number theory, too, is still a busy area, and capable of strong results.

There are also indirect approaches. There is, for example, my Theorem 15.2, concerning the M function got by accumulating Möbius μ. That is, as I said, exactly equivalent to the Hypothesis. Analytic number theorist Dennis Hejhal of the University of Minnesota actually uses this as a way to present the RH to nonmathematical audiences, to avoid having to introduce complex numbers. Here, he says (I am paraphrasing his approach, not quoting it), is the RH.

Write down all the natural numbers, starting with 2. Under each number, write its prime factors. Then, ignoring any number with a square factor (or any higher power, which will necessarily include a square), go along the line marking as "heads" any number with an even number of prime factors, "tails" any with an odd number. This gives an infinite string of heads and tails—just like a coin-tossing experiment.

2	3	4	5	6	7	8	9	10	11	12	…
2	3	2^2	5	2×3	7	2^3	3^2	2×5	11	$2^2 \times 3$	…
T	T		T	H	T			H	T		…

Now, we know very well, from classical probability theory, what to expect from a long run of N coin tosses. On average, we will get $\frac{1}{2}N$ heads and $\frac{1}{2}N$ tails. But of course, we should hardly ever get *exactly* these numbers. Suppose we subtract the number of heads from the number of tails. (Or vice versa, depending on which is larger.) What do we expect this excess to be? On average, it is $\sqrt{N}$, that is, $N^{\frac{1}{2}}$. This has been known since the time of Jakob Bernoulli, 300 years ago. If you toss a fair coin a million times, on average you have an excess of a thousand heads (or tails). You might have more or you might have less, but on average, as you keep tossing that coin—as N goes off to infinity—the size of the excess grows at a certain rate; at a rate that is less than $N^{\frac{1}{2}+\varepsilon}$, for any number ε, no matter how small. Just like my Theorem 15.2!

In fact, my Theorem 15.2, which is equivalent to the RH, says that the *M* function grows just like the excess in a coin-tossing exercise. To put it another way, it says that a square-free number is either a head or a tail—has either an even or an odd number of prime factors—with 50–50 probability. This does not seem particularly unlikely and might in fact be true. If you can prove that it *is* true, you will have proved the RH.[119]

VII. A less direct probabilistic approach concerns the so-called "Cramér model." Harald Cramér was, in spite of that accent on his name, Swedish, and yet another insurance company employee—an actuary for Svenska Livförsäkringsbolaget, but also a popular and inspiring lecturer on math and statistics.[120] In 1934 he published a paper titled "On Prime Numbers and Probability," in which he put forward the idea that the primes were distributed as randomly as they could be.

One consequence of the Prime Number Theorem (PNT), which I demonstrated in Chapter 3.ix, is that in the neighborhood of some large number *N*, the proportion of primes is ~ 1/log *N*. The log of a trillion, for example, is 27.6310211…, so in the neighborhood of a trillion, around one number in 28 is a prime. Cramér's model says that aside from this one restraint on their average frequency, the primes are utterly random.

Here is one way to see what this means.[121] Imagine a long line of earthenware jars with the natural numbers painted on them. The numbers go 2, 3, 4, 5, 6, 7, 8, 9, 10, 11, …, to infinity (or some very large number). Into each jar put a number of wooden balls. The number of balls in jar *N* should be log *N* (or the nearest whole number). So the number of balls in the first few jars are 1, 1, 1, 2, 2, 2, 2, 2, 2, 2, 3, 3, …. Furthermore, there must be exactly one black ball in each jar; all the rest of the balls in the jar are white. Jars number 2, 3, and 4, therefore, have only one black ball in them; jars number 5 to 12 have one black and one white; jars number 13 to 33 have one black and two white, and so on.

Now take a clipboard and a large (preferably infinite) sheet of paper, and take a walk along the line. Pull a ball at random from each jar. If it's black, write down the number of the jar. When you finish, you have a long list of whole numbers starting "2, 3, 4, …." The chance that 5 is on your list is 50–50, since jar 5 has one white ball and one black. The chance that 1,000,000,000,000 is on your list is 1 in 28.

Now, what can we say about this list? It is not a list of the primes, of course. There are lots of even numbers on it, for example; but only one prime, 2, is even. Well, if the Cramér model is true, the list will be *statistically* indistinguishable from the primes. Any broad statistical property the primes have—how many you expect to find in intervals of certain lengths, for instance, or the degree of clustering (what Hilbert, in stating his eighth problem, called "condensation")—this random list will have, too.

For an analogy, consider the decimal digits of π. So far as anyone knows, they are perfectly random.[122] They never repeat themselves. Digits, and pairs of digits, and triplets and quartets of digits, occur with just the frequency you would expect from pure chance. Nobody has ever been able to detect any pattern in the billions of digits of π now available for inspection. The decimal digits of π are a random sequence of digits … except that they represent π! So with the primes, on Cramér's model. They are indistinguishable from any other sequence with frequency $1/\log N$, and in that sense they are perfectly random … except, of course that they are the primes!

In 1985 Helmut Maier proved that the Cramér model in the simple form I have sketched above is not a complete picture of the primes. A modified version of the model does give accurate predictions for the distribution of primes, however, and is linked to the RH in ways subtle and indirect. There is a modest hope that further research on this topic will yield insights into the RH.[123]

VIII. Finally, I cannot resist mentioning the most indirect approach of all, the one through non-deductive logic. This is not, properly speaking, a mathematical topic. Mathematics demands rigorous logical proof before a result can be accepted. Most of the world is not like this, however. In our daily lives we work mainly from probabilities. In courts of law, in medical consultations, in drawing up insurance policies, it is the balance of probabilities that we take into account, not ironclad certainties. Sometimes, of course, we use the actual mathematical theory of probability to quantify the matters under dispute—that is why insurance companies employ actuaries. Much more often we do not, and cannot—think of a law court.

Mathematicians have often cast an interested eye at this side of life. George Pólya actually wrote a two-volume book about it,[124] in which he made the rather surprising claim that non-deductive logic is better appreciated in mathematics than in the natural sciences. This line of thought has most recently been taken up by Australian mathematician James Franklin. His 1987 paper "Non-deductive Logic in Mathematics," in *The British Journal for the Philosophy of Science*, included a section headed "Evidence for the Riemann Hypothesis and other Conjectures."

Franklin approaches the RH as if it were a courtroom case. He presents the evidence for the RH being true:

- Hardy's 1914 result that infinitely many zeros lie on the critical line.
- The RH implies the PNT, which is known to be true.
- "Denjoy's probabilistic interpretation"—that is, the coin-tossing argument given in this chapter.
- Another 1914 theorem by Landau and Harald Bohr, stating that most zeros—all but an infinitesimal proportion—are very close to the critical line. Note that since the number of zeros is infinite, one trillion counts as an infinitesimal proportion.
- The algebraic results of Artin, Weil, and Deligne, that I mentioned in Chapter 17.iii.

Then the case for the prosecution:

- Riemann himself had no sound reasons to support his statement in the 1859 paper that the RH was "very likely," and the semi-reasons that might have motivated his statement have been knocked down since.
- The zeta function exhibits some very peculiar behavior high up the critical line, as revealed by the computer-generated results of the 1970s. (Franklin seems not to have known of Odlyzko's work.)
- Littlewood's 1914 result on the error term $Li(x) - \pi(x)$. Says Franklin: "The relevance of Littlewood's discovery to Riemann's Hypothesis is far from clear. But it does give *some* reason to suspect that there may be a very large counterexample to Riemann's Hypothesis, although there are no small ones." So far as I can tell, Franklin's argument here is by analogy. "For some extremely large numbers, the error term misbehaves. It is connected with the zeros of the zeta function." [See my Chapter 21.] "So perhaps for very large T, the zeta function misbehaves, having zeros off the critical line."

This is all circumstantial, of course. It should not, however, be dismissed as mere sub-philosophical word-play. The rules of evidence can deliver very persuasive results, sometimes contrary to the strictly argued certainties of mathematics. Consider, for example, the very un-mathematical fact that a hypothesis might be seriously weakened by a confirming instance. Hypothesis: No human can possibly be more than nine feet tall. Confirming instance: A human being who is 8′11¾″ tall. The discovery of that person confirms the hypothesis … but at the same time casts a long shadow of doubt across it![125]

21

THE ERROR TERM

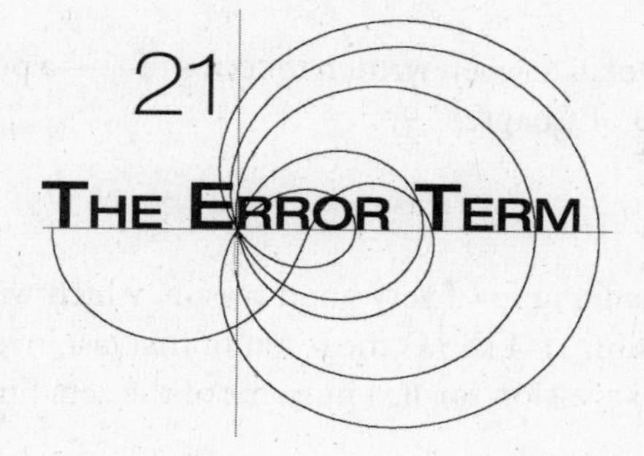

I. In Chapter 19, after defining that step function *J* in terms of the prime counting function π, I used Möbius inversion to get π in terms of *J*. Then, turning the Golden Key, I went through the steps Riemann took to express the zeta function ζ in terms of *J*. Another inversion (I said) will now give *J* in terms of ζ. The long and short of it is that:

- The prime counting function π can be written in terms of another step function, *J*.
- The function *J* can be written in terms of Riemann's zeta function ζ.

It follows that all the properties of the prime counting function π are coded, in some way, in the properties of ζ. A sufficiently close study of ζ will tell us all we want to know about π—that is, about the distribution of prime numbers.

How does this actually work? What's the code? Where do those non-trivial zeros come into it? And what does that middle-man function, J, look like when written in terms of ζ—a point I left hanging at the end of Chapter 19?

II. I left it hanging for a very good reason, which will now become clear. Expression 21-1 shows the result of that last inversion, the final and precise expression for $J(x)$ in terms of the zeta function.

$$J(x) = Li(x) - \sum_{\rho} Li(x^{\rho}) - \log 2 + \int_{x}^{\infty} \frac{dt}{t(t^2 - 1)\log t}$$

Expression 21-1

You take the point. If you're not a mathematician, that's an ugly beast of a thing (and where, by the way, is the zeta function in it?) I'm going to take it apart piece by piece, though, and show you what's going on inside it. First, I just want you to know that this equation is the main result of Riemann's 1859 paper. If you can get some kind of a handle on it, you will essentially understand Riemann's work in this area and have a clear view of all that followed.

The first thing to note is that Expression 21-1 has four parts, or terms, on the right-hand side. The first term, $Li(x)$, is generally called the "principal term." The second term, $\sum_{\rho} Li(x^{\rho})$, was referred to by Riemann in the plural as the "periodic terms" (*periodischer Glieder*) for reasons that will shortly become clear; I shall speak of it in the singular as the "secondary term." The third term is a no-brainer. It's just a number, log 2, which is 0.69314718055994....

The fourth term, though intimidating to the nonmathematician, is in fact easy to dispose of. It's an integral, that is, the area under the curve of a certain function, from argument x all the way out to infinity. The function is, of course, $1/(t(t^2 - 1) \log t)$. If you draw a graph of this function (see Figure 21-1), you will see that it is very friendly to the purpose in hand. Bear in mind that we have no interest in

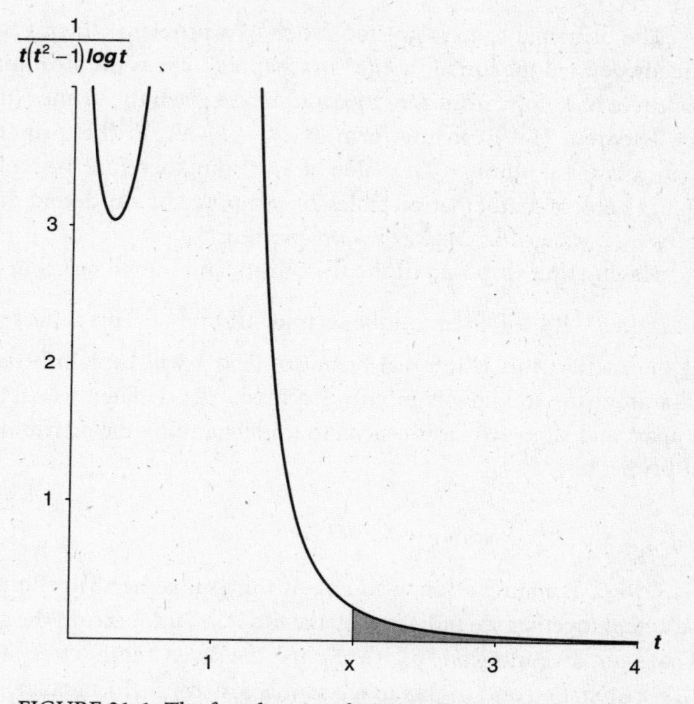

FIGURE 21-1 The fourth term of Riemann's expression for $J(x)$.

arguments x less than 2, since $J(x)$ is zero when x is less than 2. So the shaded area I've shown, corresponding to $x = 2$, is as big as this integral—this fourth term—is ever going to get. The actual value of this area, the maximum value of the fourth term for any x we might ever be interested in, is, in fact, 0.1400101011432869....

So the third and fourth terms taken together (and minding signs) are limited to the range from –0.6931... to –0.5531.... Since we are studying $\pi(x)$, which is only really interesting up in the millions and trillions, this is pretty inconsequential. I will, therefore, say almost nothing more about those last two terms and concentrate on the first two.

The principal term is not too much of a problem either. I have already defined the function $Li(x)$ in Chapter 7.viii as the area under the curve of $1/\log t$ from zero to x, and I have given the Prime Number Theorem (PNT) in the form $\pi(N) \sim Li(N)$. In this principal term, x is a real number. The value of $Li(x)$ can, therefore, be looked up in a book of mathematical tables, or computed by any decent math software package like *Maple* or *Mathematica*.[126]

Having thus disposed of the first, third, and fourth terms in Expression 21-1, I will focus on the second, $\sum_{\rho} Li(x^{\rho})$. This is the heart of the matter; this is the real business. First I will explain broadly what it means and how it got into Expression 21-1. Then I shall take it apart and show why it is crucial to understanding the distribution of primes.

III. The Σ is an invitation to add many things together. The things to be added together are indicated by the little "ρ" underneath the sign. That's not an American "p," it's a "rho," the seventeenth letter of the Greek alphabet, and in this usage stands for "root." To calculate this secondary term you must add up $Li(x^{\rho})$ for all of these roots, with ρ taking the value of one root after another. And what are these roots? Why, they are the non-trivial zeros of the Riemann zeta function!

How did these zeros turn up in the expression for $J(x)$? I can explain this, but only in outline. Recall the expression we arrived at in Chapter 19 by turning the Golden Key,

$$\frac{1}{s}\log\zeta(s) = \int_0^\infty J(x)\,x^{-s-1}\,dx$$

I said that mathematicians have a way to invert this, to turn it inside out, to get $J(x)$ in terms of the zeta function. The actual process of inversion is rather long and complex (in both senses of that word!), and most of the steps involve math beyond the level I am presenting here. That is why I have leaped straight to the final result, my Expression 21-1. I think I can explain one part of the process, though. It happens that one step in this inversion is to express the zeta function in terms of its zeros.

Expressing functions in terms of their zeros is not altogether a novel idea, if you have done high school algebra. Consider the good old quadratic equation, for example. I'll take the one I used in Chapter 17.iv, $z^2 - 11z + 28 = 0$ (but using z instead of x, since we are in the realm of complex numbers here). The left-hand side of this equation is of course a function, a polynomial function. If you feed in any argument z and do some arithmetic, it works out to some function value. If you feed in the argument 10, for example, the value is $100 - 110 + 28$, which is 18. If you feed in the argument i, the value is $27 - 11i$.

What are the solutions of the equation $z^2 - 11z + 28 = 0$? As I showed in Chapter 17, the solutions are 4 and 7. If you feed either number into the function on the left-hand side, the equation is true, the left-hand side is equal to zero. Another way to say this is that 4 and 7 are the zeros of the function $z^2 - 11z + 28$.

Now that I know the zeros, I can factorize this function. It factorizes to $(z - 4)(z - 7)$; or what, by the rule of signs, amounts to the same thing, $(4 - z)(7 - z)$. Another way to write this is $28(1 - z/4)(1 - z/7)$. And, look! Either way, I have expressed the function $z^2 - 11z + 28$ in terms of its zeros. This doesn't work only for quadratic functions, of course. The fifth-degree polynomial $z^5 - 27z^4 + 255z^3 - 1045z^2 + 1824z - 1008$ can be rewritten in terms of its zeros (which are 1, 3, 4, 7, and 12), too. Here it is: $-1008(1 - z/1)(1 - z/3)(1 - z/4)(1 - z/7)(1 - z/12)$. Any polynomial function can be rewritten in terms of its zeros.

From the point of view of complex function theory, polynomial functions have a very interesting property. The domain of a polynomial is *all* complex numbers. A polynomial never "equals infinity." There is no argument z for which its value just can't be calculated. Calculating the value of a polynomial function for any given argument just involves raising the argument to natural-number powers, multiplying it by numbers, and adding the results together. You can do that with any number.

Functions whose domain is all complex numbers and which are decently well behaved (there is a precise mathematical definition of that!) are called *entire functions.* All polynomials are entire functions, so is the exponential function. Those rational functions I showed in Chapter 17.ii, however, are not entire functions, since their denominators can be zero. The log function is not an entire function, either: it has no value at argument zero. Riemann's zeta function, likewise, has no value at argument 1, and so is not an entire function.

An entire function might have no zeros at all (like the exponential function: $e^z = 0$ is *never* true), or it might have many (like a polynomial: 4 and 7 are the zeros of $z^2 - 11z + 28$), or it might have infinitely many (like the sine function, which is zero at every integer multiple of π).[127] Now since polynomials can be rewritten in terms of their zeros, can *all* entire functions that have zeros be rewritten this way? Suppose I have some entire function, call it F, that can be defined by an infinite sum, $F(z) = a + bz + cz^2 + dz^3 + \ldots$. And suppose I happen to know that this function has infinitely many zeros; call them ρ, σ, τ .... Can I rewrite this function in terms of its zeros, as an infinite product, $F(z) = a\,(1 - z/\rho)(1 - z/\sigma)(1 - z/\tau)\ldots$? As if the infinite sum were a sort of super-polynomial?

The answer is that under certain conditions, yes, you can. When you can do it, it's often a very handy thing. This, for example, is how Euler solved the Basel problem, by applying this reasoning to the sine function.

How does this help us with the zeta function, which unfortunately is not an entire function? Well, as part of that complicated inversion process, Riemann transformed the zeta function into something slightly different—an entire function, whose zeros are exactly the non-trivial zeros of the zeta function. At that point, we can write out this slightly different function in terms of those zeros. (The trivial zeros conveniently vanished during the transformation.)

That, after some further processing, is how we end up with $\sum_{\rho} Li(x^{\rho})$, with the sum being taken over all the non-trivial zeros of the zeta function.

Now, to show the significance of this second term in Expression 21-1, and the problems it raises, I am going to take it apart. I shall do this from the inside out, first looking at x^{ρ}, then at the *Li* function, and then at the matter of summing across all possible zeros ρ.

IV. I have this number *x*, which is a real number. (The ultimate object of the exercise is to get a formula for $\pi(x)$, and $\pi(x)$ is relevant only for real numbers—only for natural numbers, to tell the truth; but we have switched from "*N*" to "*x*" so that we can apply the tools of analysis.) I raise this real number *x* to the power of ρ, which is a complex number—one having the form $\frac{1}{2} + ti$, for some real number *t*, if the Riemann Hypothesis (RH) is true. This is worth a note by itself.

If you raise a real number *x* to a complex power $a + bi$, the rules of complex arithmetic dictate the following. The *modulus* of the answer—how far it is from zero, as the crow flies—is x^a. It is not affected by *b* at all. The *amplitude* of the answer—how far round it is, which part of the complex plane it is found in—depends on *x* and *b*. It is not affected by *a*.

If you raise a real number *x* to the power $\frac{1}{2} + ti$, the modulus of the result is, therefore, *x* to the power of $\frac{1}{2}$, that is, $\sqrt{x}$. The amplitude, however, might be anything at all—the result might be anywhere in the complex plane, so long as its distance from zero is $\sqrt{x}$. To put it another way, if, for a given number x, you compute values of x^ρ for a host of different zeta zeros ρ, the numbers you get are scattered round the circumference of a circle of radius $\sqrt{x}$ in the complex plane, centered on zero. (If the RH is true!)

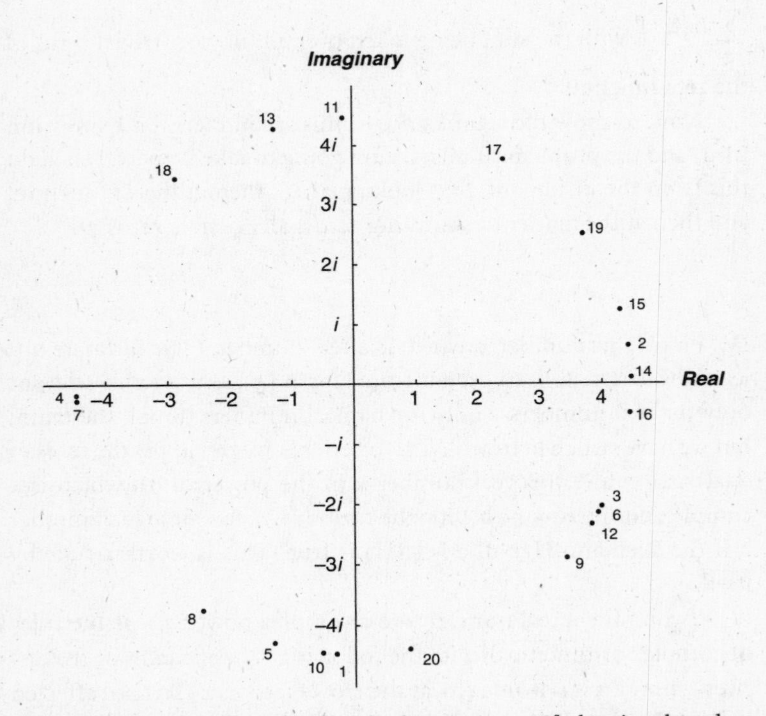

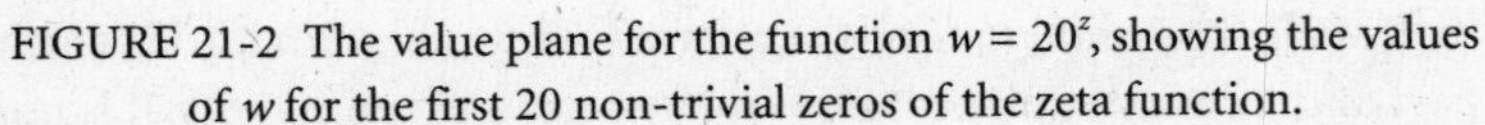
FIGURE 21-2 The value plane for the function $w = 20^z$, showing the values of w for the first 20 non-trivial zeros of the zeta function.

The points marked on Figure 21-2 are the results of raising 20 to the power of the first, second, third, …, twentieth zeros of the zeta function. You see how the results are scattered around a circle of radius $\sqrt{20}$ (which is 4.47213…) in the complex plane, in no particular order. This is because the function 20^z sends the critical line into a circle radius $\sqrt{20}$, wrapping the critical line (and all the zeta zeros speckled along it) round and round that circle an infinity of times. Mathematically speaking, that circle in the value plane is $20^{\textit{critical line}}$. If you imagine our pal the argument ant walking north up the critical line in the argument plane with her function-ometer set to the function 20^z, her twin sister, the value ant, tracing out the corresponding values in the value plane, is walking round and round and round that circle. She is proceeding counterclockwise and by the time the argument ant has reached the first zeta zero, the value ant is nearly three-quarters of the way through her seventh circuit.

V. Now, one by one, I am going to find the *Li* function of all those points—the whole infinity of them. Unfortunately, they are complex numbers. I defined the *Li* function only for real numbers, as the area under a curve. Is there a way to define *Li* for complex numbers, too? How do integrals work with complex numbers? Yes, there is a way to define it; and, yes, there is a way to develop integrals involving complex numbers. Integration is in fact a key feature of complex analysis, the subject of many of the most beautiful and powerful theorems in the topic. I shall not go into detail, only say that, yes, $Li(z)$ is defined[128] for complex numbers z.

Figure 21-3 shows where the first 10 of the points in Figure 21-2 are sent by the *Li* function. To put it another way, it shows where the critical line (to be precise, a stretch of it from $\frac{1}{2} + 14i$ to $\frac{1}{2} + 50i$) is sent by the function $Li(20^z)$. As you can see, this function maps the critical line into a counter-clockwise spiral that closes in on the number πi as the argument ascends the critical line. Where the function 20^z wrapped the critical line infinitely many times round and round the circle with radius $\sqrt{20}$, applying the *Li* function unwraps it into this elegant spiral, with the zeros still dotted along it.

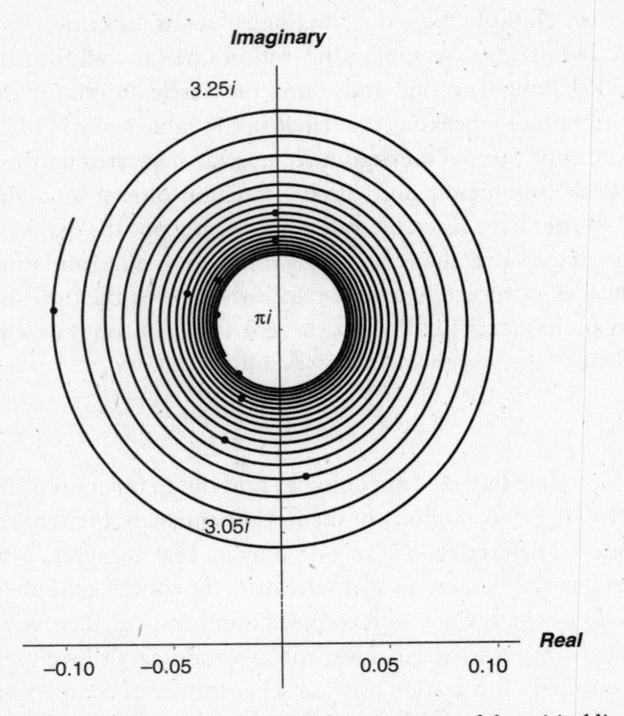

FIGURE 21-3 The function $Li(20^z)$ for a segment of the critical line.

VI. Now I shall tackle the sigma sign—the business of summing those dots (each of which is just a complex number) across all possible non-trivial zeros of the zeta function. To do this, let me first bring up a point I have mostly been ignoring so far. For any non-trivial zero on the north half of the critical line, there is a corresponding one in the south half. That is, if $\frac{1}{2} + 14.134725i$ is a zero of the zeta function, so must $\frac{1}{2} - 14.134725i$ be. In proper math language, if z is a zero, then so is its complex conjugate $\bar{z}$. (Remember that "$\bar{z}$" is pronounced "*z*-bar." At this point you might want to check back with Figure 11-1 to refresh your memory on complex number basics.)

In carrying out this summation, the south half of the critical strip plays a key role. Figures 21-2 and 21-3 were concerned only with the first few zeros along the northern half of the critical line. For a fuller picture, including the southern half of the line, Figure 21-4 shows, at the far left, a plane of complex numbers with the critical strip marked in from $\frac{1}{2} - 15i$ to $\frac{1}{2} + 15i$. This is enough to show the first zero at $\frac{1}{2} + 14.134725i$, and also its complex conjugate at $\frac{1}{2} - 14.134725i$. I have marked them as "ρ" and "$\bar{\rho}$."

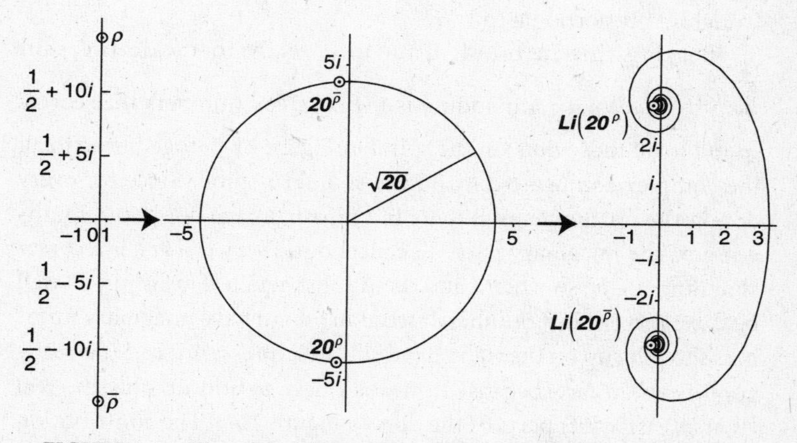

FIGURE 21-4 The critical line, out to the first pair of non-trivial zeros, mapped first by the function 20^z, then by the function $Li(20^z)$.

Taking this as the argument plane for the function 20^z, the middle part of the diagram shows the "comes from" figure in the value plane, a circle of radius $\sqrt{20}$, with 20^ρ marked as in Figure 21-2, and now also $20^{\bar{\rho}}$. Notice that the values are complex conjugates just as the arguments are. This doesn't happen with all functions, but fortunately it does with 20^z. If we apply the *Li* function, using that middle figure this time as the argument plane for *Li*, we see that the critical line, which got wound round that circle an infinity of times by 20^z, now unwinds into that pleasing double spiral at the right. (Figure 21-3 was a close-up of the top spiral.) And again, when the arguments are complex conjugates, so are the values.

There is just one more thing to notice before I actually evaluate the sum $\sum_{\rho} Li(20^{\rho})$. That spiral—Figure 21-3 shows it best—isn't closing in on its target very fast. Its closing-in rate is, in fact, harmonic. That is, if you imagine the argument ant walking north up the critical line with her function-ometer set to $Li(20^{z})$, the value ant traversing the spiral in the value plane is getting closer and closer to πi at a rate inversely proportional to the argument ant's height. If the argument ant's height is T, the value ant's distance from πi is (roughly) proportional to $1/T$.

Bearing this in mind, I am now ready to tackle the sum $\sum_{\rho} Li(20^{\rho})$. What I am adding is the complex numbers that correspond to all those dots on the spiral in Figure 21-3, together with all the complex conjugate dots on the southern spiral. Since for every dot on the northern spiral there is a mirror-image dot in the southern one, the imaginary parts all cancel out. Every $a + bi$ has a corresponding $a - bi$, so when I add them I just get $2a$. This is just as well because $J(x)$ is a real number. It wouldn't do to have imaginary numbers showing up on the right-hand side of Expression 21-1! It is really good news, in fact, because it means I have to add up only the real (that is, east-west) parts of the dots in Figure 21-3. The contribution of the southern hemisphere is merely to double the answer, $(a + bi) + (a - bi) = 2a$.

The rest of the news is not so good. The dots scattered along that spiral in Figure 21-3 are, as I observed, closing in on πi—their real parts, therefore, closing in on zero—at a harmonic rate. Adding up the real parts of all those dots, therefore, offers the danger that I might be adding up something like the harmonic series, which, we recall from Chapter 1, is divergent. How do I know this sum $\sum_{\rho} Li(20^{\rho})$ converges?

It helps that the dots can have real parts that are either positive or negative. In fact, the sum this one resembles is not the harmonic sum but its cousin, which I introduced briefly in Chapter 9.vii:

$$1-\frac{1}{2}+\frac{1}{3}-\frac{1}{4}+\frac{1}{5}-\frac{1}{6}+\frac{1}{7}-\frac{1}{8}+\cdots$$

Here the terms approach zero harmonically: $1, \frac{1}{2}, \frac{1}{3}, \frac{1}{4}, \frac{1}{5}, \ldots$, but the alternating plus and minus signs mean that each term to some degree cancels out the term before, and convergence is possible. However, the convergence is, in the terminology I introduced in Chapter 9.vii, only conditional. It depends on adding up the terms *in the correct order.*

Just so with $\sum_{\rho} Li(x^{\rho})$. We need to be careful about the order in which we do the addition if we want to be sure of convergence to the correct number. So what is the proper order? It is just what you would think it should be. Take the zeros one by one, heading north up the critical line, pairing off each zero with its complex-conjugate zero down south.

VII. So to evaluate $\sum_{\rho} Li(x^{\rho})$ we first pair off each zeta zero with its mirror image (i.e., complex conjugate) in the south half of the argument plane. Then these pairs must be taken in ascending order of the positive imaginary parts. So we take the zeros in this order,

$\frac{1}{2}+14.134725i$ and $\frac{1}{2}-14.134725i$; then
$\frac{1}{2}+21.022040i$ and $\frac{1}{2}-21.022040i$; then
$\frac{1}{2}+25.010858i$ and $\frac{1}{2}-25.010858i$; then....

To see how this process actually works out, and to get an insight into why Riemann called this secondary term the "periodic terms," let me work through the arithmetic for an actual value of x. I'll take $x = 20$ as before, so we are engaged in computing $J(20)$—which, you can easily verify from the original definition of J, is actually $9\frac{7}{12}$, that is, 9.5833333…. Here goes.

First, I must raise 20 to the power $\frac{1}{2} + 14.134725i$. The result is $-0.302303 - 4.46191i$, which is the dot marked "1" in Figure 21-2. Take the logarithmic integral—the *Li* function—of that to get the answer $-0.105384 + 3.14749i$, which is the western-most dot in Figure 21-3. Now do the conjugate member of this pair of zeros. Raise 20 to the power $\frac{1}{2} - 14.134725i$. The result is $-0.302303 + 4.46191i$. That's shown in the middle picture of Figure 21-4. It's the mirror image in the real axis of the "1" dot in Figure 21-2. Take the logarithmic integral to get answer $-0.105384 - 3.14749i$, which is the one way down south on the right of Figure 21-4. Add the two answers: -0.210768. The imaginary parts have, of course, canceled out. So much for the first matching pair of zeros.

Repeat for the second pair, $\frac{1}{2} + 21.022040i$ and $\frac{1}{2} - 21.022040i$. The final answer this time is 0.0215632. For the third pair it's −0.0535991. Three down, an infinity to go!

After 50 of these calculations, you have the following answers (read down the columns):

−0.210768	0.0563226	−0.0332852	0.00801349	0.0240114
0.0215632	−0.0274298	−0.00692417	0.0279464	−0.0223427
−0.0535991	0.0481966	0.0205354	0.0159041	−0.0225924
−0.00432174	0.00127986	−0.0312052	−0.0102871	−0.000132221
−0.0868451	0.0128283	0.0280167	0.0224912	−0.0180932
−0.037716	−0.00472225	0.0188243	−0.00106082	0.0221559
−0.0046281	0.0361164	0.0228139	0.0130158	−0.017333
−0.0577894	0.0317626	−0.0301646	−0.0191586	−0.0150514
−0.0400277	0.0222196	0.0208943	−0.018169	0.0206192
−0.0595976	−0.037927	0.0275883	−0.0165671	0.0207551

That first is a bit of an anomaly, because that westernmost dot in Figure 21-3 is more than twice as far from the vertical axis as any of the others. After that, though, they get smaller as the values corresponding to the north half of the critical line spiral in toward πi.

And look at the signs—there are about as many positives as negatives.[129] That's good news, because, though the answers are getting smaller, they're not getting smaller very fast, and we need all the help we can get from positives and negatives canceling each other out on addition. This is all happening under the sigma sign, remember—those 50 numbers have to be added up. (The sum is –0.343864, which is, as a matter of fact, within 8 percent of the infinite sum. Not bad for just 50 terms.)

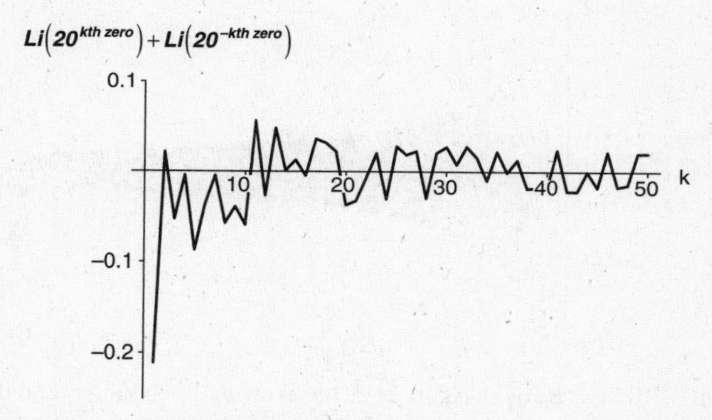

FIGURE 21-5 The first 50 values got by taking a non-trivial zero and its complex conjugate, computing the function values $Li\ (20^z)$, and then adding them.

You can see from Figure 21-5 why Riemann referred to these components of the secondary term as "periodic." They vary irregularly (which means, if you want to be finicky about it, that they are not strictly "periodic," only "oscillatory") up and down, from positive to negative and back.[130] The reason for this is plain in Figure 21-3.

The oscillatory quality of these secondary terms arises because, as Figure 21-3 shows, the function $Li(x^\rho)$ winds the critical line round and round in an ever-tighter spiral. The function values for the zeros are likely to end up anywhere on this spiral; especially since, for large x, the critical line gets immensely stretched before being wound. The winding is so tight that a high segment of the critical line maps into something very close to a circle. The values for the zeros, therefore, resemble points scattered around the circumference of a circle again. If you know some trigonometry, you know that this brings us into the world of sines and cosines, of wave functions, oscillations, vibrations ... of music. This is the ultimate root of Sir Michael Berry's notion of the "music of the primes."

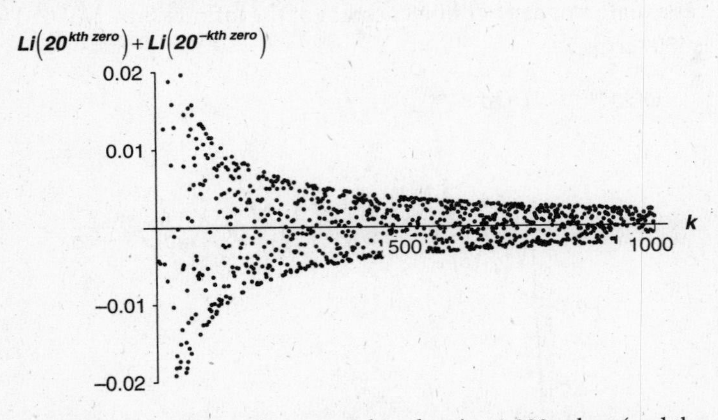

FIGURE 21-6 Same as Figure 21-5, but showing 1,000 values (and the points not joined up).

As you add them up, the terms are gradually decreasing, and the positives and negatives are canceling out, and you get convergence, though it's awfully slow. For three-digit accuracy you need to add up over 7,000 terms; for four-digit accuracy, more than 86,000. In Figure 21-6 I have plotted the first 1,000 results (though some over at the left got lost in the scaling) without attempting to join the dots this time. You can see that they really do get smaller, though at a leisurely pace.

The final result is −0.370816425.... This is, to remind you, the second term in Expression 21-1. The first is, in this case, *Li*(20), which is 9.90529997763.... The third term is log 2, which is 0.69314718055994.... The fourth term, that nuisance integral, delivers a piddling 0.000364111.... Feed those into Expression 21-1 and—ker-ching! $J(20) = 9.58333333$, which we knew all along.

VIII. I'll round off with a complete calculation of $\pi(1{,}000{,}000)$, the number of primes up to one million, using Riemann's formula—not for the fun of it, though it is of course great fun, but to make some important points about the error term.

Remember from Chapter 19.iv that

$$\pi(1{,}000{,}000) = J(1{,}000{,}000) - \frac{1}{2}J\left(\sqrt{1{,}000{,}000}\right) - \frac{1}{3}J\left(\sqrt[3]{1{,}000{,}000}\right) - \cdots$$

How far do I need to take that right-hand side? Until the number inside the parenthesis is less than 2, because $J(x)$ is zero when x is less than 2. The nineteenth root of 1,000,000 is 2.069138...; the twentieth root is 1.995262.... We can, therefore, stop at 19. Since 19 is square-free and has only one prime factor—itself—the Möbius function $\mu(19)$ has value -1. The last item on the right-hand side is, therefore, $-\frac{1}{19}J\left(\sqrt[19]{1{,}000{,}000}\right)$. Altogether there are 13 items on that right-hand side since there are 13 numbers from 1 to 19 whose Möbius function is not zero: 1, 2, 3, 5, 6, 7, 10, 11, 13, 14, 15, 17, 19. The Möbius function is zero, remember, for any number that is divisible by a perfect square like 4 or 9.

Each of those 13 items has four terms: the principal term, the secondary term (which involves the zeros of the zeta function), the log 2 term, and the integral term. If I add up all 52 of these fragments, I have $\pi(1{,}000{,}000)$—which, we know in advance from Chapter 3.iii, is 78,498.

I have laid out all this arithmetic in Table 21-1 (omitting row N when $\mu(N)$ is zero). Reading across row N, and using y to stand for the N-th root of one million, the principal term is $(\mu(N)/N)\ Li(y)$, the secondary term is $-\frac{\mu(N)}{N}\sum_{\rho} Li(y^{\rho})$, the log 2 term is

$-\frac{\mu(N)}{N}\log 2$, and the integral term is $\frac{\mu(N)}{N}\int_{y}^{\infty}\frac{dt}{t(t^2-1)\log t}$.

TABLE 21-1 Calculation of π (1,000,000).

N	Principal term	Secondary term	Log 2 term	Integral term	Row totals
1	78627.54916	−29.74435	−0.69315	0.00000	78597.11166
2	−88.80483	0.11044	0.34657	0.00000	−88.34782
3	−10.04205	0.29989	0.23105	0.00000	−9.51111
5	−1.69303	0.08786	0.13863	−0.00012	−1.46667
6	1.02760	−0.02349	−0.11552	0.00031	0.88889
7	−0.69393	−0.04737	0.09902	−0.00058	−0.64286
10	0.29539	−0.02791	−0.06931	0.00183	0.20000
11	−0.23615	−0.00634	0.06301	−0.00234	−0.18182
13	−0.15890	0.03206	0.05332	−0.00340	−0.07692
14	0.13281	−0.01581	−0.04951	0.00394	0.07143
15	0.11202	−0.00362	−0.04621	0.00448	0.06667
17	−0.08133	−0.01272	0.04077	−0.00554	−0.05882
19	−0.06013	−0.02241	0.03648	−0.00657	−0.05263
Column totals	78527.34662	−29.37378	0.03515	−0.00799	78498.00000

The row totals should, and in fact do, work out to $(\mu(N)/N)\,J(y)$. For an easy check, look at the row $N=6$. Since a million is 10^6, the sixth root of a million is 10. The value of $J(10)$ is easy to work out, it comes to $\frac{16}{3}$. Because 10 is square-free and the product of two primes, its Möbius function $\mu(10)$ has the value +1. For the row $N=6$, that last column should therefore work out to $(+1)\times(\frac{1}{6})\times(\frac{16}{3})$. That is $\frac{8}{9}$, which is just what we have for the row total when $N=6$.

The principal term when $N = 1$ is of course just $Li(1{,}000{,}000)$, the approximation given by the PNT. What is the difference between that and $\pi(1{,}000{,}000)$? A quick subtraction gives the answer. The difference, taking it as $\pi(1{,}000{,}000)$ minus $Li(1{,}000{,}000)$ to preserve the signs in my table, is –129.54916. How is that difference made up? As follows.

From principal terms:	–100.20254
From secondary terms:	–29.37378
From log 2 terms:	0.03515
From integral terms:	–0.00799

The largest difference arises from the principal terms. However, these are pretty predictable. They decline steadily and rapidly.

The difference arising from the secondary terms is of the same order of magnitude, and its components, those secondary terms, are much more worrisome. The first secondary term is quite large and negative; but there is no obvious reason why this should be so. Even the others do not look helpful. If you just read down the column of secondary terms, ignoring minus signs, and noting whether each term is bigger or smaller than the one above it, they read: smaller, bigger, smaller, smaller, bigger, smaller, smaller, bigger, smaller, smaller, bigger, bigger. The one for $N = 19$ is almost as big as the one for $N = 6$. Those secondary terms, the terms that involve zeros of the zeta function, are the wild cards in this calculation. The log 2 and integral terms are, as I promised, negligible.

Think of Littlewood's 1914 paper (Chapter 14.vii), in which he proved it is not true that $Li(x)$ is always greater than $\pi(x)$. That means that the difference will eventually be positive. Since the principal terms decrease very fast in size, and since the Möbius function makes most of the first few, including the really big ones ($N = 2$, $N = 3$, $N = 5$), negative, it's hard to see how those primary terms can ever contribute anything to the difference but a big negative number. If the net difference is going to be positive, as Littlewood proved it eventually will be, then that number will have to be swamped by bigger, positive, secondary terms. For that to happen, the secondary terms—the zeros of the zeta function—are going to have to seriously misbehave. Apparently they do.

IX. For further insights into the meaning of the error term, look back at that double spiral on the right of Figure 21-4. It is $Li(x^{\text{critical line}})$ when $x = 20$. The critical line—with, if the RH is true, all the zeta zeros speckled along it—is sent to that spiral by the function $Li(20^z)$. What happens if, instead of 20, we choose some larger value of x? What will the corresponding spirals look like?

Figure 21-7 gives the general idea. It shows $Li(10^{\text{critical line}})$, $Li(100^{\text{critical line}})$ and $Li(1{,}000^{\text{critical line}})$. In all three cases I have mapped the same segment of the critical line, the segment from $\frac{1}{2} - 5i$ to $\frac{1}{2} + 5i$. Notice the following things that happen as x goes from 10, to 100, to 1,000.

- The spirals get bigger. They still, however, converge on the same two points, $-\pi i$ and πi.
- The segment of the critical line that we are mapping, which has length 10 units, gets more and more stretched, winding more and more times round the result points at $-\pi i$ and πi.
- The top spiral and the bottom spiral approach each other, "kiss" at some value of x between 100 and 1,000, and thereafter overlap. (The spirals actually kiss when $x = 399.6202933538\ldots$.)

The segment of the critical line I have mapped here is too short to reach to the first pair of zeros at $\frac{1}{2} + 14.134725i$. Because the line is getting stretched, wound more and more round the result points, even as the spirals get bigger, an interesting question arises. Does the stretching and winding perhaps keep the zeta zeros close in to $-\pi i$ and πi, regardless of how big the spirals become? Answer: No, for bigger and bigger x, the zeta zeros map into points that get arbitrarily large. When ρ is the first zeta zero, the one at $\frac{1}{2} + 14.134725i$, for

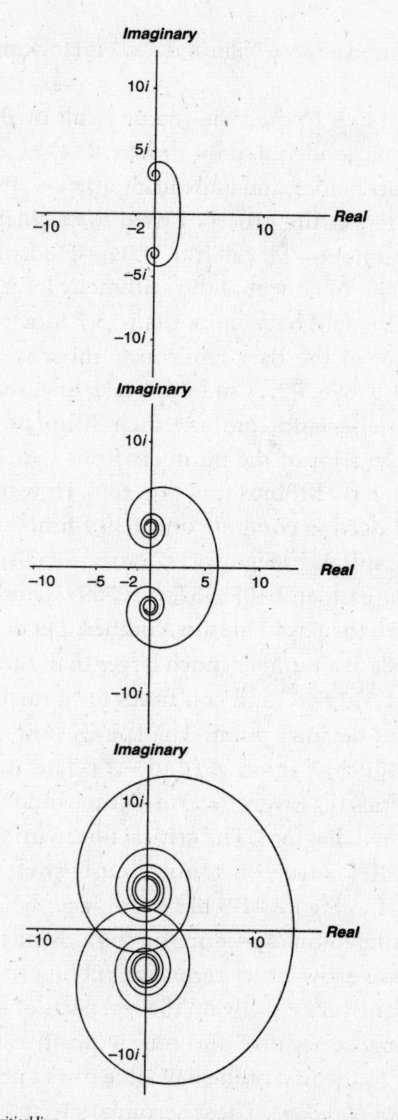

FIGURE 21-7 $Li(x^{\text{critical line}})$, for $x = 10$, 100, and 1,000. The part of the critical line being mapped here is the segment from $\frac{1}{2} - 5i$ to $\frac{1}{2} + 5i$.

arguments x around a mere trillion, $Li(x^\rho)$ is clocking up real parts of more than 2,200.

In Chapter 14.vii I noted the recent result by Bays and Hudson that the first Littlewood violation—when $\pi(x)$ exceeds $Li(x)$ for the first time—occurs before, and quite likely at, $x = 1.39822 \times 10^{316}$. Suppose I were to repeat the process I used to calculate $\pi(1{,}000{,}000)$, but using this number—I'll call it the "Bays-Hudson Number"—instead of 1,000,000. What would the arithmetic look like?

Obviously I would have more than 13 J-functions to work out. The 1,050th root of the Bays-Hudson number is 2.0028106..., the 1,051st root is 1.99896202..., so I must take first, second, ..., 1,050th roots of the number and compute their J-functions. It's not quite that bad, because a lot of the numbers from 1 to 1,050 are square-divisible, and so have Möbius function zero. How many? As a matter of fact 411, so I need to compute only 639 J-functions.[131]

The double spirals in Figure 21-7 cross the positive real axis successively further east, at 2.3078382, 6.1655995, and 13.4960622. If I was working with the Bays-Hudson Number, that double spiral would cross the real axis at a number much larger than these, a number that begins "325,771,513,660" and continues for a further 144 digits before reaching its decimal point. The spirals now are inconceivably vast. Yet they still close in on πi and $-\pi i$. This means that the top and bottom spirals massively overlap—you would not be able to distinguish them in a diagram. The critical line, with the zeros speckled along it (if the RH is true!) is tremendously stretched out. The diagram equivalent to Figure 21-3 has a far bigger hole in the middle—though still centered on πi—and the spiral winds trillions of times between successive low-order zeros, scrambling their coordinates in the complex plane very effectively, the real parts of the values oscillating between hugely negative and hugely positive numbers. And all this refers only to the first of the 639 table rows I need for computing π (*Bays-Hudson Number*). Those secondary terms are very unruly.

For the calculations and diagrams in this chapter, I have assumed the truth of the RH. If the RH is *not* true, then these elegant circles and spirals are mere approximations, and at some unknown height up the critical line—for some zero ρ far out along the infinite sum in that secondary term—the logic of this chapter falls apart.

However, it is important to understand that while I have assumed the truth of the RH in order the more neatly to illustrate the meaning of Riemann's formula for $J(x)$(Expression 21-1), that formula does not itself depend on the truth of the RH. The formula is absolutely, indisputably true: von Mangoldt proved that in 1895. Expression 21-1, combined with Expression 19-2, gives you a perfectly precise formula for $\pi(x)$. That was the main result of Riemann's 1859 paper. As he himself noted (see Chapter 10.i), a proof of the RH was "not necessary for the immediate objective of my investigation"—not necessary, that is, in order to get the formula for $\pi(x)$.

Riemann's formula for $\pi(x)$ shows us that the distribution of the primes depends intimately on the zeta zeros. The RH is a statement about those zeros. If the RH is true, certain things about the distribution of primes follow from Riemann's formula for $\pi(x)$. If the RH is false, different things follow from that formula. The truth of the formula, however, does not depend on the truth of the RH. Only *the nature of the results delivered by the formula* depends on the truth of the RH.

X. I have attained the main object of the mathematics in this book, to show the intimate connection between the distribution of prime numbers, as embodied in $\pi(x)$, and the non-trivial zeros of the zeta function, which make up a large—and, by Littlewood's result, sometimes dominant—component of the difference between $\pi(x)$ and $Li(x)$, that is, of the error term in the PNT.

All this was revealed to us by Bernhard Riemann's dazzling 1859 paper. We know much more today, of course, than we did in 1859. Yet the great conundrum first set out in that paper still stands unresolved, as resistant to assaults by the world's finest minds as when Riemann recorded his own "fleeting vain attempts" to prove it, back when analytic number theory had just been born. What are the prospects now, in the fifteenth decade of our efforts to crack the RH?

22

Either It's True, or Else It Isn't

I. There is a satisfying symmetry about the fact that the Riemann Hypothesis (RH), after 120 years among the mathematicians, has got the attention of the physicists. Riemann's own imagination was, as I noted in Chapter 10.i, very much that of a physical scientist. "Four of the nine papers that he himself managed to publish must be viewed as belonging to physics" (Laugwitz). And in fact, number theorist Ulrike Vorhauer[132] reminds me, the distinction between mathematician and physicist was not much made in Riemann's time. Shortly before that it was not made at all. Gauss was a first-rank physicist as well as a first-rank mathematician and would have been puzzled to hear the two disciplines spoken of as separate spheres of interest.

Jonathan Keating[133] tells the following anecdote, which I must say I find rather eerie.

> I was vacationing in the Harz Mountains with some colleagues. Two of us decided to drive the 30 miles or so to Göttingen to look at Riemann's working notes, which are kept in the library there. I my-

self wanted to look at his notes from around the time of the 1859 zeta function paper.

My colleague, however, an applied mathematician with no interest in number theory, was interested in some completely different work Riemann had done, relating to perturbations. Imagine a large blob of gas in empty space, held together by the gravitational attraction between its particles. What happens if you give it a good kick? Well, there are basically two things that might happen: it might fly apart, or it might just start wobbling at some frequency. It depends on the size, direction, and location of the kick, the shape and size of the original blob, and so on.

We got to the library, and I asked to see the notes on number theory, and my colleague asked to see the notes on perturbation theory. The librarian did some checking, then she came back and told us that a single set of Riemann's notes would do for both of us. *He had been working on both these problems at the same time.*

Of course, Jonathan adds, Riemann didn't have twentieth-century operator algebra to help him with the perturbation problem, to give him the set of all possible wobble frequencies as a spectrum of eigenvalues. He'd just slogged through the differential equations, creating a sort of ad hoc, embryonic operator theory for himself. Still, it's hard to believe that a mind as acute and penetrating as Riemann's would have missed the analogy between the zeta zeros strung out on the critical line, and his spectrum of perturbation frequencies—the analogy that was so dramatically paralleled over afternoon tea in Fuld Hall 113 years later!

II. It was at New York University's Courant Institute that I heard Keating tell that anecdote, in the early summer of 2002. The occasion was a four-day series of lectures and discussions organized by the American Institute of Mathematics (AIM). The title of the thing was "Workshop on Zeta Functions and Associated Riemann Hypotheses."

There were many famous names at the Courant conference. Atle Selberg himself showed up, 84 years old and still sharp as a tack. (He pulled up Peter Sarnak on a point of historical-mathematical fact in the very first lecture. During lunch break I went up to the Courant's excellent library and checked the point. Selberg was right.) Many of the other names mentioned in these last few chapters were present, too, including both halves of the Montgomery-Odlyzko Law. Other attendees included the current superstar of math, Andrew Wiles, famous for having proved Fermat's Last Theorem; Harold Edwards, whose definitive book on the zeta function I have mentioned several times in these pages; and Daniel Bump, one of the two names attached to the most euphonious of all RH-related results, the Bump-Ng Theorem.[134]

The AIM has been a considerable force in assaults on the RH during recent years. The Courant conference was the third they had sponsored on RH-related topics. The first, at the University of Washington in Seattle, in August 1996, was inspired by a wish to commemorate the proof of the Prime Number Theorem by Hadamard and de la Vallée Poussin 100 years earlier. The second was held in 1998 at the Erwin Schrödinger Institute in Vienna. The AIM by no means restricts its activities to RH studies—nor even just to number theory. They currently have a project on general relativity, for example. They have, though, done great work in bringing together scholars from different fields, pursuing all the different approaches I have mentioned: algebraic, analytic, computational, and physical.

AIM was established in 1994 by Gerald Alexanderson, a senior figure in American mathematics (and author of a very good book about George Pólya), and John Fry, a California businessman. Fry comes from a family of entrepreneurs. His parents owned a successful chain of supermarkets in California. John fell in love with math early on and in the 1970s he majored in the subject at Santa Clara University, where Alexanderson was on the faculty. After graduation John faced the choice of following the family tradition into business or going to graduate school. John opted for business and with his two brothers started the Fry's Electronics chain of stores, originally just in California, but at the time of writing going nationwide.

John Fry and Jerry Alexanderson stayed in touch. They shared a common interest, collecting rare math books and original papers. In the early 1990s they kicked around the idea of establishing a math library to house their collections. This developed into a plan for a math institute. They called in Brian Conrey, an old classmate of John's at Santa Clara, a number theorist of some repute, and a very successful head of department at Oklahoma State University.

For the first few years of its existence, AIM was funded almost entirely by personal donations from John Fry, to the tune of around $300,000 a year. This was a case of doing good by stealth. John is a reserved and private man who does not publicize his activities. When I first learned about AIM I went looking for a picture of him on the internet; there weren't any. In his element, though, that is, among mathematicians and people who love math, John is perfectly accessible. He took a party of us to lunch at the Courant conference in New York. A tall, boyish man, his face lights up when he talks math. I quietly wondered whether he had ever regretted the decision to go into business rather than the academy, but thought it might be impertinent to ask, and so missed the opportunity.

Visiting AIM headquarters a few days before the Courant conference, I found it occupying a utilitarian suite of rooms attached to the Fry's store in Palo Alto, California. In 2001, however, AIM applied for National Science Foundation funding to help establish a conference center on a leafy 200-acre property south of San Jose, California. The funding was approved, and research programs at the new location will begin in December 2002.

Another privately-funded enterprise similar to AIM began on the East Coast of the United States in 1998, when Boston businessman Landon T. Clay and Harvard mathematician Arthur Jaffe established the Clay Mathematics Institute (CMI). While AIM's first major initiative was to commemorate the proof of the Prime Number Theorem, CMI's was to mark the anniversary of Hilbert's speech at the 1900 Paris Congress.

For that purpose, the Clays held a two-day Millennium Event, also in Paris, at the Collège de France, in May 2000, during the course of which a $7 million fund was unveiled, $1 million to be awarded for the solution to each of seven great mathematical problems. The RH was naturally included, as problem number 4. (The order was based on the lengths of the problems' titles, to give the announcement an attractive appearance.)Whatever may be the case with the other six problems, $1 million is very little extra incentive to prove, or disprove, the Hypothesis. It is sufficiently established as *the* open problem in math at the beginning of the twenty-first century that whoever can resolve it will attain, in addition to everlasting fame, financial success—in lecture, interview, and royalty fees alone—far in excess of $1 million.[135]

III. What are the prospects for a proof or disproof of the RH? Delivering prognostications about this sort of thing is a very good way to make a fool of yourself. This is true even if you are a great mathematician, which of course I am not. Seventy-five years ago, lecturing to a lay audience, David Hilbert ranked three problems in ascending order of difficulty:

- The RH.
- Fermat's Last Theorem.
- "The Seventh"—that is, number 7 in the list of 23 problems Hilbert presented at the 1900 congress. In its more explicit form: If *a* and *b* are algebraic numbers, then a^b is transcendental (see Chapter 11.ii) except when it trivially isn't.

Hilbert said that the RH would be resolved in his lifetime, and Fermat's Last Theorem within the lifetime of younger audience members; but "no-one in this room will live to see a proof of the Seventh." In fact the Seventh was proved less than 10 years later, by Alexander Gel'fond and Theodor Schneider working independently. Hilbert was right, at a stretch, about Fermat's Last Theorem, proved by Andrew Wiles in 1994, when younger members of Hilbert's audience would have been in their nineties. He was drastically wrong about the RH, though. Should the RH make a fool of me, too—should the words I am about to write be rendered null and void by a proof of the RH turning up while this book is at the casebinder—I shall at least be able to console myself that I am in excellent company.

I am, therefore, going to stick my neck out and say that I believe a proof of the RH to be a long way beyond our present grasp. Surveying the modern history of attempts on the RH is something like reading an account of a long and difficult war. There are sudden surprising advances, tremendous battles, and heartbreaking reverses. There are lulls—times of exhaustion, when each side, "fought out," does little but conduct small-unit probes of the enemy defenses. There are breakthroughs followed by outbursts of enthusiasm; and there are stalemates followed by spells of apathy.

My impression of the current (mid-2002) state of affairs—though, to be sure, it is only the impression of a noncombatant—is that researchers are stalemated. We are in a lull. The great burst of interest generated by Deligne's proof of the Weil Conjectures in 1973 and by the Montgomery-Odlyzko developments of 1972–1987 seems to me to have spent itself.

In May 2002 I spent three days at the AIM office in Palo Alto, reviewing the videotaped record of the 1996 Seattle conference. The following month I attended the Courant Institute workshop. If you subtract 1996 from 2002, you get six years. If you "subtract" the contents of the Seattle conference from those of the Courant workshop, the mathematicians assembled at the Courant had little new to show. That is not a very surprising statement, to be sure, and I certainly do not mean it in a disparaging sense. This is work of the utmost difficulty. Progress is naturally slow, and six years is a very short time in the history of mathematics. (It took 357 years to prove Fermat's Last Theorem!) And there *were* some striking presentations at the Courant by younger mathematicians like Ivan Fesenko.

Still, the overriding impression was of stalemate. It is as if the RH were a mountain to be climbed, but from whichever direction one approaches it, one sooner or later finds oneself stuck at the rim of a wide, bottomless crevasse. I lost count of the number of times, in both 1996 and 2002, a lecturer ended his presentation with a verbal throwing up of hands: "This is of course a very important advance. However, it is not clear how we can proceed from here to a proof of the classical RH...."

Sir Michael Berry, who has a way with words, has coined the concept of the "clariton," which he defines to be "the elementary particle of sudden understanding." In the realm of the RH, claritons are currently in short supply.

Andrew Odlyzko: "It was said that whoever proved the Prime Number Theorem would attain immortality. Sure enough, both Hadamard and de la Vallée Poussin lived into their late nineties. It may be that there is a corollary here. It may be that the RH is false; but, should anyone manage to actually *prove* its falsehood—to find a zero off the critical line—he will be struck dead on the spot, and his result will never become known."

IV. Setting aside the search for a proof, how do mathematicians *feel* about the RH? What does their intuition tell them? Is the RH true, or not? What do they think? I made a point of asking every mathematician I spoke with, very directly, whether he or she believed the Hypothesis to be true. The answers formed a wide spectrum, with a full range of eigenvalues.

Among that majority of mathematicians who believe it true (Hugh Montgomery, for example), it is the sheer weight of evidence that tells. Now, all professional mathematicians are aware that weight of evidence can be a very treacherous measure. There was a good weight of evidence for $Li(x)$ being always greater than $\pi(x)$, until Littlewood's 1914 result disproved it. Ah, yes, RH believers will tell you, but that was merely one line of evidence, numerical evidence, together with the unsupported assumption that the second log-integral term $-\frac{1}{2}Li(x^{\frac{1}{2}})$ would continue to dominate the difference, which would therefore always be negative. For the Hypothesis we have far more lines. The RH underpins an enormous body of results, most of them very reasonable and—to bring in a word mathematicians are especially fond of—"elegant." There are now hundreds of theorems that begin, "Assuming the truth of the Riemann Hypothesis…." They would all come crashing down if the RH were false. That is undesirable, of course, so the believers might be accused of wishful thinking, but it's not the undesirability of losing those results, it's the fact of their existence. Weight of evidence.

Other mathematicians believe, as Alan Turing did, that the RH is probably false. Martin Huxley[136] is a current nonbeliever. He justifies his nonbelief on entirely intuitive grounds, citing an argument first put forward by Littlewood: "A long-open conjecture in analysis generally turns out to be false. A long-open conjecture in algebra generally turns out to be true."

The answer I liked best was Andrew Odlyzko's. He was actually the first person to whom I posed the question—the first mathematician I approached, when I was preparing the proposal for this book. We went for dinner at a restaurant in Summit, New Jersey. Andrew was at that time working for Bell Labs; he is now at the University of Minnesota.

I was fairly new to the RH at this point and had been learning a lot. With an excellent Italian meal under our belts and two hours of solid math talk behind us, having finally run out of things to ask, I said this:

JD: Andrew, you have gazed on more non-trivial zeros of the Riemann zeta function than any person alive. What do you think about this darn Hypothesis? Is it true, or not?

AO: Either it's true, or else it isn't.

JD: Oh, come on, Andrew. You must have some *feeling* for an answer. Give me a probability. Eighty percent it's true, twenty percent it's false? Or what?

AO: Either it's true, or else it isn't.

I could get no more from him than that. He simply would not commit himself. In a later conversation, in another place, I asked Andrew if there are any good mathematical reasons to believe the Hypothesis false. Yes, he said, there are some. You can, for example, decompose the zeta function into different parts, each of which tells you something different about zeta's behavior. One of these parts is the so-called *S* function. (This has *no connection at all* with the function I called $S(x)$ in Chapter 9.ii.) For the entire range for which zeta has so far been studied—which is to say, for arguments on the critical line up to a height of around 10^{23}—*S* mainly hovers between –1 and +1. The largest value known is around 3.2. There are strong reasons to think that if *S* were ever to get up to around 100, then the RH might be in trouble. The operative word there is "might"; *S* attaining a value near 100 is a *necessary* condition for the RH to be in trouble, but not a *sufficient* one.

Could values of the *S* function ever get that big? Why, yes. As a matter of fact, Atle Selberg proved in 1946 that *S* is unbounded; that is to say, it will *eventually*, if you go high enough up the critical line, exceed any number you name! The rate of growth of *S* is so creepingly slow that the heights involved are beyond imagining; but certainly *S* will eventually get up to 100. Just how far would we have to explore up the critical line for *S* to be that big? Andrew: "Probably around *T* equals $10^{10^{10,000}}$." Way beyond the range of our current computational abilities, then? "Oh, yes. *Way* beyond."

V. A thing that nonmathematical readers want to know, a question that is always asked when mathematicians address lay audiences, is, *What use is it?* Suppose the RH were proved true, or false. What practical consequences would follow? Would our health, our convenience, our safety be improved? Would new devices be invented? Would we travel faster? Have more devastating weapons? Colonize Mars?

I had better unmask myself at this point as a pure mathematician *sans mélange*, having no interest in such questions at all. Most mathematicians—and most theoretical physicists, too—are motivated not by any thought of advancing the health or convenience of the human race, but by the sheer joy of discovery and the challenge of tackling difficult problems. Mathematicians are generally pleased when their work turns out to have some practical result (at any rate if the result is peaceful), but they rarely think about such things in their working lives. At the Courant conference I sat through four days of solid lectures and discussions on topics related to the RH, from 9:30 A.M. to 6:00 P.M. every day, without ever hearing a mathematician mention practical consequences.

Here is what Jacques Hadamard had to say on this point in *The Psychology of Invention in the Mathematical Field.*

> [T]he answer appears to us before the question.... Practical application is found by not looking for it, and one can say that the whole progress of civilization rests on that principle.... [P]ractical questions are most often solved by means of existing theories.... It seldom happens that important mathematical researches are *directly* undertaken in view of a given practical use: they are inspired by the desire which is the common motive of every scientific work, the desire to know and to understand.

G.H. Hardy, in the concluding pages of his strange little *Apology*, was more blunt and more personal about it.

> I have never done anything "useful." No discovery of mine has made, or is likely to make, directly or indirectly, for good or ill, the least difference to the amenity of the world.... Judged by all practical standards, the value of my mathematical life is nil.

In the case of prime number theory, Hadamard's "the answer appears to us before the question" applies, and Hardy's claim is no longer true. Beginning in the late 1970s, prime numbers began to attain great importance in the design of encryption methods for both military and civilian use. Ways to test a large number for primality, ways to resolve large numbers into their prime factors, ways to manufacture gigantic primes; these all became very practical matters indeed in the last two decades of the twentieth century. Theoretical results, including some of Hardy's, were essential in these developments, which, among other things, allow you to use your credit card to order goods over the internet. A resolution of the RH would undoubtedly have further consequences in this field, validating all those countless theorems about primes that begin, "Assuming the truth of the RH..." and acting as a spur to further discoveries.

And of course, if the physicists really do succeed in identifying a "Riemann dynamics," our understanding of the physical world will be transformed thereby.

Unfortunately, it is impossible to predict what things will follow from that transformation. Not even the cleverest people can make such predictions, and those who do should not be trusted. Here is a mathematician at work, not quite 100 years ago.

> Every morning I would sit down before a blank sheet of paper. Throughout the day, with a brief interval for lunch, I would stare at the blank sheet. Often when evening came it was still empty.... [T]he two summers of 1903 and 1904 remain in my mind as a period of complete intellectual deadlock.... [I]t seemed quite likely that the whole of the rest of my life might be consumed in looking at that blank sheet of paper.

That is from Bertrand Russell's autobiography. What was stumping him was the attempt to find a definition of "number" in terms of pure logic. What does "three," for example, actually mean? The German logician Gottlob Frege had come up with an answer; but Russell had found a flaw in Frege's reasoning and was searching for a way to plug the leak.

If you had asked Russell, during those summers of frustration, whether his perplexities were likely to lead to any practical application, he would have hooted with laughter. This was the purest of pure intellection, to the degree that even Russell, a pure mathematician by training, found himself wondering what the point was. "It seemed unworthy of a grown man to spend his time on such trivialities...," he remarked. In fact, Russell's work eventually brought forth *Principia Mathematica*, a key development in the modern study of the foundations of mathematics. Among the fruits of that study have been, so far, victory in World War II (or at any rate, victory at a lower cost than would otherwise have been possible) and machines like the one on which I am writing this book.[137]

The RH should therefore be approached in the spirit of Hadamard and Hardy, though preferably without the overlay of melancholy Hardy put on his disclaimer. As Andrew Odlyzko told me, "Either it is true, or else it isn't." One day we shall know. I have no idea what the consequences will be, and I don't believe anyone else has, either. I am certain, though, that they will be tremendous. At the end of the hunt, our understanding will be transformed. Until then, the joy and fascination is in the hunt itself, and—for those of us not equipped to ride—in observing the energy, resolution, and ingenuity of the hunters. *Wir müssen wissen, wir werden wissen.*

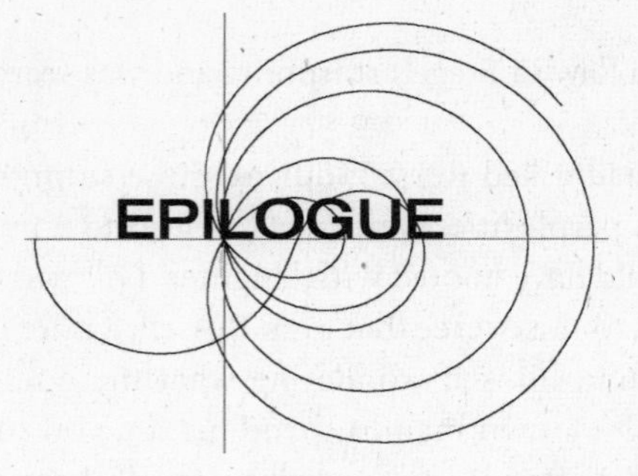

EPILOGUE

Bernhard Riemann died on Friday, July 20, 1866, a few weeks short of his 40th birthday. He had caught a heavy cold in the fall of 1862, and this had accelerated the tuberculosis from which he had probably suffered since childhood.[138] The efforts of Göttingen colleagues had secured a series of government grants to enable Riemann to travel to a better climate, this being the only way known for a TB sufferer to obtain relief from the disease and slow its progress.

Thus, Riemann's last four years had been spent almost entirely in Italy. When he died he was staying in Selasca, on the western shore of Lago Maggiore in the Piedmontese Alps. His wife, Elise, and their three-year-old daughter, Ida, were with him. Richard Dedekind recorded the event in the brief biography of his friend that he appended to the *Collected Works.*

> On June 28 he arrived at Lago Maggiore, where he lived at the Villa Pisoni in Selasca, near Intra.[139] Swiftly his strength ebbed away, and he himself perceived with full clarity that his end was approaching. Still, on the day before his death, resting in the shade of a fig tree, full of joy at the beautiful scenery laid out before him, he was at

work on those papers that, sad to say, he left unfinished. His end was very peaceful, with no struggle or death-spasm. It seemed as though he watched with interest the separation of the soul from the body. His wife brought bread and wine to him. He asked her to take his greetings to those at home, and said to her: "Kiss our child." She recited the Lord's Prayer for him, but he himself could no longer speak. At the words "forgive us our trespasses" he directed his eyes devoutly upward. She felt his hand become colder in hers, and after a few breaths his pure, noble heart ceased beating. That pious sense that was planted in him under his father's roof, stayed with him all his life, and he served God faithfully, in his own way. With devotion of the highest kind, he never interfered with the faith of others: the main thing in religion was, in his opinion, daily self-examination before the face of God.

He rests in the churchyard of Biganzolo, in the parish of Selasca. His gravestone carries the inscription:

HERE RESTS IN GOD
GEORG FRIEDRICH BERNHARD RIEMANN
PROFESSOR AT GÖTTINGEN
BORN IN BRESELENZ, SEPT. 17, 1826
DIED IN SELASCA, JULY 20, 1866

ALL THINGS WORK TOGETHER FOR GOOD
TO THEM THAT LOVE GOD

The inscription is all in German. The epitaph is from St. Paul's epistle to the Romans, 8:28. (In German, *Denen die Gott lieben müssen alle Dinge zum Besten dienen.*) Riemann's grave site no longer exists. It was destroyed in a later reorganization of the church property. The inscribed stone survived, though, and has been set in a nearby wall.

Elise Riemann returned to Göttingen with her daughter. They lived there with Bernhard Riemann's one surviving sister, also named Ida, at Weender Chaussee 17. The next door house, No. 17A, was occupied by Hermann Schwartz, a professor of mathematics at the University.[140] Riemann's chair at the University was taken by Alfred Clebsch, who wrote the founding text of modern algebraic geometry.

In 1884, Riemann's daughter Ida, then 20 years old, married Carl David Schilling, who had taken his doctor's degree under Schwartz in 1880 and remained friendly with him. Soon after this, Schilling took up a position as director of the marine academy in Bremen. In September 1890, Riemann's widow and his sister went to live with the Schillings in Bremen. Riemann's daughter lived until 1929, her husband until 1932. They seem to have produced a large family, but the precise number of their children has eluded me. The descendants of Bernhard Riemann are, at any rate, now merged into the general mass of humanity.

> Few as were the years of work allotted to him, and few as are the printed pages covered by the record of his researches, his name is, and will remain, a household word among mathematicians. Most of his memoirs are masterpieces—full of original methods, profound ideas and far-reaching imagination.
>
> —George Chrystal, from the article headed "Riemann" in the 1911 *Encyclopædia Britannica*

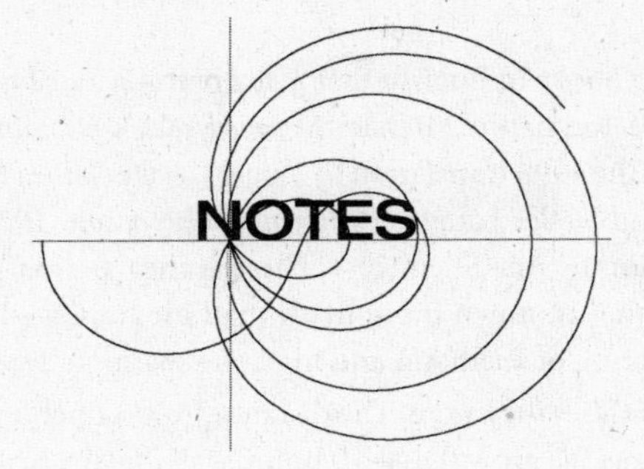

CHAPTER 2

1. A fact I learned at school in England by means of the following Victorian ditty:

   George the First was always reckoned
   Vile; but viler George the Second.
   No one ever said or heard
   A decent thing of George the Third.
   When to heaven the Fourth ascended,
   God be praised!—the Georges ended.

   In fact, they did not end; the twentieth century brought forth two more Georges.

2. There was another great Elbe flood in 1962, causing many deaths and much destruction in the Wendland district. Following that, a system of major dikes was built. In August 2002, as I was finishing this book, the Elbe flooded again. However, the post-1962 dikes appear to have held, and the region has suffered less than those further upriver.

3. Erwin Neuenschwander is professor of the history of mathematics at the University of Zürich. He is the leading authority on the life and work of Bernhard Riemann and has edited Riemann's letters. I have made use of his researches in this book. I have also relied heavily on the

only two books in English that give anything like a comprehensive account of Riemann: Michael Monastyrsky's *Riemann, Topology and Physics* (the 1998 translation by Roger Cooke, James King, and Victoria King) and Detlef Laugwitz's *Bernhard Riemann, 1826–1866* (the 1999 translation by Abe Shenitzer). Though they are mathematical biographies—that is, much more math than biography—both books give a good picture of Riemann and his times, with many valuable insights.

4. I should think they were. The distance from Lüneburg to Quickborn is 38 miles as the crow flies—10 hours walking at a brisk pace.
5. Hanover did not become a kingdom until 1814. Before that, its rulers were titled "Elector"—that is, they had the right to participate in electing the Holy Roman Emperor. The Holy Roman Empire was wound up in 1806.
6. Ernest Augustus was the last but one king of Hanover. The kingdom was incorporated into the Prussian Empire in 1866, a key moment in the creation of modern Germany.
7. Rankings vary, but he is almost always in the top three, usually with Newton and either Euler or Archimedes.
8. Heinrich Weber and Richard Dedekind published that first edition in 1876. The most recent edition of the *Collected Works* was compiled by Raghavan Narasimhan and published in 1990. The German for "Collected Works" is *Gesammelte Werke*, by the way; and this is a phrase so often encountered in mathematical research that English-speaking mathematicians, in my experience, say it in German quite unselfconsciously.
9. An Abelian function is a multivalued function obtained by inverting certain kinds of integrals. The term is hardly used nowadays. I shall mention multivalued functions in Chapter 3, complex function theory in Chapter 13, and the inverting of integrals in Chapter 21.

CHAPTER 3

10. Here is an example of *e* turning up unexpectedly. Select a random number between 0 and 1. Now select another and add it to the first. Keep doing this, piling on random numbers. How many random numbers,

on average, do you need to make the total greater than 1? Answer: 2.71828….

11. One of the great mathematical discoveries of antiquity, made by Pythagoras or one of his followers around 600 B.C.E., was that not every number is either a whole number or a fraction. The square root of 2, for example, is obviously not a whole number. Brute arithmetic shows that it is between 1.4 (whose square is 1.96) and 1.5 (whose square is 2.25). It's not a fraction either, though. Here is a proof. Let S be the set of all positive whole numbers n for which the following thing is true: $n\sqrt{2}$ is also a positive whole number. If S is not empty, it has a least member. (*Any* non-empty set of positive whole numbers has a least member.) Call this least member k. Now form the number $u = (\sqrt{2} - 1)k$. It is easy to show that (i) u is less than k, (ii) u is a positive whole number, (iii) $u\sqrt{2}$ is also a positive whole number, so that (iv) u is a member of S. This is a contradiction, since k was defined to be the least member of S, and therefore the founding assumption—that S is not empty—must be false. Therefore, S is empty. Therefore, there is no positive whole number n for which $n\sqrt{2}$ is a positive whole number. Therefore, $\sqrt{2}$ is not a fraction. A number that is neither whole nor fractional is called "irrational," because it is not the ratio of any two whole numbers.

12. Rule of signs: a minus times a minus is a plus. This is a major sticking point in arithmetic for a lot of people. "What does it mean to multiply a negative by a negative?" they ask. The best explanation I have seen is one of Martin Gardner's, as follows. Consider a large auditorium filled with two kinds of people, good people, and bad people. I define "addition" to mean "sending people into the auditorium." I define "subtraction" to mean "calling people out of the auditorium." I define "positive" to mean "good" (as in "good people") and "negative" to mean "bad." Adding a positive number means sending some good people into the auditorium, which obviously increases the net quantity of goodness in there. Adding a negative number means sending some bad people in, which decreases the net goodness. Subtracting a positive number means calling out some good people—net goodness in the auditorium

decreases. Subtracting a negative number means calling out some bad people—net goodness increases. Thus, adding a negative number is just like subtracting a positive, while subtracting a negative is like adding a positive. Multiplication is just repeated addition. Minus three times minus five? Call out five bad people. Do this three times. Result? Net goodness increases by 15.... (When I tried this out on 6-year-old Daniel Derbyshire, he said, "What if you call for the bad people to come out and *they won't come*?" A moral philosopher in the making.)

13. One reader of this book's manuscript thought that "twiddle" sounds like a Britishism. (I was educated in England.) I agree, it does. American mathematicians certainly use it, though. I have heard, for example, Nicholas Katz of Princeton University use it in a lecture. Prof. Katz is from Baltimore and was educated entirely in the U.S.A.

CHAPTER 4

14. George was the last king of Hanover. The kingdom was swallowed by Prussia in 1866, after taking the wrong side in the Austro-Prussian war of that year. The medal seems not to have actually been struck until the Gauss centenary in 1877.

15. Among the Duke's claims to fame, perhaps it is worth noting that he was the father of Caroline of Brunswick, who was married off to the Prince Regent of England. The marriage was a disaster and Caroline left England; but when the Prince ascended the English throne as King George IV, she returned to claim her rights as Queen. This caused a constitutional crisis of the minor sort, as well as much public merriment over the unpopular king's discomfiture, his queen's rather bumptious personality, her peculiar personal habits, and her flagrant *liaisons.* The following ditty was widely circulated.

    Gracious Queen, we thee implore
    To go away and sin no more;
    But if this effort be too great,
    To go away, at any rate.

    One of the Duke's maternal aunts married a Holy Roman Emperor and begat Maria Theresa, the great Hapsburg empress. Another married Alexis Romanov and was the mother of Peter II, nominal Tsar when Leonhard Euler came ashore in St. Petersburg (Section VI of this chapter). Once you start following the genealogies of these petty German rulers, there is no end to it.

16. Did I mention that as well as being a towering mathematical genius and a physicist of the first rank, Gauss was also a brilliant astronomer, the first person to correctly compute the orbit of an asteroid?

17. To find out if some number N is prime, you just keep dividing it by primes 2, 3, 5, 7, ... one after another until either one of them divides exactly, in which case you have shown that N is *not* prime, or ... what? How do you know when to stop? Answer: you stop when the prime you are about to divide by is bigger than $\sqrt{N}$. Suppose N is 47, for example; $\sqrt{47}$ is 6.85565..., so I only need to try division by 2, 3, and 5. If none of them works, 47 must be prime. Why don't I need to try 7? Because $7 \times 7 = 49$, so if 7 divided exactly into 47, the quotient would be *some number less than 7.* Likewise, $\sqrt{701{,}000}$ is 837.2574.... The last prime below this is 829; the next prime above it is 839. If 839 divided into 701,000, the quotient would be a number less than 839; either some prime less than 839 (which I would therefore already have tried), or a composite number made up of *even smaller* prime factors....

18. Legendre died in poverty, having offended his political superiors by taking a stand on principle. His dates are 1752–1833. I am sorry that I have presented him here as a disgruntled and slightly comical figure. Legendre was a fine mathematician, at the top of the second division, and did valuable work over many years. His *Elements of Geometry* was the leading elementary textbook on the subject for over a century. It is said to have inspired the tragic Évariste Galois—the narrator in Tom Petsinis's novel *The French Mathematician*—to take up a career in mathematics. More relevant to the present narrative, his book *Theory of Numbers*—the renamed third edition of the *Essay* mentioned in the text—was lent by a schoolmaster to the adolescent Bernhard Riemann, who returned it in less than a week with the comment, "This is truly a wonderful book; I know it by heart." The book has 900 pages.

19. For a full account of the Euler-Mascheroni number, see Julian Havil's book *Gamma: Exploring Euler's Constant.* Though I have not described

it properly in this book, the very observant reader will glimpse the Euler-Mascheroni number in Chapter 5.ix.

20. In the mathematics department of my English university, all undergraduates were expected to take a first-year course in German. Those like myself who had studied German in secondary school were shipped off to the nearby School of Slavonic and East European Studies to learn Russian, which our instructors considered to be the language of most importance to mathematicians, after German. There you have the legacy of Peter.
21. I have taken this story from a hilarious account of Frederick's relations with Voltaire written by the English wit and satirist Lytton Strachey in 1915 and found in his *Books and Characters: French and English.*
22. Euler's Latin is a stripped-down, racing version of the language, designed not to show off the writer's superb grasp of Augustan style (which Euler probably could have done if he had wanted to—he knew the *Aeneid* by heart) but to communicate ideas as plainly as possible with a minimum of verbiage to readers much less concerned with form than with content. I shall give some actual examples in Chapter 7.v.
23. The President of the Berlin Academy, Pierre Maupertuis, was accused by Swiss mathematician Samuel König, probably correctly, of having plagiarized Leibnitz's work. Maupertuis called on the Academy to pronounce König a liar, which they duly did. Writes Strachey: "The members of the Academy were frightened; their pensions depended on the President's good will; and even the illustrious Euler was not ashamed to take part in this absurd and disgraceful condemnation."
24. First English edition 1795; first American, 1833. For some reason this book can now be found only in expensive collector's editions.

CHAPTER 5

25. It had been posed by Pietro Mengoli in 1644. Mengoli was a professor at the University of Bologna at the time, so we really should say "the Bologna problem." It was Jakob Bernoulli who first brought the problem to the attention of a wide audience, though, and "the Basel problem" has stuck.

26. If the shape of the curve looks oddly familiar, that's because if you add up N terms of the harmonic series (Chapter 1.iii), you get a number close to log N. In fact,

$$1+\frac{1}{2}+\frac{1}{3}+\frac{1}{4}+\frac{1}{5}+\frac{1}{6}+\frac{1}{7}+\cdots+\frac{1}{N}\sim\log N$$

and the profile of that tottering stack of cards, if you rotate it clockwise through 90 degrees then reflect it in a vertical mirror, is the graph of $\log x$.
27. Note: It is a convention in math to use ε—that's epsilon, the fifth letter of the Greek alphabet—to mean "some very tiny positive number."
28. The proof was devised by Greek-French mathematician Roger Apéry, who was 61 years old at the time—so much for the notion that no mathematician ever does anything worthwhile after the age of 30. In honor of this achievement, the sum—its actual value is 1.2020569031 595942854...—is now known as "Apéry's number." It actually has some use in number theory. Take three positive whole numbers at random. What is the chance they have no proper factor in common? Answer: around 83 percent—to be precise, 0.831907372580707468683..., the reciprocal of Apéry's number.

CHAPTER 6

29. English edition published by Bloomsbury USA, 2000. The novel was first published in Greek in 1992. As Doxiadis points out, the conjecture was first framed in proper mathematical form by Euler.
30. Of topics like the Goldbach Conjecture and Fermat's Last Theorem, you might want to say "Oh, that's not arithmetic, that's number theory." These two terms have had an interesting relationship. The phrase "number theory," or at any rate "theory of numbers," goes back to at least Pascal (1654, in a letter to Fermat), but was not clearly distinct from "arithmetic" until the nineteenth century. Gauss's great classic on number theory was titled *Disquisitiones Arithmeticae* (1801). It seems to have been sometime in the later nineteenth century that "arithmetic" was definitely reserved for the basic manipulations learned in elementary school, with "number theory" used for the deeper researches of

professional mathematicians. Then, around the middle of the twentieth century, there began to be a road back. Perhaps it all began with Harold Davenport's 1952 book *The Higher Arithmetic*, an excellent popular presentation of serious number theory, whose title echoed an occasional synonym for "number theory" going back at least as far as the 1840s. Then, some time in the 1970s (I am working from personal impressions here) it began to be thought cute for number theorists to refer to their work as just "arithmetic." Jean-Pierre Serre's *A Course in Arithmetic* (1973) is a text for graduate students of number theory, covering such topics as modular forms, *p*-adic fields, Hecke operators, and, yes! the zeta function. I smile to think of some doting mother picking it out for her third-grader, to help him master long multiplication.

31. The pronunciation of Dirichlet's name gives a lot of trouble. Since he was German, the pronunciation should be "Dee-REECH-let," with the hard German "ch." English-speakers hardly ever say this. They either use the French pronunciation "Dee-REESH-lay," or half-and-half it: "Dee-REECH-lay."

32. Constantin Carathéodory, though of Greek ancestry, was born, was educated, and died in Germany. Cantor was born in Russia and had a Russian mother, but he moved to Germany at age 11 and lived there practically all his life. Mittag-Leffler was the Swede. According to mathematical folklore, he was the cause of there being no Nobel Prize in mathematics. The story goes that he had an affair with Nobel's wife, and Nobel found out. It's a nice story, but Nobel was not married.

33. Felix's first cousin, Ottilie, married the great German mathematician Eduard Kummer; their grandson, Roland Percival Sprague, was co-creator of "Sprague-Grundy Theory," in twentieth-century Game Theory.... I have to resist the temptation to take this further; it's like tracing the genealogies of those German princes. Another Mendelssohn link will show up in Chapter 20.v.

CHAPTER 7

34. "Eratosthenes" is pronounced—at any rate by mathematicians—"era-TOSS-the-niece."

35. Mathematics allows infinite products, just as it allows infinite sums. As with infinite sums, some of them converge to a definite value, some diverge to infinity. This one converges when s is greater than 1. When s is 3, for example, it is

$$\frac{8}{7}\times\frac{27}{26}\times\frac{125}{124}\times\frac{343}{342}\times\frac{1331}{1330}\times\frac{2197}{2196}\times\frac{4913}{4912}\times\frac{6859}{6858}\times\cdots$$

The terms get closer and closer to 1 really fast, so at each step in the multiplication you are multiplying by something a teeny bit bigger than 1 ... which, of course, hardly changes the result. Add 0 to something: no effect. Multiply something by 1: no effect. In an infinite sum, the terms have to get close to 0 really fast, so that adding them has very little effect; in an infinite product, they have to get close to 1 really fast, so that multiplying by them has very little effect.

36. "Golden Key" is strictly my nomenclature. "Euler product formula" is standard. So are the following terms for the two parts, "the Dirichlet series" for the infinite sum, and "the Euler product" for the infinite product. Strictly speaking, the left-hand side is *a* Dirichlet series and the right-hand side is *an* Euler product. In the narrow context of this book, though, "the" is fine.

37. There are two ways to define $Li(x)$, both, unfortunately, in common use. In this book I shall use the "American" definition given in Abramowitz and Stegun's classic *Handbook of Mathematical Functions*, published in 1964 by the National Bureau of Standards. This definition takes the integral from 0 to x, and this is also the sense in which Riemann used $Li(x)$. Many mathematicians—including the great Landau (see Chapter 14.iv)—have preferred the "European" definition, which takes the integral from 2 to x, avoiding the nasty stuff at $x = 1$. The two definitions differ by 1.04516378011749278.... The *Mathematica* software package uses the American definition.

38. You can get a good approximation for $Li(N)$ by just adding up $1/\log 2$, $1/\log 3$, $1/\log 4$, ..., $1/\log N$. If you do this for N equal to a million, for example, you get 78,627.2697299..., while $Li(N)$ is equal to 78,627.5491594.... So the sum gives an approximation that is low by 0.0004 percent. That integral sign sure does look like an "S" for "sum."

CHAPTER 8

39. Mostly. Prussia and Austria also held parts of historic Poland.
40. He worked for a year and a half as an assistant in Weber's physics lab and might have earned some spare change thereby, so perhaps was not utterly without income.
41. Topology is "rubber-sheet" geometry—the study of those properties of figures left unaffected by stretching, without tearing or cutting. The surface of a sphere is topologically equivalent to that of a cube, but not to that of a doughnut or a pretzel. The word "topology" was coined by Johann Listing in 1836, in a letter to his old schoolmaster. In 1847 Listing wrote a short book titled *Preliminary Sketch of Topology.* He was a professor of mathematical physics at Göttingen during Riemann's time there, and Riemann certainly knew him and his work. However, Riemann seems never to have used the word "topology," always referring to the topic by the Latin term favored by Gauss, *analysis situs*—"the analysis of position."
42. *Eugene Onegin*, 1833; *A Hero of Our Times*, 1840; *Dead Souls*, 1842.
43. He was also the subject of a 1959 comic song, *Lobachevsky*, by mathematician/musician Tom Lehrer.
44. Atle Selberg, now the Grand Old Man of number theory, is still at the Institute at the time of writing (June 2002) and still mathematically active. There is a story about this in Chapter 22. He was born June 14, 1917, in Langesund, Norway.
45. Riemann, Gauss, Dirichlet, and Euler also enjoy this distinction. Riemann's crater is at 87°E 39°N.
46. I should perhaps explain that mathematicians have their own particular approach to the learning of foreign languages. To be able to read mathematical papers in a language not one's own, it is by no means necessary to master that language thoroughly. You need to learn only the few dozen words, phrases, and constructions that are common in mathematical exposition: "it follows that…," "it is sufficient to prove that…," "without loss of generality…," and so on. The rest is symbols like $\surd$ and Σ, that are common to all languages (though there are some minor national dialects in their usage). Some mathematicians, of

course, are fine linguists. André Weil (see Chapter 17.iii) spoke and read English, German, Portuguese, Latin, Greek, and Sanskrit, besides his native French. I am speaking of ordinary mathematicians.

47. Two of Gauss's six children emigrated to the United States, where they helped populate the state of Missouri.

CHAPTER 9

48. "Heck of a formula...." It is not actually so daunting, unless you have forgotten all your high school math. Other than the zeta function, there is nothing in there that high school math doesn't cover, at least in part. The sine and factorial functions are, as mathematicians say, "elementary," so this formula "elementarily" relates the value of zeta at argument $1 - s$ to its value at s. This formula, by the way, is called "the functional equation."

49. A fact first proved by Bernhard Riemann, incidentally.

CHAPTER 10

50. *Riemann's Zeta Function*, by H.M. Edwards (1974). Reprinted by Dover in 2001.

51. A few unfortunate cases like Riemann notwithstanding, higher mathematics is wonderfully healthful. In writing this book, I have been struck by the number of mathematicians who lived to advanced ages, active to near the end. "Mathematics is very hard work, and dons tend to be above the average in health and vigor. Below a certain threshold a man cracks up, but above it hard mental work *makes* for health and vigor (also—on much historical evidence through the ages—for longevity)."—*The Mathematician's Art of Work* by J.E. Littlewood, 1967. Littlewood, of whom I shall have much more to say in Chapter 14, was an illustration of his own argument. He lived to be 92. A colleague, H.A. Hollond, recorded the following note about him in 1972: "In his 87th year he is still working long hours at a stretch, writing papers for publication and helping mathematicians who send their problems to

him."—Quoted by J.C. Burkill in *Mathematics: People, Problems, Results* (Brigham Young University, 1984).

52. I cannot restrain myself. "If f is an analytic function in the annulus $0 < r_1 < |z| < r_2 < \infty$, r is some number between r_1 and r_2 exclusive, and M_1, M_2, and M are the maxima of f on the three circles corresponding to r_1, r_2, and r, respectively, then $M^{\log(r_2/r_1)} \leq M_1^{\log(r_2/r)} M_2^{\log(r/r_1)}$."
53. Stieltjes's dates are 1856–1894. The most popular pronunciation of his name among English-speaking mathematicians is "STEEL-ches."
54. "Reports Received." This term is so common in scholarly bibliographies, it is often abbreviated to "*C.R.*"
55. He did not join the Communist Party, though his daughter Jacqueline did.
56. Though the glory of proving the PNT belongs to Hadamard and de la Vallée Poussin equally, I have written a great deal about the former and next to nothing about the latter. This is only in part because I find Hadamard an interesting and sympathetic character. It is also because there is much less material on de la Vallée Poussin. Though a fine mathematician, he appears to have been active in no other sphere. I mentioned this to Atle Selberg, the only mathematician I have spoken with who might have known both men. Hadamard? "Oh, yes. I met him at the Cambridge Congress" (i.e., in 1950). De la Vallée Poussin? "No. I never met him, and I don't know anyone who did. I think he did not travel much."

CHAPTER 11

57. Nowadays it is more often called "the argument" and denoted by *Arg*(z). I have used the older term, partly out of loyalty to G.H. Hardy (see Chapter 14.ii) and partly to avoid confusion with my use of "argument" to mean "the number to which a function is applied."

CHAPTER 12

58. I do not mean to dismiss Kronecker as a crackpot. The case he made was, though I disagree with it, subtle and mathematically sophisticated.

For a spirited defense of Kronecker, see Harold Edwards' article in the *Mathematical Intelligencer*, Vol. 9, No. 1. Kronecker was, says Prof. Edwards, "reasonable, not vitriolic."

59. In German, *Wer von uns würde nicht gern den Schleier lüften, unter dem die Zukunft verborgen liegt, um einen Blick zu werfen auf die bevorstehenden Fortschritte unserer Wissenschaft und in die Geheimnisse ihrer Entwickelung während der künftigen Jahrhunderte?*

60. Hilbert actually only presented 10 of the problems to his audience, having been urged by those who had read the printed form of his address to shorten it for delivery. All 23 problems are listed in the printed address, and they are generally referred to by their numbers in that paper. The ones he actually read out to his audience at the Sorbonne were numbers 1, 2, 6, 7, 8, 13, 16, 19, 21, and 22. A further confusion arises from the fact that some of Hilbert's 23 bullet points just single out areas for investigation, and are only arguably problems. Typical is number 2, "To investigate the consistency of the axioms of arithmetic." This accounts for the different numbering schemes you will sometimes see. Andrew Hodges, for example, in his biography of Alan Turing, counts 17 Hilbert problems, not 23, with the proof of the Riemann Hypothesis at number 4, not 8. Those of Hilbert's items that were actual well-defined problems have now all been solved, with the single exception of the Riemann Hypothesis.

61. The best such book-length account that I know of is Jeremy J. Gray's *The Hilbert Challenge* (Oxford University Press, 2000).

62. For a good popular account, see John L. Casti's book *Mathematical Mountaintops* (Oxford University Press, 2001).

63. Most mathematicians of the time would have given that title to Henri Poincaré (1854–1912). The Hungarian Academy of Sciences in fact did so in 1905, awarding Poincaré its first Bolyai Prize as "that mathematician whose achievement during the past 25 years have most greatly contributed to the progress of mathematics." The second Bolyai Prize was awarded to Hilbert in 1910.

64. George Pólya (1887–1985). Look at those dates—another immortal. Pólya was Hungarian. Even more striking than the rise of the Germans in the early nineteenth century was the rise of Hungarians in the early twentieth. While the German states (excluding Austria and Switzer-

land) in 1800 had about 24 million people, the Hungarian-speaking population of Hungary was around 8.7 million in 1900, and I believe never rose above 10 million. This small and obscure nation produced an astonishing proportion of the world's finest mathematicians: Bollobás, Erdélyi, Erdős, Fejér, Haar, Kerékjártó, two Kőnigs, Kürschák, Lakatos, Radó, Rényi, two Rieszes, Szász, Szegő, Szokefalvi-Nagy, Turán, von Neumann, and I have probably missed a few. There is a modest literature attempting to explain this phenomenon. Pólya himself thought that the major factor was Fejér (1880–1959), an inspiring teacher and gifted administrator, who attracted and encouraged mathematical talent. A high proportion of the great Hungarian mathematicians (including Fejér) were Jewish—or, like Pólya's parents, "social" converts to Christianity, of originally Jewish stock.

65. "The vertex figures of a regular polytope are all equal." A polytope is the n-dimensional equivalent of a polygon in two dimensions, or a polyhedron in three. It is regular if all its "cells"—its $(n-1)$-dimensional "faces"—are regular and all its vertex figures regular. The cells of a cube are squares; the vertex figures are equilateral triangles. Longevity watch: "Donald" Coxeter was born February 9, 1907, and died March 31, 2003, after sixty-seven years as professor at the University of Toronto. He was publishing in math journals at least as late as 2001. Of the famously prolific Coxeter, a mathematician remarked to me in 2002: "Donald seems to have slowed down some recently."

66. Theory assures us, by the way, that the real part is precisely and mathematically $\frac{1}{2}$, not 0.4999999, or 0.5000001. I shall say more about this in Chapter 16.

CHAPTER 13

67. Note incidentally, that the "unknown" complex number is most commonly represented by "z," not "x." Mathematicians customarily use "n" and "m" for whole numbers, "x" and "y" for real numbers, and "z" and "w" for complex numbers. We can, of course, use any other letters we feel like using—this is just a custom. (For the argument of the zeta function, I shall persist in that other custom of calling it "s," as all math-

ematicians do.) Pólya used to tell his students that the common use of "z" for the argument and "w" for the value in complex function theory derived from the German words *Zahl*, which means "number," and *Wert*, which means "value." I don't know if this is true, though.

68. Estermann (1902–1991) made his mark in mathematics by proving, in 1929, that the Goldbach Conjecture, which asserts that every even number greater than 2 is the sum of two primes, is almost always true. He was also the originator of my proof for the irrationality of $\sqrt{2}$ in Note 11—"the first new proof since Pythagoras," he used to boast.
69. Mathematicians working with functions of a complex variable generally say "the z plane" and "the w plane," it being understood that "z" is the generic argument and "w" the generic value in complex function theory.
70. And both kinds of illustration have really come into their own only with the advent of fast computer workstations and PCs. Before then, constructing pictures like my Figures 13-6 through 13-8 was an awfully painstaking business.

CHAPTER 14

71. E.W. Barnes, Littlewood's director of studies. He later became an Anglican bishop.
72. Author of *Calcul des Résidus*, a textbook of complex function theory. Ernst Lindelöf (1870–1946) was a great hero of Scandinavian mathematics, which he worked hard to advance through teaching, research, and writing textbooks. Born in Helsinki, he began his life a subject of the Russian Tsar—Finland did not get independence from Russia until 1917. Lindelöf was, however, a Finnish patriot (one of only two Finns in this book), and participated enthusiastically in the life of the new nation. He was the originator of the Lindelöf Hypothesis, a famous conjecture about the Riemann zeta function, concerning its rate of growth in the critical strip. I describe this conjecture in the Appendix.
73. A fellowship at Trinity was a lecturing position, with a regular stipend, and the right to take rooms in the college and eat dinner in the "hall" (refectory). It was not necessarily tenured.

74. In the mid-1930s, the Soviet intelligence services recruited five young Cambridge undergraduates. Their names were Guy Burgess, Donald Maclean, Kim Philby, Anthony Blunt, and John Cairncross. This "Ring of Five," as the Soviets referred to them, all went on to attain high positions in the British political and intelligence establishments during the 1940s and 1950s and passed vital information to the U.S.S.R. through World War II and the Cold War. Four of the five were at Trinity; Maclean was at Trinity Hall, a separate, smaller college.
75. Lytton Strachey, Leonard Woolf, Clive Bell, Desmond MacCarthy, Saxon Sydney-Turner, and both Stephen brothers (Thoby and Adrian) were Trinity men. John Maynard Keynes, Roger Fry, and E.M. Forster, however, were at King's.
76. So it is always said. In his book on George Pólya, though, Jerry Alexanderson claims that the Pólya estate holds many more.
77. Though the spine of my copy, a first edition, says simply "*Primzahlen.*"
78. There are also *lower* bounds in problems of this sort. A lower bound is a number *N* for which we could prove that whatever the precise answer may be, it is certainly greater than *N*. In the case of the Littlewood violations, there seems to have been less work done here, presumably because everyone knew that the precise value of the first violation was extremely large. Deléglise and Rivat established 10^{18} as a lower bound in 1996 and have since extended the lower bound to 10^{20}, but in view of the Bays and Hudson result, these lower bounds are almost nugatory.
79. If the names Bays and Hudson ring a bell, that is because I mentioned them in Chapter 8.iv in connection with the Chebyshev bias. There is in fact a deep level, too deep to explore further here, at which the tendency of $Li(x)$ to be greater than $\pi(x)$ is kin to the Chebyshev biases. These two issues are generally dealt with as one by analytic number theorists. In fact, Littlewood's 1914 paper showed not only that the tendency of $Li(x)$ to be greater than $\pi(x)$ is violated infinitely many times, but that this is also true of Chebyshev biases. For some very fascinating recent insights on this topic, see the paper "Chebyshev's Bias," by Michael Rubinstein and Peter Sarnak, in *Experimental Mathematics*, Vol.3, 1994 (pp. 173–197).

80. Von Koch is better known to readers of pop-math books for the "Koch snowflake curve." The "von" always gets dropped in that context, I don't know why.

CHAPTER 15

81. Either unaware of Bachmann's book, or (more likely) just choosing not to use the new big oh notation, von Koch actually expressed his result in a more traditional form

$$|f(x) - Li(x)| < K \cdot \sqrt{x} \cdot \log x$$

82. There has been a vast amount of research in this area. It is quite probably the case, in fact, that $\pi(x) = Li(x) + O(\sqrt{x})$, which may be what Riemann meant by his "order of magnitude" remark. However, we are nowhere near being able to prove this. Some researchers, by the way, prefer the notation $O_\varepsilon(x^{\frac{1}{2}+\varepsilon})$, to emphasize that the constant implied by the definition of big oh depends on ε. If you use this notation, the logic of Section 15.iii changes slightly. Note that the square root of *N* is about half as long (I mean, has about half as many digits) as *N*. It follows, though I shall not pause to prove it in detail, that $Li^{-1}(N)$ gives the *N*-th prime, correct to about half-way along, that is, roughly the first half of the digits are correct. The expression "$Li^{-1}(N)$" here is to be understood in the inverse-function sense of Chapter 13.ix, with this meaning: "The number *K* for which $Li(K) = N$." The billionth prime, for instance, is 22,801,763,489; Li^{-1}(1,000,000,000) is 22,801,627,415—five digits, very nearly six, out of eleven.

83. Möbius is best remembered for the Möbius strip, shown in Figure 15-4, which he discovered for himself in 1858. (It had previously been described by another mathematician, Johann Listing, also in 1858. Listing published, and Möbius didn't, so according to the academic rules it should really be called "the Listing strip." There is no justice in this world.) To create a Möbius strip, take a strip of paper, hold the two ends together (one in your right hand, one in your left), twist one end through 180°, and glue the ends together. You now have a one-sided

strip—an ant can walk from any point on the strip to any other point without going over the edge.

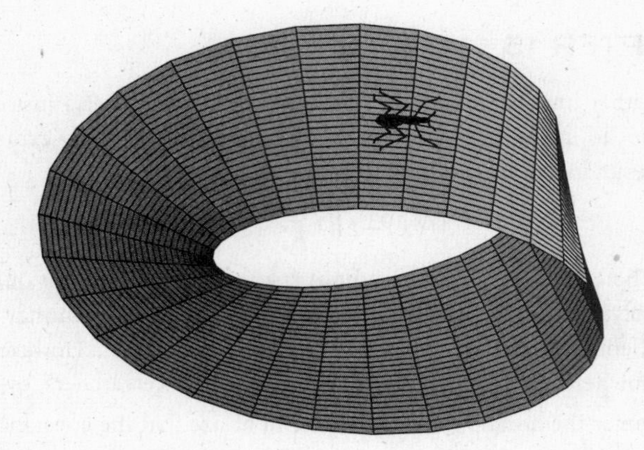

FIGURE 15-4 A Möbius strip, with ant.

84. In case you think it was somewhat vainglorious of Möbius to pick a symbol equivalent to his own initial, let it be known that Möbius himself did not use μ when he first described the function in 1832; the μ is due to Franz Mertens in 1874, and Mertens was honoring Möbius, by then dead, not himself.
85. If the logic there escapes you, here's a parallel case. Imagine that Theorem 15-1 said "All human beings are less than 10 feet tall," while the Riemann Hypothesis said "All U.S. citizens are less than 10 feet tall." If the first is true, the second must be true, since every U.S. citizen is a human being. The weaker result follows from the stronger one. If a human being 11 feet tall were discovered living in the remote highlands of New Guinea, then the existence of that person would prove Theorem 15-1 to be false. The Riemann Hypothesis, however, would still be open, since the giant is not a U.S. citizen. (Though I suspect he soon would be....)

CHAPTER 16

86. Bernstein became a professor only in 1921. I have seen it written that he was also technically exempt under the Hindenburg modifications, but I do not know the basis for this statement. Bernstein (1874–1956) fled to the U.S. during the Hitler period but returned to Göttingen in 1948.
87. Carl Siegel told Harold Davenport the following story. "In 1954, to celebrate the 1,000th anniversary of Göttingen's founding, the city fathers decided to give the freedom of the city to three of those professors who had been dismissed in 1933. The *Tageblatt* sent a reporter to Rellich (i.e., Franz Rellich, then director of the university's Mathematics Institute) to ask if he could write an article on the three. Rellich replied, 'Why don't you just look up what you wrote back in '33?'"
88. There is actually a branch of geometric function theory known, not altogether accurately, as "Teichmüller Theory." It deals with the properties of Riemann surfaces. Teichmüller volunteered for active duty in World War II. He disappeared in fighting along the Dnieper in September 1943.
89. In the world of mathematics another instance was Ludwig Bieberbach, author of a famous conjecture in complex function theory (proved in 1984 by Louis de Branges). In 1933 at Berlin University, Bieberbach was conducting spoken examinations of doctoral candidates in full Nazi uniform.
90. I can think of no satisfactory English translation for *Nachlass*. Neither, to judge from the word's frequent appearance in English-language materials, can anyone else. "Literary remains," says my German dictionary. In this context the meaning is "unpublished papers found among a scholar's effects after his death."
91. Recall from my explanation of big oh that it involves some fixed constant multiplier. Thus, $O(\log T)$ means "This term never exceeds some fixed multiple of $\log T$." To describe the formula as "very good" is to say that the fixed multiplier is small. In this case it is less than 0.14.
92. This particular piece of theory deals with zeros actually, precisely, mathematically *on* the critical line. It is important to grasp the logic here. Theory *A* tells you: "There are *n* zeros in the rectangle from T_1 to

T_2." (See Figure 16-1.) Theory *B* tells you: "There are *m* zeros on the critical line from T_1 to T_2." If it turns out that $m = n$, then you have verified the Riemann Hypothesis between T_1 and T_2. If, on the other hand, *m* is less than *n*, you have disproved the Riemann Hypothesis! (It is, of course, logically impossible for *m* to be greater than *n*.) Theory *B* deals with matters *on* the critical line. There is no possibility that the zeros being discussed here might have real parts 0.4999999999 or 0.5000000001. Compare the note on this in Chapter 12.vii.

93. All the zeros computed so far appear to be irrational numbers, by the way. It would be astonishing and wonderful if an integer showed up among them, or even a repeating decimal (indicating a rational number). I know no reason this should not happen, but it hasn't.
94. The Fields Medal, first awarded 1936, was the idea of Canadian mathematician John Charles Fields (1863–1932). Now given at four-year intervals, its main purpose is to encourage promising younger mathematicians. Therefore, it is given only to those under 40. Several of the mathematicians named in this book have been Fields medalists: Atle Selberg (1950), Jean-Pierre Serre (1954), Pierre Deligne (1978), and Alain Connes (1982). The Fields Medal is held in high esteem by mathematicians. If you are a Fields winner, every mathematician knows it, and speaks your name with great respect.
95. Not "104," as Hodges says.
96. *The Theory of the Riemann Zeta-function* (1951). Still in print.
97. Just one more biographical note. Josef Backlund (1888–1949) is the other Finn in this book, born into a working-class family in Jakobstad on the Gulf of Bothnia. "The family was gifted but seems to have been mentally unstable; three brothers of Josef committed suicide." (*The History of Mathematics in Finland, 1828–1918*, by Gustav Elfving; Helsinki, 1981.) A student of Lindelöf's, Backlund became an actuary after taking his doctorate and made a career in insurance, like Gram. Human knowledge owes a great deal to the insurance business. Gram, by the way, died an absurd death—struck and killed by a bicycle.
98. Professor Edwards's book includes some photographs of pages from the *Nachlass*, illustrating the scale of the task Siegel undertook.

CHAPTER 17

99. For example, S.J. Patterson, in his book, *An Introduction to the Theory of the Riemann Zeta-Function*, §5.11, wrote: "The most convincing reason that has so far been evinced for the validity of the Riemann Hypothesis is that an analogous statement is valid for the zeta-functions attached to curves over finite fields. The formal similarities are so striking that *it is difficult to believe that they do not lead to even more far-reaching coincidences*." (My italics.)

100. To coin an apothegm, algebraists do not care so much about what things are, as about what you can do with them. They are verb people, not noun people. Another interesting conceptual perspective on algebra was offered by Sir Michael Atiyah at a Fields Lecture in Toronto in June 2000. While geometry is obviously about space (said Sir Michael, a Fields Medal winner), algebra is about time. "[G]eometry is essentially static. I can just sit here and see, and nothing may change, but I can still see. Algebra, however, is concerned with time, because you have operations that are performed sequentially...." (Shenitzer, A. and M.F. Atiyah. "Mathematics in the 20th century," *American Mathematical Monthly*, Vol. 108, No. 7.)

101. Pronounced "Vay" by most English-speaking mathematicians. The main thing is to avoid listeners' confusing him with Hermann Weyl ("Vail"). Weil, one of the most illustrious names in twentieth-century mathematics, was the brother of the mystic and French Resistance heroine Simone Weil. He had been a student of Hadamard's at the Collège de France.

102. It might be better to say "from 1 to N zeros," because zeros sometimes repeat. The zeros of the polynomial $x^2 - 6x + 9$ are 3 and 3. It factorizes as $(x-3)(x-3)$. You might, therefore, prefer to say that this polynomial has only one zero, namely 3. In strict mathematical terms, this is "a zero of order 2." There is a way to assign a similar order to any zero of any function, by the way. So far as we know, all the non-trivial zeros of the zeta function have order 1; but this has not been proved. Should a non-trivial zero of the zeta function show up with order 2 or greater, it would not disprove the Hypothesis, but it would create havoc with some of the computational theory.

CHAPTER 18

103. I am really speaking here about *operators*, of course. Operators provide a mathematical model for describing dynamical systems. "Ensemble" (the originator of this usage seems to have been Willard Gibbs, in his 1902 book *Elementary Principles in Statistical Mechanics*) refers to a collection of such operators that share some common statistical properties.
104. To be more precise, Montgomery's area of interest was the so-called "class number problem," of which there is a very accessible account in Keith Devlin's book, *Mathematics: The New Golden Age* (Columbia University Press, 1999).
105. Harold Diamond is a number theorist. He is currently Professor of Mathematics in the University of Illinois at Urbana-Champaign.
106. Sarvadaman Chowla, 1907–1995. A fine number theorist, mainly at the University of Colorado.
107. The standard introductory text on random matrix theory is Madan Lal Mehta's *Random Matrices and the Statistical Theory of Energy Levels* (Academic Press, 1991).
108. Dyson was in fact another Trinity man, having attended that college in the early 1940s. He recalls that Hardy, at that time slipping into his terminal depression, was "not encouraging."
109. This raises the interesting question of the degree to which these are really theorems. A result that assumes the truth of the RH is technically, it seems to me, a hypothesis itself—or perhaps a sub-hypothesis, but at any rate not a proper theorem. Considering, in fact, that mathematics is supposed to be the most precise of disciplines, mathematicians are not very consistent about the use of terms like "conjecture," "hypothesis," and "theorem." Why, for example, is the RH a "hypothesis," not a "conjecture"? I don't know, and I haven't found anyone who can tell me. These remarks seem, on a cursory examination, to apply in languages other than English, too. The German for "the Riemann Hypothesis," by the way, is *Die Riemannsche Vermutung*, from the verb *vermuten*—"to surmise."
110. Professor of Physics at Bristol University in England. Berry was elevated to the knightage in the Queen's Birthday Honors of June, 1996, becom-

ing Sir Michael Berry. I have done my best to refer to him as "Berry" in writing of his activities up to 1996, and "Sir Michael" thereafter; but I don't guarantee consistency.

111. The Cray-1 was supplemented by a Cray X-MP at some point in the late 1980s.

112. The earliest reference I have been able to track down to the Montgomery-Odlyzko Law thus named is in a paper by Nicholas Katz and Peter Sarnak published in 1999. The word "Law" is of course to be understood in a physical, rather than a mathematical sense. That is, it is a fact established by empirical evidence, like Kepler's laws for the motions of the planets. It is not a mathematical principle, like the rule of signs. The Sarnak-Katz paper actually proved the law for zeta-like functions over finite fields (see Chapter17.iii), thus establishing a bridge between the algebraic and physical approaches to the RH.

113. The answer is not "a half." That would be to confuse the median with the average. The average of these four numbers: "1, 2, 3, 8510294," is 2127575; but half of them are less than 3.

114. Known to mathematicians as a "Poisson distribution." The number *e*, by the way, is all over here. That 6,321, for example, is $10{,}000(1-1/e)$.

115. The equation I used for the curve in Figure 18-5 is $y = (320000/\pi^2)\,x^2 e^{-4x^2/\pi}$. It is a skewed distribution, not (like the Gaussian-normal) a symmetrical one. Its peak is at argument $\frac{1}{2}\sqrt{\pi}$, i.e., 0.8862269.... This was the curve surmised by Eugene Wigner for the GUE consecutive-spacings distribution. His surmise was based on the small amounts of data that can be gathered from experiments on the nucleus. It later turned out that this is not precisely the correct curve, though it is accurate to about a 1% error. The true curve, found by Michel Gaudin, has a more difficult equation. Andrew Odlyzko had to write a program to draw it.

CHAPTER 20

116. Though the word "chaos" was not applied to these theories until 1976, when physicist James Yorke first coined it. James Gleick's 1987 best-seller *Chaos: Making a New Science* remains the best guide to chaos

theory for the layman ... unless you count Tom Stoppard's 1993 play *Arcadia.*

117. Hensel (1861–1941) was yet another branch of the Mendelssohn tree. His grandmother, Fanny, was the sister of the composer; his father, Sebastian Hensel, was her only son. Sebastian was 16 when Fanny died, and he was sent to live with the Dirichlets (Chapter 6.vii), with whom he remained until his marriage. Kurt spent most of his career as a professor at the University of Marburg, in central Germany, retiring in 1930. In spite of the Jewish lineage, he seems not to have suffered under the Nazis. "In general, the Mendelssohns did not feel the full brunt of the Nuremberg anti-Semitic laws because most of the family had undergone conversion several generations back." (H. Kupferberg, *The Mendelssohns.*) In 1942, Hensel's daughter-in-law donated his large mathematical library to the newly Nazified University of Strasbourg in occupied Alsace, reopened in November that year as the Reichsuniversität Straßburg (but nowadays back in France once more).

118. And at least one mathematician has expressed guarded skepticism in print. Reviewing Connes 1999 paper "Trace Formulae in Noncommutative Geometry and the Zeros of the Riemann Zeta Function," Peter Sarnak (who is neither of my mathematicians X and Y) noted: "The analogies and calculations in the paper and its appendices are suggestive, pleasing and intricate and for these reasons this appears to offer more than just another equivalence of RH. Whether in fact these ideas and in particular the space X can be used to say anything new about the zeroes of $L(s, \lambda)$ is not clear to this reviewer." The $L(s, \lambda)$ Sarnak refers to is one of those analogues of the Riemann zeta function I mentioned in 17.iii.

119. The official name for this approach is "Denjoy's Probabilistic Interpretation," after the French analyst Arnaud Denjoy (1884–1974). Denjoy was Professor of Mathematics at the University of Paris, 1922–1955.

120. "Touching the dull formulas with his wand, he turned them into poetry."—Gunnar Blom, from the memorial essay included in Cramér's collected works. Cramér (1893–1985) was yet another immortal. He died a few days after his 92nd birthday.

121. I have borrowed this thought experiment from Chapter 3 of *The Prime Numbers and Their Distribution*, by Gérald Tenenbaum and Michel Mendès France (American Mathematical Society publications, 2000).

122. A good article on this topic is "Is π Normal?" by Stan Wagon, in *Mathematical Intelligencer*, Vol. 7, No. 3.

123. I have a preprint copy of a very recent paper by Hugh Montgomery and Kannan Soundararajan, titled "Beyond Pair Correlation," and delivering another blow to the Cramér model. The last words in the paper are, "...it seems that there is something going on here that remains to be understood."

124. *Mathematics and Plausible Reasoning* (1954).

125. Franklin has written a very good book about nonmathematical probability theory, *The Science of Conjecture* (2001). I reviewed this book for *The New Criterion*, June 2001.

CHAPTER 21

126. I should perhaps say, for the benefit of any reader so fired up by my exposition as to be on the point of running out and buying a math software package, that very strong opinions are held about the relative merits of the different packages, along the lines of the evergreen PC/Macintosh debate, with Stephen Wolfram, who created *Mathematica*, playing the part of Bill Gates. As a mere journalist, I consider myself *hors de combat* in this war. I am certainly not propagandizing on behalf of *Mathematica*. It was the first math software package that came to my attention, and it is the only one I have ever used. It has always done what I asked it to do. Sometimes, to be sure, I had to tweak it a little (see Note 128), but I never knew a software package that didn't need tweaking now and again.

127. It has no direct bearing on the argument here, but I can't resist adding, as a matter of interest, that one of the most famous theorems in complex function theory concerns entire functions. The theorem was stated and proved by Émile Picard (1856–1941). Picard's Theorem says that if an entire function takes more than one value—if, that is, it is not merely

a flat constant function—then it takes *every* value, with at most one exception. For e^z, the exception is 0.

128. Though the definition involves some ambiguities, on the resolution of which there is no general agreement. The *Mathematica 4* software package, for example, provides $Li(x)$ as one of its built-in functions—it calls it LogIntegral[x]. For real numbers, it is just as I described it—in fact, I used it to draw the graph of $Li(x)$ in Chapter 7.viii. For complex numbers, however, *Mathematica*'s definition of the integral is slightly different from Riemann's. Therefore, I didn't use *Mathematica*'s LogIntegral[z] for these complex calculations. I actually set up $Li(x^{\frac{1}{2}+ir})$ in *Mathematica* as ExpIntegralEi[($\frac{1}{2}$ + ir) Log[x]].

129. Looking at this list with one eye and Figure 21-3 with the other, you can see that the tendency of the first few zeros to be sent to numbers with negative real parts is just a chance effect, and soon rights itself.

130. In Figures 21-5 and 21-6, I have referred to the complex conjugate of the kth zero as the $-k$th zero. This is just a handy way of enumerating the zeros. It is, of course, *not* the case that $\overline{\rho} = -\rho$.

131. Note that $639 \div 1050 = 0.6085714\ldots$. For large numbers N, the probability that N is square-free is $\sim 6/\pi^2$, that is, 0.60792710.... Recalling Euler's solution of the Basel problem in Chapter 5, you might notice that this probability is $1/\zeta(2)$. This is generally true. The chance that a positive whole number N chosen at random is not divisible by any nth power is indeed $\sim 1/\zeta(n)$. Of all the numbers up to and including 1,000,000, for example, 982,954 are not divisible by any sixth power. $1/\zeta(6)$ is 0.98295259226458....

CHAPTER 22

132. Ulrike's pages on the University of Ulm website have a photograph of her standing next to Bernhard Riemann's memorial stone in Selasca, Italy.

133. Professor of Applied Mathematics at the University of Bristol, England. Keating has worked closely with Sir Michael Berry on the physical aspects of the RH.

134. "The zeros of Mellin transforms of Hermite functions have real part one-half," (1986). Bump's collaborator in the proof was one Eugene K.-S. Ng.

135. So it seems to me. One of the professional mathematicians who looked over my manuscript expressed frank disbelief at this, though. The idea that one might be able to make money by doing mathematics is extremely difficult for mathematicians to take seriously.

136. Professor of Pure Mathematics, University of Wales, Cardiff.

137. Here is the chain of events in barest outline. The method adopted for *Principia Mathematica* offered no guarantee against flaws, like the flaw Russell had spotted in Frege's work. Hilbert's "metamathematics" program tried to encompass both logic and mathematics in a more waterproof symbolism. This inspired the work of Kurt Gödel and Alan Turing. Gödel proved important theorems by attaching numbers to Hilbert-type symbols; Turing coded both instructions and data as arbitrary numbers in his "Turing machine" concept. Picking up on this idea, John von Neumann developed the stored-program concept on which all modern software is based, that code and data can be represented in the same way in a computer's memory....

EPILOGUE

138. In a letter to his brother dated June 26, 1854, he mentioned a recurrence of *mein altes Übel*—"my old malady"—brought on by a spell of bad weather.

139. In the modern municipality of Verbania.

140. Weender Chaussee has since been renamed Bertheaustrasse.

Appendix
The Riemann Hypothesis in Song

Tom Apostol, Professor of Mathematics Emeritus at Caltech, wrote the following tribute to the Riemann Hypothesis (RH) in 1955 and performed it at the Caltech Number Theory conference held in June of that year. Tom's original lyrics went only as far as Line 32; the last two stanzas were posted on a bulletin board at Cambridge University in 1973 by algebraic topologist Saunders MacLane.

The song mentions the Lindelöf Hypothesis (LH), a younger cousin of the RH. Dating as it does from 1908, the LH really belongs in Chapter 14 somewhere; but because it is peripheral to the main story, and because it involves the "big oh" notation from Chapter 15, and because I felt that my book already had too much math at that point, I left it out. Tom's lyrics can't be understood without it, though, and I couldn't bear to omit them; so you get a song and a bonus hypothesis!

Where are the zeros of zeta of *s*?

by Tom M. Apostol

(To the tune of *Sweet Betsy from Pike.*)

Where are the zeros of zeta of *s*? 1
G.F.B. Riemann has made a good guess:
"They're all on the critical line," stated he,
"And their density's one over two pi log *T*."

This statement of Riemann's has been like a trigger, 5
And many good men, with vim and with vigor,
Have attempted to find, with mathematical rigor,
What happens to zeta as mod *t* gets bigger.

The efforts of Landau and Bohr and Cramér,
Hardy and Littlewood and Titchmarsh are there. 10
In spite of their effort and skill and finesse,
In locating the zeros there's been no success.

In 1914 G.H. Hardy did find,
An infinite number that lie on the line.
His theorem, however, won't rule out the case, 15
That there might be a zero at some other place.

Let *P* be the function pi minus *Li*;
The order of *P* is not known for *x* high.
If square root of *x* times log *x* we could show,
Then Riemann's conjecture would surely be so. 20

Related to this is another enigma,
Concerning the Lindelöf function mu sigma,
Which measures the growth in the critical strip;
On the number of zeros it gives us a grip.

But nobody knows how this function behaves. 25
Convexity tells us it can have no waves.

Lindelöf said that the shape of its graph
Is constant when sigma is more than one-half.

Oh, where are the zeros of zeta of *s*?
We must know exactly. It won't do to guess. 30
In order to strengthen the prime number theorem,
The integral's contour must never go near 'em.

André Weil has improved on old Riemann's fine guess
By using a fancier zeta of *s*.
He proves that the zeros are where they should be, 35
Provided the characteristic is *p*.

There's a moral to draw from this long tale of woe
That every young genius among you must know:
If you tackle a problem and seem to get stuck,
Just take it mod *p* and you'll have better luck. 40

Notes.

Tune. "Sweet Betsy from Pike" is the song Americans sing to this tune. The tune is older than those lyrics, though. It first showed up attached to an English song popular in the mid-nineteenth century, "Villikens and his Dinah." (From which, by the way, the cat in Lewis Carroll's *Alice* books got its name. "Villikens and his Dinah" was a favorite with Alice Liddell, the girl who inspired the books, and she actually did have a cat named Dinah.) If you had a British education that included membership in a school rugby club, you will most likely recognize the tune as that of the melancholy ballad beginning, "O Father, O Father, I've come to confess. I've left some poor girl in a hell of a mess...."

Line 1. See Chapter 5.vii.

Line 2. Riemann's full name was Georg Friedrich Bernhard Riemann (Chapter 2.iii). He seems only ever to have used the "Bernhard."

Line 3. "Critical line", see Chapter 12.iii, Figure 12-1.

Line 4. Compare my statement in Chapter 13.viii that at height T up the critical line, the average spacing of zeros is ~ $2\pi / \log(T/2\pi)$. That means that in a unit length of the line, there are ~ $(1/2\pi)\log(T/2\pi)$ zeros. That's what the songwriter means by "density." Note that by the rules for logs, $\log(T/2\pi)$ is equal to $\log T - \log(2\pi)$, that is, $\log T -$ 1.83787706.... If you multiply that by $1/2\pi$, you get $(1/2\pi)\log T -$ 0.29250721.... As T gets larger and larger, so does $\log T$ (albeit much more slowly), and the significance of the 0.29250721... term dwindles to nothing. The density is, therefore, ~ "one over two pi log T."

Line 8. "Mod t" refers to the *modulus* of t, as defined in Chapter 11.v. When, as here, t is understood to be a real number, "mod t"—in proper symbols, "$|t|$"—just means "the size of t," that is, t without its sign. $|5|$ is 5; $|-5|$ is also 5. As I pointed out in Chapter 16.iv, "t" (or "T") is pretty standard in zeta-function theory for referring to height up the critical line; or, more generally, as in the discussion of the LH in the notes on Lines 21–28, to the imaginary part of a zeta-function argument.

Line 9. Harald Bohr (Chapter 14.iii) and Edmund Landau proved an important theorem about the S function (see Chapter 22.iv) in 1913. The theorem states that, so long as there is only a finite number of zeta zeros off the critical line, $S(t)$ is unbounded when t goes to infinity. Selberg's 1946 proof that $S(t)$ is unbounded, which I mentioned in Chapter 22.iv, is stronger, as it does not need that initial condition. For Cramér, see Chapter 20.vii. As well as developing that "probabilistic model" for the prime numbers, Cramér also proved a minor result about the S function: If the LH (see the notes for Lines 21–28) is true, then $S(t)/\log t$ dwindles to zero as t goes to infinity. For Littlewood and Hardy, see Chapter 14; for Titchmarsh, see Chapter 16.v.

Lines 13–16. Chapter 14.v.

Line 17. The term "*Li*" here should be pronounced "ell-eye," to preserve the meter. The songwriter is here discussing the error term $\pi(x) - Li(x)$, which I cover extensively in Chapter 21.

Line 18. "The order of P is not known" means: "P is 'big oh' of... what? We don't know." For big oh, see Chapter 15.ii-iii. By "x high," the songwriter means, "large values of x."

Lines 19–20. If we could show that $\pi(x)-Li(x)=O(\sqrt{x}\log x)$, the RH would follow. That is the converse of von Koch's 1901 result in Chapter14.viii. I didn't mention it at the time, but if von Koch's result is true, the RH follows. Each implies the other.

Lines 21–28. These next few lines are all about the Lindelöf Hypothesis (LH), a famous conjecture in the theory of the zeta function. For Lindelöf the man, see Note 72. His hypothesis concerns the growth of the zeta function in a vertical direction—that is, up a vertical line in the complex plane.

Lindelöf, writing the argument of the zeta function as $\sigma+it$, asked: For any given real part σ (that's a lowercase Greek "sigma," by the way), what can be said about the size of $\zeta(\sigma+it)$ as t, the imaginary part, goes from zero to infinity? "Size" here means the modulus, as defined in Chapter 11.v; in other words, it means $|\zeta(\sigma+it)|$, the distance of the value from zero. This is a real number, so that for any given σ, both the argument t and the value $|\zeta(\sigma+it)|$ are real numbers. We can therefore draw a graph. Figures A-1 through A-8 show these graphs, for some representative values of σ, and explain the issue better than any number of words.

Note the non-trivial zeros of the zeta function in Figure A-5. Note, in fact, the *busyness* of Figures A-4 through A-6, compared to the others. With the zeta function, all the interesting action is in the critical strip.

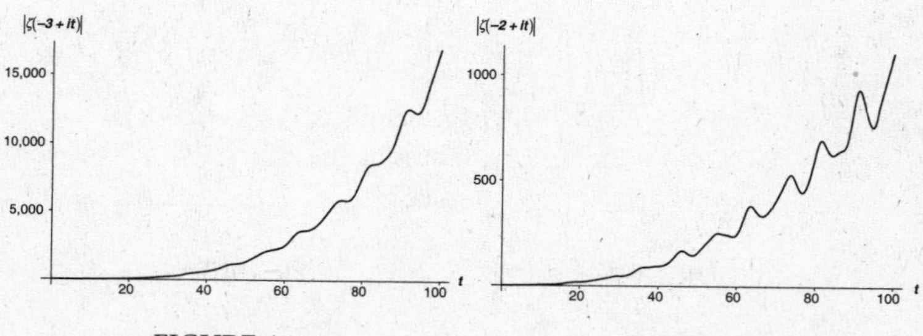

FIGURE A-1

FIGURE A-2

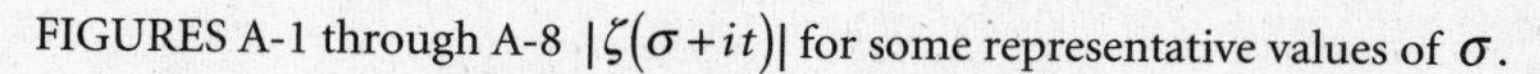
FIGURES A-1 through A-8 $|\zeta(\sigma+it)|$ for some representative values of σ.

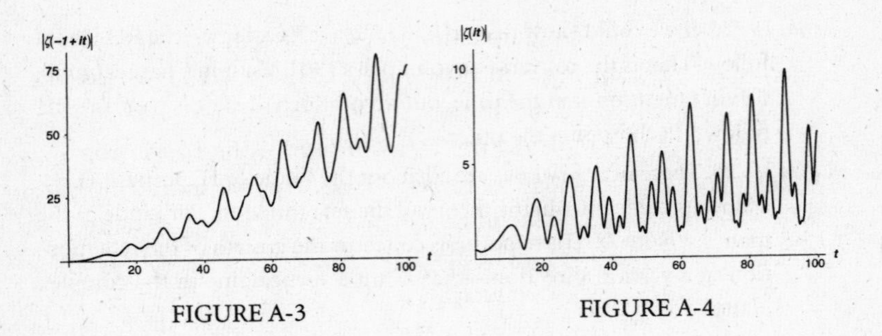

FIGURE A-3

FIGURE A-4

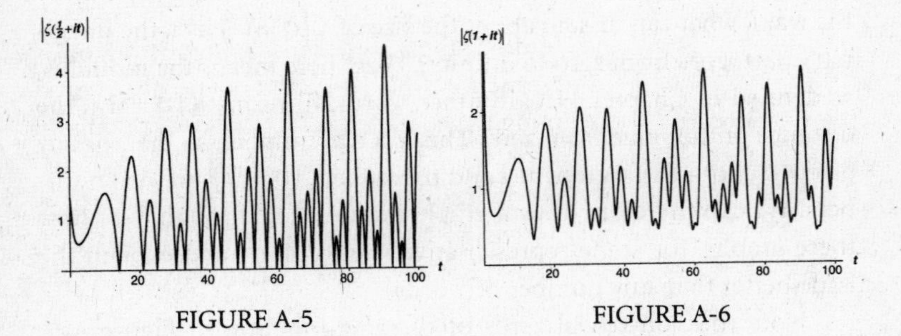

FIGURE A-5

FIGURE A-6

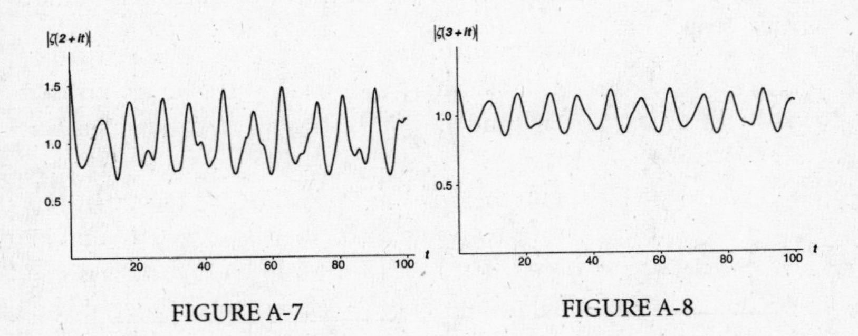

FIGURE A-7

FIGURE A-8

Note also some familiar values when $t = 0$: $\frac{1}{2}$ in Figure A-4 (corresponding to $\zeta(0) = -\frac{1}{2}$ in Figure 9-3, since of course $\left|-\frac{1}{2}\right|$ is just $\frac{1}{2}$); infinity, in Figure A-6 (divergence of the harmonic series, Chapter 1.iii); 1.644934 … in Figure A-7 (solution of the Basel problem, Chapter 5.i); and 1.202056 … in Figure A-8 (Apéry's number, Chapter 5.vi). The function value of zero at $t = 0$ in Figure A-2 is genuine, a trivial zero (Chapter 9.vi). The apparent zeros in Figures A-1 and A-3 are false; the actual $t = 0$ values are just too small to register. (They are, respectively, 0.0083333…, and 0.0833333….)

The LH is about finding a big oh (Chapter 15.ii) for these graphs. Just from looking at them, you can guess the following:

- For $\sigma = -1$, -2, and -3, the graph looks as if it is big oh of some accelerating function of t, perhaps a power like t^2 or t^5, those powers seeming to get bigger as σ heads west along the negative real axis.
- For $\sigma = 2$ and 3, it looks as though we are in the world of $O(1)$, or in other words, of $O(t^0)$.
- In the critical strip, that is, for $\sigma = 0$, $\frac{1}{2}$, and 1, it is not easy to say what an appropriate big oh might be.

Could it be that, for any value of σ, there is a definite number μ for which $\left|\zeta(\sigma + it)\right| = O(t^{\mu})$? With $\mu = 0$ when σ is bigger than 1, and μ some increasing positive number when σ goes west from zero? That's how things look. But then, what happens in the critical strip when σ is between 0 and 1? And in particular, what happens on the critical line when $\sigma = \frac{1}{2}$?

Well, here (see Figure A-9) is what we know for certain at the time of writing. For any given value of σ, there is indeed a number μ for which $\left|\zeta(\sigma + it)\right| = O(t^{\mu+\varepsilon})$, for arbitrarily small ε. This is not quite the same as my suggestion in the previous paragraph, but you could be forgiven for ignoring the difference. (If you compare the ε that showed up in Chapter 15.iii, though, you will understand its significance here.) Clearly, this number μ is a function of σ. Hence

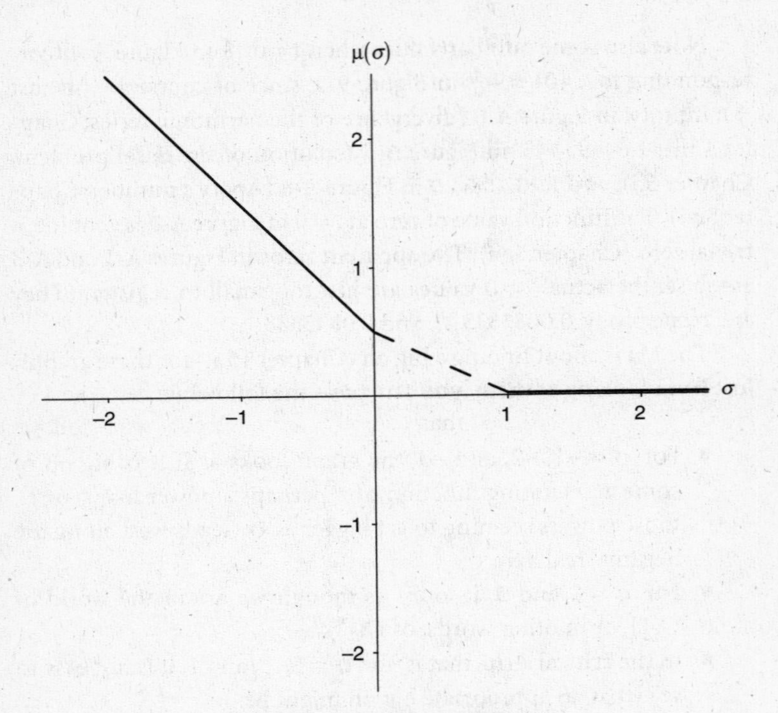

FIGURE A-9 Lindelöf's function.

"the Lindelöf function mu sigma" in Line 22. This is nothing to do with the Möbius μ function of Chapter 15, of course. Here we have another unfortunate case of overloaded symbols.

We also know the following, with mathematical certainty.

- When σ is less than or equal to zero, $\mu(\sigma) = \frac{1}{2} - \sigma$.
- When σ is greater than or equal to 1, $\mu(\sigma) = 0$.
- In the critical strip (that is, when σ is between 0 and 1 exclusive), $\mu(\sigma) < \frac{1}{2}(1 - \sigma)$. In other words, it lies below the dotted line in Figure A-9.

- For all values of σ, $\mu(\sigma)$ is convex downward. That is, if you join any two points of the graph with a straight line, the arc you cut off lies entirely below, or on, the line. This is true everywhere, including the critical strip; and it implies that for σ between 0 and 1, $\mu(\sigma)$ must be positive or zero. (Line 26 of the song.)
- The truth of the RH would imply the truth of the LH (which I am just about to state), but not vice versa. The LH is the weaker result.

That, I repeat, is the limit of our current knowledge. The LH, shown in Figure A-10, says that $\mu(\frac{1}{2}) = 0$, from which it easily follows

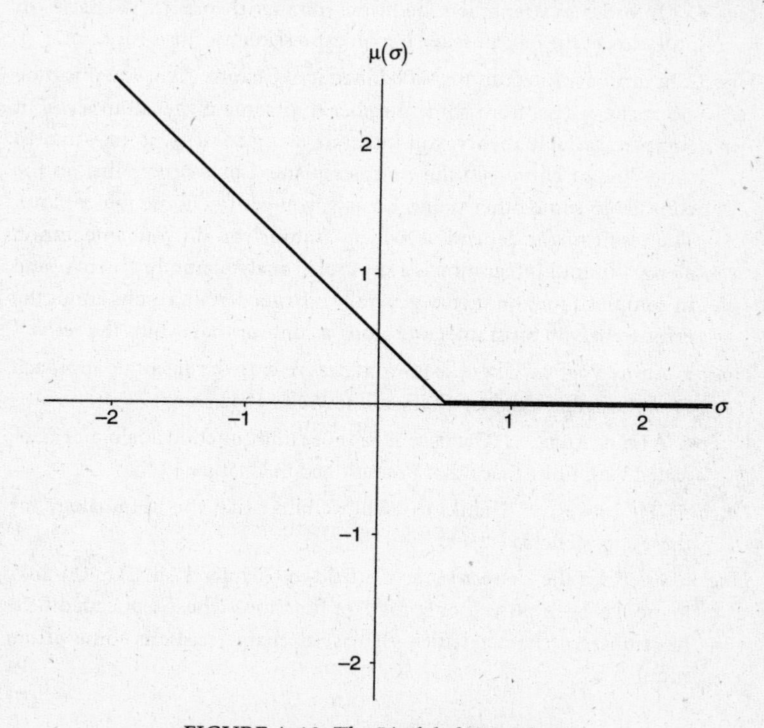

FIGURE A-10 The Lindelöf Hypothesis.

that $\mu(\sigma) = \frac{1}{2} - \sigma$ all the way from negative infinity to $\sigma = \frac{1}{2}$, and then is zero for every argument east of that. Compare Lines 27 and 28 of the song. This is an open hypothesis, still unproven. As a matter of fact, we don't know a single value for $\mu(\sigma)$ when σ is between 0 and 1, exclusive. The LH is the greatest challenge in zeta-function theory after the RH and has been the subject of intense interest and investigation since Lindelöf stated it in 1908.

Line 24. It can be proved that the LH is equivalent to a statement limiting the number of zeta zeros off the critical line. If the RH is true, of course, there should be no such zeros; but then, as I have already pointed out, if the RH is proved, the LH follows.

Line 31. "In order to strengthen the prime number theorem. . . ." That is, in order to get the best possible big oh expression for the error term.

Line 32. In ordinary integration, as I defined it in Chapter 7.vii, you integrate along the *x*-axis, from some number *a* to some bigger number *b*. In complex variable theory, you integrate along some contour—that is, some line or curve—in the complex plane, from some point on the contour to some other point. Usually, you get to choose the contour. The result might depend on which contour you do your integration along. Contour integration is a key tool in analytic number theory (and in complex function theory generally). To get certain results about the error term, you must integrate along a contour that avoids the zeros.

Line 33. "André Weil. . . ." These last two stanzas refer to the algebraic approach I mentioned in Chapter 17.iii, and to Weil's 1942 result.

Line 34. "A fancier zeta. . . ." That is, one of those zeta-function analogues associated with finite fields that I mentioned in Chapter 17.iii.

Line 35. "He proves. . . ." Thanks to Weil, we know that the RH-analogy for these special fields is true.

Line 36. I defined the *characteristic* of a field in Chapter 17.ii. The RH-analogue has been proved only for zeta functions whose associated field has non-zero characteristic—that is, its characteristic is some prime number *p*.

Line 40. The word "mod" here is being used in the clock arithmetic sense of Chapter 6.viii; as I remarked in Chapter 17.ii, this has connections with field theory.

Among the many alternative versions of Tom's lyrics to be found on the internet, I note that one ends with the line, "Use R.M.T., and you'll have better luck." This is a good-natured dig at the "physical" approach. "R.M.T." stands for "Random Matrix Theory."

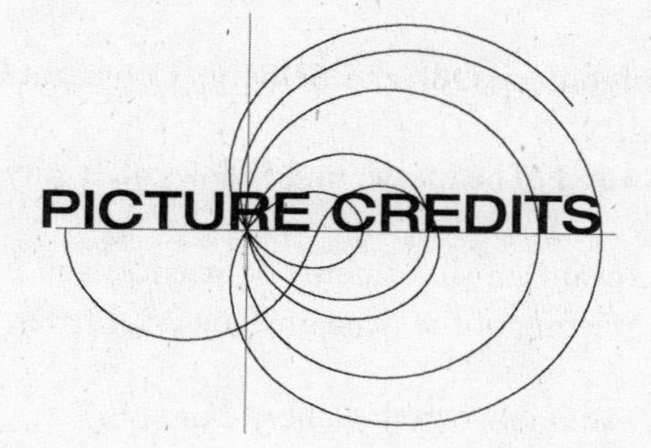

PICTURE CREDITS

Euler, Pólya: By permission of Gerald Alexanderson. The portion of George Pólya's letter in Chapter 17 is reproduced by permission of Andrew Odlyzko.

Peter the Great: The State Hermitage Museum, St. Petersburg. Artist: Jean Marc Nattier (1717).

Dirichlet, Gauss, Hilbert: Deutsches Museum.

Duke of Brunswick: Braunschweigisches Landesmuseum.

Riemann: The younger Riemann, by permission of Michael Monastyrsky; the older, courtesy of the Staatsbibliothek zu Berlin, Preussischer Kulturbesitz.

Dedekind, Landau, Siegel: Niedersächsische Staats- und Universitätsbibliothek, Göttingen; Abteilung für Handschriften und seltene Drucke.

de la Vallée Poussin: Louvain-la-Neuve, Archives de l'Université Catholique de Louvain, CHUL.

Hadamard: Archives of the Woodson Research Center, Fondren Library, Rice University.

Chebyshev: Scientific Library named after M. Gorky, St. Petersburg State University.

Connes, Montgomery, Odlyzko, Selberg: Copyright C.J. Mozzochi, Princeton, NJ.

Hardy, Littlewood: The Master and Fellows of Trinity College Cambridge.

Gram: The Royal Danish Academy of Sciences and Letters. (Detail from "A Meeting of the Academy," by P.S. Krøyer, painted 1895-1897.)

Turing: The National Portrait Gallery, London.

Artin: Princeton University Library.

Weil, Deligne: Archives of the Institute for Advanced Study, Princeton. Photographers Herman Landshoff (Weil), Randall Hagadorn (Deligne).

Dyson: By permission of Freeman Dyson.

Berry: By permission of Sir Michael Berry.

Lindelöf: Helsinki University Museum. Artist: W. Sjörström (1930).

Cramér: By permission of Professor Anders Martin-Löf, Department of Mathematical Statistics, Stockholm University.

Taiye: By the author.

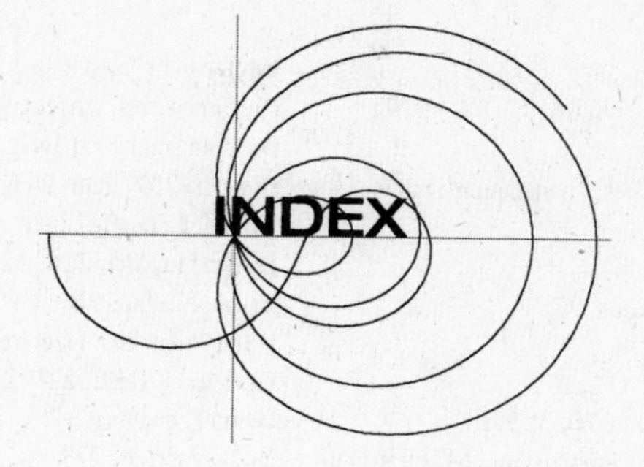

A

G